Lecture Notes in Artificial Intelligence   2969

Edited by J. G. Carbonell and J. Siekmann

Subseries of Lecture Notes in Computer Science

Matthias Nickles   Michael Rovatsos
Gerhard Weiss (Eds.)

Agents and Computational Autonomy

Potential, Risks, and Solutions

 Springer

Series Editors

Jaime G. Carbonell, Carnegie Mellon University, Pittsburgh, PA, USA
Jörg Siekmann, University of Saarland, Saarbrücken, Germany

Volume Editors

Matthias Nickles
Michael Rovatsos
Gerhard Weiss
Technical University of Munich
Department of Informatics
Boltzmannstr. 3, 85748 Garching bei München, Germany
E-mail: {nickles, rovatsos, weissg}@cs.tum.edu

Library of Congress Control Number: 2004109920

CR Subject Classification (1998): I.2.11, C.2.4, I.2, D.2

ISSN 0302-9743
ISBN 3-540-22477-7 Springer Berlin Heidelberg New York

Springer is a part of Springer Science+Business Media

springeronline.com

© Springer-Verlag Berlin Heidelberg 2004
Printed in Germany

Typesetting: Camera-ready by author, data conversion by Scientific Publishing Services, Chennai, India
Printed on acid-free paper   SPIN: 11301097   06/3142   5 4 3 2 1 0

Preface

This volume contains the postproceedings of the 1st International Workshop on Computational Autonomy – Potential, Risks, Solutions (AUTONOMY 2003), held at the 2nd International Joint Conference on Autonomous Agents and Multi-agent Systems (AAMAS 2003), July 14, 2003, Melbourne, Australia. Apart from revised versions of the accepted workshop papers, we have included invited contributions from leading experts in the field. With this, the present volume represents the first comprehensive survey of the state-of-the-art of research on autonomy, capturing different theories of autonomy, perspectives on autonomy in different kinds of agent-based systems, and practical approaches to dealing with agent autonomy.

Agent orientation refers to a software development perspective that has evolved in the past 25 years in the fields of computational agents and multiagent systems. The basic notion underlying this perspective is that of a computational agent, that is, an entity whose behavior deserves to be called flexible, social, and autonomous. As an autonomous entity, an agent possesses action choice and is at least to some extent capable of deciding and acting under self-control. Through its emphasis on autonomy, agent orientation significantly differs from traditional engineering perspectives such as structure orientation or object orientation. These perspectives are targeted on the development of systems whose behavior is fully determined and controlled by external units (e.g., by a programmer at design time and/or a user at run time), and thus inherently fail to capture the notion of autonomy.

To date autonomy is still a poorly understood property of computational systems, both in theoretical and practical terms, and among all properties usually associated with agent orientation it is this property that is being most controversially discussed. On the one hand, it is argued that there is a broad range of applications in complex domains such as e/m-commerce, ubiquitous computing, and supply chain management which can hardly be realized without taking autonomy as a key ingredient, and that it is first of all agent autonomy which enables the decisive features of agent-oriented software, namely robustness, flexibility and the emergence of novel solutions of problems at run time. On the other hand, it is argued that autonomy is mainly a source of undesirable and chaotic system behavior. Obviously, without a clarification of these two positions, it is unlikely that agent orientation and agent-oriented systems (having "autonomy" as a real property and not just as a catchy label) will become broadly accepted in real-world, industrial and commercial applications.

The AUTONOMY 2003 workshop was the first workshop organized to discuss different definitions and views of autonomy, to analyze the potential and the risks of computational autonomy, and to suggest solutions for the various issues raised by computational autonomy.

Acknowledgements

We would like to express our gratitude to the authors of this volume for their contributions and to the workshop participants for inspiring discussions, as well as to the members of the Steering Committee, the Programme Committee, and the additional reviewers of AUTONOMY 2003 for their reviews and for their overall support for the workshop.

Special thanks also go to Simon Parsons, the AAMAS 2003 Workshops Chair, for his great support.

April 2004   Matthias Nickles, Michael Rovatsos, Gerhard Weiß

Table of Contents

Models and Typologies

Design and Applications

Agency, Learning and Animal-Based Reinforcement Learning

Eduardo Alonso[1] and Esther Mondragón[2]

[1] Department of Computing, City University
London EC1V 0HB, United Kingdom
eduardo@soi.city.ac.uk
[2] Department of Psychology, University College London
London WC1H 0AP, United Kingdom
e.mondragon@ucl.ac.uk

Abstract. In this paper we contend that adaptation and learning are essential in designing and building autonomous software systems for real-life applications. In particular, we will argue that in dynamic, complex domains autonomy and adaptability go hand by hand, that is, that agents cannot make their own decisions if they are not provided with the ability to adapt to the changes occurring in the environment they are situated. In the second part, we maintain the need for taking up animal learning models and theories to overcome some serious problems in reinforcement learning.

1 Agency and Learning

Agents (and thus agency) have been defined in many different ways according to various research interests. It is universally accepted though that an agent is a software system capable of flexible, autonomous behaviour in dynamic, unpredictable, typically multi-agent domains. We can build up on this fundamental definition and state which characteristics an agent should display following the traditional distinction between strong and weak agency. The later prescribes autonomy, social ability, reactivity, and pro-activeness, to which the former adds various mental states, emotions, and rationality. How these features are reflected in the systems architecture will depend on the nature of the environment in which it is embedded and the degree of control the designer has over this environment, the state of the agent, and the effect of its actions on the environment.

It can be said, therefore, that, at first glance, learning does not seem to be an essential part of agency. Quoting Michael Wooldridge, "learning is an important agent capability, but it is not central to agency" [11]. Following the same line of argumentation, agent research has moved from investigating agents components to multi-agent systems organization and performance. For instance, the CfP for the International Joint Conference on Autonomous Agents and Multi-Agent Systems has discouraged papers that address isolated agent capabilities *per se* such as learning. In addition, AgentLinkII's (Europe's Network of Excellence in Agent-Based Computing) Special Interest Group on Agents that Adapt, Learn

M. Nickles, M. Rovatsos, and G. Weiss (Eds.): AUTONOMY 2003, LNAI 2969, pp. 1–6, 2004.

and Discover (ALAD) may disappear under the Sixth EU Framework. However generous AgentLinkII was in supporting activities directed towards the increase awareness and interest in adaptive and learning agent research (sponsoring, for example, the Symposia Series on Adaptive Agents and Multi-Agent Systems), and although there are plenty of recent references to learning algorithms and techniques (*e.g.*,[2],[10]) and that a considerable effort has been done in providing a suitable infrastructure for the development of close collaboration between machine learning experts and agent systems experts with the creation of the Adaptive and Learning Agents and Multi-Agent Systems (ALAMAS) Consortium, the truth is that learning agents does not seem to be a priority any longer.

It is our understanding that one of the main arguments against considering learning as a requisite for agency is that there are scenarios in which agents can be used and learning is not needed. For example, little can be learned in accessible domains where agents can obtain complete, accurate, up-to-date information about the environment's states, or in deterministic domains where any action has a single guaranteed effect, or in static domains where the environment remains unchanged unless an action is executed.

It is our contention though that in such domains agents are not strictly necessary anyway and that applying object-oriented technology would suit best the requirements and constraints designers must meet. Put roughly, if you can use objects, do not use agents. Unlike agent-oriented technology, object-oriented technology is well-established and understood, with clear modeling and specification languages (UML) and programming languages (Java, C++). On the other hand, as denounced in [1], an Unified Agent Modeling Language is still under development, and although some object-oriented features such as abstraction, inheritance and modularity make it easier to manage increasingly more complex systems, JAVA (or its distributed extensions JINI and RMI) and other OO programming languages cannot provide a direct solution to agent development.

Agents are ideal for uncertain, dynamic systems. Lehman and Belady's Laws of Software Evolution [6], particularly, those referring to continuing change and increasing complexity have proven true with the growth of the Internet and the arrival of the Grid computing. Certainly, it has become increasingly complicated to model and control the way software systems interact and get co-ordinated. Perhaps, many claim, we should focus on observing their emergent behaviours. Perhaps we should move from software engineering to software phenomenology and study the performance of multi-agent systems the same way we study biological or chemical systems (for example, by using minimalist Multi-Agent System platforms such as BTExact's DIET [4]). On the one hand, by distributing the tasks among different autonomous entities we gain in both speed and quality. On the other hand, such systems seem to work as if guided by an "invisible hand". Designers cannot foresee in which situations the systems will encounter themselves or with whom they will interact. Consequently, such systems must adapt to and learn from the environment so that they can make their own decisions when information comes. To sum up, agents need to learn in real-life domains. Therefore, in real-life domains, learning is essential to agency.

Several initiatives have been launched recently to investigate adaptive intelligent systems. In an attempt to improve our understanding of the mechanisms and organisational principles involved in the generation of adaptive behaviour in intelligent machines or adaptive systems inspired by the study of animals the BBSRC and EPSRC have announced a call for Adaptive and Interactive Behaviour of Animal and Computational Systems (AIBACS). Separately, a Special Interest Group on Computation and Animal Learning has been formed. Moreover, the Society for the Study of Artificial Intelligence and the Simulation of Behaviour (SSAISB) and the International Society for Adaptive Behaviour (ISAB) are organising symposia and workshops on this research area in 2004.

Our contribution to this trend of thought is briefly explained in the next section. We propose to use animal learning theories to overcome some drawbacks in the quintessential machine learning technique for agents, namely, reinforcement learning.

2   Animal-Based Reinforcement Learning

Reinforcement learning has been defined as learning what to do —how to map situations to actions— so as to maximise a numerical reward signal. Unlike supervised learning such as pattern recognition or artificial neural networks, the learner is not told which actions to take, but instead must discover which actions yield the most reward by trying them. Typically, actions may affect not only the immediate reward but also the next situation and, through that, all subsequent rewards. These two characteristics, trial-and-error search and delayed reward, are the two most important features of reinforcement learning.

The reinforcement learning problem is the problem of finding an optimal policy, *i.e.*, the optimal way of behaving at a given time defined as a mapping from perceived states of the environment to actions.

Among the different techniques used to solve the reinforcement learning problem (see [7] for a detailed account) temporal-difference methods are the most widely used due to their great simplicity: They can be applied on-line, with minimal computation, to experience generated from interaction with an environment; besides, they can be expressed nearly completely by single equations that can be implemented with small computer programs.

Despite the success of these methods, for most practical tasks, reinforcement learning does fail to converge even if a generalising function approximation is introduced. It turns out to generate extreme computational cost when not dealing with small state-action pairs, which are, in practice, very rare in any real learning scenarios. For example, since all state-action pairs must be repeatedly visited, Q-learning does not address generalisation over large state-action spaces. In theory, it may converge quite slowly to a good policy. In practice however, Q-learning can require many thousands of training iterations to converge even in modest-sized problems.

To overcome these problems we propose to go back to animal learning psychology and update reinforcement learning algorithms with contemporary associative and instrumental learning models.

One historical thread of reinforcement learning concerns learning by trial and error and started in the psychology of animal learning. In particular, Thorndike [8] described the effect of reinforcing events on the tendency to select actions in what he called the Law of Effect. This Law includes the two most important aspects of trial-and-error learning and, in turn, of reinforcement learning, namely, it is selectional (it involves trying alternatives and selecting among them by comparing the consequences) and associative (the alternatives are associated with particular situations).

There is however plenty of evidence that suggest that Thorndike's Law is wrong. For example, it has been established experimentally that rewards are not essential for learning and that the assumption that learning consists of the gradual strengthening of a connection between neural centres concerned with the perception of a stimulus and the performance of a response is far too simple.

Reinforcement learning has tried to keep abreast with new animal learning theories by, for example, incorporating Tolman's findings on instrumental learning [9]. Yet, these new paradigms have, in turn, proved to be inaccurate. As a consequence, reinforcement learning techniques based on out-of-date animal learning models may be conceptually incorrect. Indeed, it could be argued that this failure to understand recent developments in animal psychology is of no consequence for reinforcement learning. On the contrary, we claim that the poor results so far encountered may be at least in part due to this gap between theory and practice.

To start with, an analysis of the terminology allegedly taken from animal psychology shows that most of the concepts used in reinforcement learning do not match with their counterparts in animal learning. States (or events) are compounds of stimuli, actions are responses, and rewards are values associated to actions that can be understood as reinforcers on stimulus-response associations. Of course, changing our vocabulary should not be a problem. The real problem arises when such changes are ontological and thus prevent reinforcement learning from studying the nature of reinforcers, the sort of associations formed and the conditions for their formation.

What else are we missing? First of all, associations are formed among different stimuli (classical learning) and between the response and the reinforcer (instrumental learning), not ony between a stimulus and a response. Secondly, reinforcers are stimuli and, therefore, elements of the associations, not mere values assigned to stimulus-response associations. Moreover, reinforcers have not only specific but also affective characteristics, that is, characteristics that reflect their motivational quality. Finally, learning depends on contiguity, contingency, associative competition, attention and surprisingness, none of which is considered in reinforcement learning.

One more example. Unlike other machine learning paradigms, reinforcement learning assumes that, for optimal performance, animals do explore (state-action pairs which outcomes are unknown) and exploit (those state-action pairs which rewards are known to be high). Numerous functions have been presented in the literature to control the balance between these two opposite processes (*e.g.*, soft-max action selection). Either way, behaviour as such is neutral in reinforcement learning, *i.e.*, behaviours acquire values only when rewarded. On the contrary, animals explore by default: Exploratory behaviours act as (internal) reinforcers *per se*. The strength of the association between these behaviours and other re-inforcers will, thus, depend on the behaviours intrinsic value. Indeed, Kaelbling *et al.* [5] have already suggested that, in order to solve highly complex problems, we must give up *tabula rasa* learning techniques and begin to incorporate bias that will give leverage to the learning process. The very nature of animal learning may well be such a bias.

Of course, there have been several attempts to bring together the machine learning community and the animal learning community. For example, Peter Dayan and L.F. Abbott [3] have successfully modelled several phenomenon in classical conditioning and instrumental learning. This trend has been directed towards the identification of mathematical techniques, mainly temporal difference algorithms, that psychologists might use to analyse and predict animal behaviour.

Our goal is complementary. We are not trying to bridge gaps in the analytical skeleton of animal learning using reinforcement learning techniques. Instead, we intend to use animal learning models to improve reinforcement learning performance. Our contention is that convergence and generalization problems so common to reinforcement learning can be corrected by re-defining some of its more fundamental basis according to animal learning models.

References

[1] E. Alonso. AI and Agents: State of the Art. *AI Magazine*, 23(3):25–29, 2002.

[2] E. Alonso, D. Kudenko, and D. Kazakov, editors. *Adaptive Agents and Multi-Agent Systems*. LNAI 2636, Springer-Verlag, Berlin, 2003.

[3] P. Dayan and L.F. Abbott, *Theoretical Neuroscience: Computational and Mathematical Modeling of Neural Systems*. The MIT Press, Cambridge: MA, 2001.

[4] C. Hoile, F. Wang, and P. Marrow. Core specification and experiments in DIET: A decentralised ecosystem-inspired mobile agent system. In C. Castelfranchi and W. Lewis Johnson, editors, *Proc. 1st Int. Conf. on Autonomous Agents and Multi-Agent Systems (AAMAS-2002)*, pages 623–630, ACM Press, New York, 2002.

[5] L.P. Kaelbling, M.L. Littman, and A.W. Moore. Reinforcement Learning: A Survey. *Journal of Artificial Intelligence Research* 4:237–285, 1996.

[6] M.M. Lehman, and L. Belady. *Program Evolution: Processes of Software Change*. Academic Press, London, 1985.

[7] R.S. Sutton and A.G. Barto. *Reinforcement Learning: An Introduction*. The MIT Press, Cambridge: MA, 2002.

[8] E.L. Thorndike. *Animal Intelligence: Experimental Studies*. Macmillan, New York, 1911.

[9] E.C. Tolman. *Purposive Behavior in Animals and Men*. Century, New York, 1932.

[10] G. Weiss, editor. *Distributed Artificial Intelligence Meets Machine Learning*. LNAI 1221, Springer-Verlag, 1997.

[11] M. Wooldridge. *An Introduction to Multiagent Systems*. John Wiley & Sons, Chichester, 2002.

Agent Belief Autonomy in Open Multi-agent Systems

K. Suzanne Barber and Jisun Park

The Laboratory for Intelligent Processes and Systems
The University of Texas at Austin
{barber, jisun}@lips.utexas.edu

Abstract. The goals and the beliefs of an agent are not independent of each other. In order for an agent to be autonomous an agent must control its beliefs as well as tasks/goals. The agent's beliefs about itself, others and the environment are based on its derived models of perceived and communicated information. The degree of an agent's belief autonomy is its degree of dependence on others to build its belief models. We propose source trustworthiness, coverage and cost as factors an agent should use to determine on which sources to rely. Trustworthiness represents how reliable other agents are. Coverage is a measure of an information source's contribution to an agent's information needs. Cost of getting information from a source is defined by the timeliness of information delivery from the source. Since a respective agent only knows about a limited amount of information sources which can communicate with the agent, it must also rely on other agents to share their models describing sources they know about. These factors along with the degree to which a respective agent shares its knowledge about its neighbors are represented in this paper and proposed as contributions to agent's decisions regarding its belief autonomy for respective goals.

1 Introduction

Agents are situated in the environment, and proactively pursue their goals [1]. Consequently, autonomy is relational and limited, meaning that an agent's autonomy is meaningful in relation to the influence of other agents while the autonomy is necessarily limited by the relationship [2]. Information represented as the agent's internal beliefs is essential for an agent to reason about when deciding how to accomplish its goals. Thus, an agent's goal achievement is certainly a function of its modeled beliefs.

We define the degree of an agent's belief autonomy for a respective goal as:

- The degree of dependence on others to build its belief models; or
- The degree of control over its beliefs.

Since an agent's beliefs may be constructed from information acquired from others (e.g. agents), an agent's belief autonomy is key for accurate belief formulation and goal achievement.

We argue that an agent sets its degree of belief autonomy by establishing its radius of awareness, which is the degree to which it uses the beliefs of others to construct its

M. Nickles, M. Rovatsos, and G. Weiss (Eds.): AUTONOMY 2003, LNAI 2969, pp. 7–16, 2004.

own beliefs (e.g. the degree to which its beliefs are constructed based on beliefs shared by others). An agent's internal beliefs consist of its derived models of perceived and communicated information about itself, others and environment. We propose three factors to be considered for guaranteeing an agent's appropriate degree of belief autonomy: trustworthiness, coverage, and cost. Informally, trustworthiness represents how reliable other agents are. Coverage is a measure of an information source's contribution to an agent's information needs. Cost of getting information from a source is defined by the timeliness of information delivery from the source.

2   Overview

An agent must select the information and information sources so that the achievability of the goal which is depending on the respective sources and information is maximized (or near-maximized). Thus, the degree of belief autonomy (e.g. how much the agent depends on the beliefs of others) as well as the actual dependencies (i.e. dependencies on which beliefs from which information source) is paramount.

S is the set of information sources. Assuming $\Phi_a(s)$ is the abstract representation of an information source's $(s \in S)$ evaluation from agent a's perspective as well as the degree to which the source, s, is sharing evaluations of other sources it knows about (dms). Let $< r_k, s, \Phi_a(s) >$ be a tuple representation of the information and the corresponding information source, where s is a provider of r_k, r_k is an information with index k.[1] Also, let $X_j = \{< r_1, s_1, \Phi_a(s_1) >, < r_m, s_k, \Phi_a(s_k) >, ..., < r_n, s_n, \Phi_a(s_n) >\}$, which is a set of tuple combinations in which the union of all r_k in each tuple satisfies the information requirements ($\bigcup_{\forall m, < r_m, s_k, \Phi_a(s_k) > \in X_j} r_m$ is equal to a set of information required by an agent a). An agent must seek to establish the appropriate belief autonomy which means finding the best combination of sources (X_j) to request and receive information for the best goal achievability.

The agents are assumed to be symmetric and distributed. Agents are symmetric if the rationality of an agent does not subsume the other agents' rationality. Agents are distributed if there is no central point of control in a system. In addition, the agents are not assumed to be self-sufficient in both information acquisition and communication capability. Therefore, the agents may need to get information from other agents or may not be able to directly communicate with the potential sources. The lack of information self-sufficiency causes each agent to model the external system as a set of distributed information sources.

Figure 1 shows an instance of an assumed multi-agent system, where the nodes represent the agents and the edges represent the communication links. The graph is not complete because an agent can be either physically limited in its communication radius or unaware of all neighboring agents. The degree of the model sharing (dms) can also be clarified in this context as the distance of neighbors sharing their evaluations of their respective neighbors. For example, dms is 1 if a_2 shares its

---

[1] k is used as a simple identification index for both information and sources.

evaluations of its neighbors with a_1 so that a_1 has access to how a_2 evaluates its neighbor(s) (a_8), and the same for the other neighbors (a_3, a_5, a_6). If *dms* is 2, a_1 has access not only to how its neighbors evaluate their neighbors, but also to how agents 2-hops away evaluate their neighbors. Therefore, a_1 has access to a_8's evaluation of a_9 and a_4's evaluation of a_{10}.

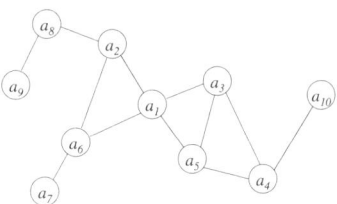

Fig. 1. Agents are the distributed information sources with limited information acquisition and communication

The system can be formally described as follows.

$$N(a_i) = \{a_k \mid a_k \text{ is a neighbor of } a_i\}$$
$$KB(a_i) = \{r_k \mid r_k \text{ is required by } a_i\}$$
$$PROV(a_i) = \{r_k \mid r_k \text{ is provided by } a_i\},$$

where $N(a_i)$ is a set of agents which are the neighbors of a_i, $KB(a_i)$ is a set of information which is required by a_i, and $PROV(a_i)$ is a set of information which is provided by a_i.

A neighbor of an agent is the agent who can be reached by direct communication. An agent's KB represents the Knowledge Base comprised of a set of information which is required by one of the agent's goals. When an agent requires information it does not have, the agent needs to find whom to request and consequently receive the information. Instead of assuming an oracle which resolves the location of information, we assume multicast of requests and information to and from neighbors. For example, assuming $KB(a_1) = \{r_1\}, PROV(a_8) = \{r_1\}$ in Figure 1, a_1 multicasts a request for r_1 to its neighbors ($\{a_2, a_3, a_5, a_6\}$). Then, $\{a_2, a_3, a_5, a_6\}$ also request the information to their neighbors. Intuitively, there are 2 paths for the delivery of the request and information with a_2, a_6 willing to provide the information to a_1, which are $(a_1, a_2, a_8, a_2, a_1)$ and $(a_1, a_6, a_2, a_8, a_2, a_6, a_1)$. A set of paths for the delivery of the requests and information is called the Information Supply Chain. Adding $PROV(a_4) = \{r_1\}$ to the assumption will generate more options for the delivery paths of request and information, resulting in $(a_1, a_2, a_8, a_2, a_1)$, $(a_1, a_6, a_2, a_8, a_2, a_6, a_1)$, $(a_1, a_6, a_2, a_8, a_2, a_6, a_1)$, $(a_1, a_3, a_4, a_3, a_1)$, $(a_1, a_3, a_5, a_4, a_5, a_3, a_1)$, $(a_1, a_5, a_3, a_4, a_3, a_5, a_1)$,

$(a_1, a_5, a_4, a_5, a_1)$. However, since a_1 communicates only with the neighbors it is neither efficient nor necessary to know the complete paths to the sources. Especially, in an open system, it is not always possible for an agent even to know the existence of all agents in the system or the IDs of those agents. When an agent learns which neighbors can provide the respective information, the agent can exercise the control over the selection of information sources, and thus, the control over the acquisition of beliefs (i.e. belief autonomy), deciding from whom to get information the "next time." There are two main issues for an agent constructing the most reliable paths to the information sources. The first is the evaluation metrics for information and information sources (Section 3), and the second is the effect of different degrees of the model sharing (i.e. the degree to which others share their evaluations of information sources).

There are two kinds of evaluation metrics for information and information sources – collective and respective. The evaluation metrics for collective information measure the expected quality and efficiency of information as a set. For example, given a set of information, the coverage of the information can be a collective metric for information evaluation, where the coverage can be defined as how much the set of information contributes to the goals. The evaluation metrics for respective information represent the quality and efficiency of individual information from a specific source. Although the models which are shared among agents in the Information Supply Chain can be domain dependent, coverage, cost and trustworthiness are proposed as the key factors to be considered for information and source evaluations when determining the appropriate degree of belief autonomy. Therefore the relationship between an appropriate degree of belief autonomy and the evaluation metrics can be formally described by the following equation. It should be noted that the appropriate degree of belief autonomy is defined in the context of a respective *dms* where degree to which an agent knows about others will surely affect it decisions about how much to depend on others to form beliefs.

Appropriate degree of belief autonomy

$$= \underset{\text{degree of dependence}}{\arg} \max \Psi(\text{trustworthiness,coverage,cost})\,|_{dms}$$

, where Ψ is an integration function of the arguments

3   Evaluation Metrics

In order for an agent to determine its most appropriate degree of belief autonomy for a goal, it must determine (1) the potential sources that might deliver those beliefs and (2) the quality and efficiency of those sources. The following metrics, trustworthiness, coverage and cost, serve to describe an agent's potential information sources. A respective agent evaluates the neighbors with regard to those metrics, and selects the most appropriate sources from the potential sources. The dependences on the selected sources are imposed by an agent itself (i.e. an agent determine from whom it will gather information to form its beliefs), resulting in an agent's control over its own beliefs by determining an appropriate degree of belief autonomy.

3.1   Trustworthiness

Trustworthiness of an information provider from an agent's point of view can be represented by the notion of reputation. Reputation $R(a_i, a_j)$ can be defined as a probability that the information a_j provides is true from a_i's point of view [3]. There also exist many different representations and management schemes for reputation [4–7], but basically what they represent is how reliable other agents are. The only assumption about the reputation in this paper is that an agent keeps track of the reputations of only its own neighbors. This assumption makes a difference in both selecting the potential sources and suggesting the request paths. In Figure 1, if a_1 does not know about a_4 and the degree of the model sharing (*dms*) is 1, a_1's evaluation of the path to a_4 is dependent on both a_3's reputation evaluation on a_4 and a_1's reputation evaluation on a_3. However, if a_1 knows about a_4, a_1 must add its own reputation evaluation of a_4 to the decision-making. Reputation of the sources as a trustworthiness evaluation metric is one of the factors for guaranteeing autonomy on beliefs [2] because it is desirable for an agent to be completely independent from the untrustworthy sources. Consequently, if an agent knows the reputations of potential sources, the agent can identify who to depend on for information; thus, it is a significant factor for finding an appropriate degree of belief autonomy (degree of dependence). In addition, the degree of the model sharing (i.e. others telling the agent about sources and source reputations they know about) affects the radius of awareness for source reputations an agent may not have.

3.2   Coverage

Coverage of an information source is a measure for representing the contribution of the source to an agent's information needs. Tasks are necessitated by an agent's goals, and each task requires some amount of information. The union of the required information for all the tasks of an agent constitutes a knowledge base for the goal achievement. Each portion of information in the knowledge base can be assigned a weight so that the weight implies the priority of information. Assume an agent a_i has a goal (G) with three tasks (T_1, T_2, T_3). T_1 requires r_1, r_3, T_2 requires r_2, r_3, r_4, and T_3 requires r_4 resulting in $KB(a_i) = \{r_1, r_2, r_3, r_4\}$. Figure 2 depicts the relationships between tasks and information, and priorities assigned to the information. One way to assign a priority to each portion of information is by calculating the number of satisfied information requirements for each task. For example, in Figure 2, satisfying T_2's requirements (r_2, r_3, r_4) satisfies T_3's requirement as well as a half of T_1's requirements. The satisfaction of any other task's requirement does not result in higher satisfaction ratio than T_2. Therefore, T_2's requirements (r_2, r_3, r_4) are assigned a higher priority than the rest of the requirements (r_1).

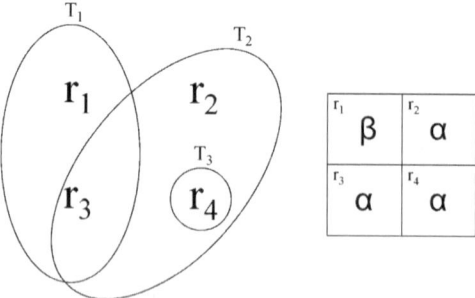

Fig. 2. Relationship between required information (KB) and tasks. α and β are priorities assigned to each portion of information.

With this notion of priority, coverage of an information source (a_i) from an information requester (a_j)'s perspective is defined as the following equation. We can assume all the priority to be 1 when the prioritization is impossible.

$$Coverage(a_i, a_j) = \frac{\sum_{r_k \in PROV(a_i, a_j)} PRIO(a_j, r_k)}{\sum_{r_k \in KB(a_j)} PRIO(a_j, r_k)}, \text{ where } PRIO(a_j, r_k) \text{ is a priority assigned to } r_k \text{ from } a_j.$$

Also, given a combination of information sources we can measure the total coverage of the information source combination.

$$TotalCoverage(a_j) = \sum_{a_i \in N(a_j)} Coverage(a_i, a_j)$$

Coverage is a metric for representing the contribution of information sources to a requestor's goal achievement. The degree of the distribution of information source changes depending on an agent's decision preference. A high degree of information source distribution means that an agent prefers (or must select) a larger number of information sources for a given number of information. This results in the distribution of dependence to a larger number of sources while the dependence on each source is a smaller portion of information than the opposite case. A Low degree of information source distribution means that an agent prefers (or must select) a smaller number of information sources for the same case. This results in a decreased dependence with respect to the number of sources while the amount of dependence on each source is larger than the former case.

Total coverage measures the quality of an information source combination. Obviously, the total coverage increases if the selected combination of sources provides a higher degree of goal achievability. Therefore, an agent's decisions regarding the degree of belief autonomy should also take the total coverage into account for building a reliable knowledge base.

3.3   Cost

The cost of getting information affects the performance and efficiency of an agent's decision-making. Even if a source provides the most reliable information, high cost of information acquisition degrades the system. Timeliness of the information is closely related to the cost because the acquisition of the information is accomplished by a multi-hop request and delivery protocol. Untimely delivery of information can be caused by either delay due to agents or delay due to communication links. As an information consumer, the requesting agent cannot resolve the causes of the untimely reception of the information but can measure the delay for cost quantification. The cost can also be summed up to estimate the total cost of receiving all the information in the KB from a combination of sources.

$$Cost(a_i, a_j) = \frac{k}{Timeliness(a_i, a_j)}, \text{ where } k \text{ is a scaling factor.}$$

$$TotalCost(a_j) = \sum_{a_i \in N(a_j) \wedge PROV(a_i) \neq \varnothing} Cost(a_i, a_j)$$

Since a respective agent only knows about a limited amount of information sources (known in this paper as the agent's neighbors), it must also rely on other agents to share their models describing sources they know about.   The degree to which a respective agent shares its knowledge about its neighbors is referred to as the degree of model sharing.

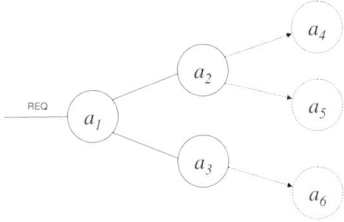

Fig. 3. Information request multicasting to neighbors with and without the awareness of the indirect neighbors

4   Model Sharing

An agent in an Information Supply Chain receives a request, if it is not an initiator of the request, from an agent, and if it cannot provide the information it will request the information from other neighbors. The request is directed to the neighbor which looks like the most reliable provider based on the requestor's models about the providers. The degree of the model sharing refers to the degree to which a respective agent shares its knowledge about it neighbors (i.e. shares information about the source's trustworthiness, coverage and cost, see Section 3). For example, in Figure 3, a_1 decides who to request information from based on its own model of neighbors

(a_2, a_3) without knowing how a_2, a_3 model and evaluate their neighbors. The information providing agents beyond the neighborhood (a_4, a_5, a_6) are invisible to a_1. However, if a_2, a_3 share their models of their neighbors with a_1, a_1 can have increased awareness about potential sources, potential dependencies for information and thus, different belief autonomy assignments. Specifically, the internal models which can be shared by the adjacent neighbors include perceived information, evaluation of neighbors, and suggestion of request paths. They are exchanged between the agents as a form of messages as follows.

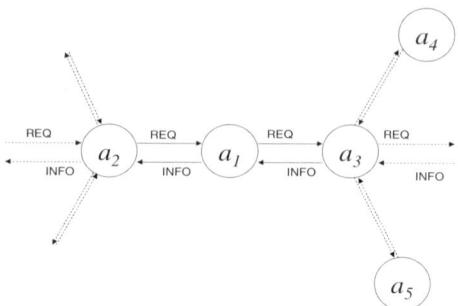

Fig. 4. Information supply chain with request and information

Figure 4 depicts an example Information Supply Chain. Focusing on the agent in the middle (a_1), a_1 takes care of two messages – request and information. A request message (*REQ*) includes the information identifications which are required either by a_1 itself or other agents, and suggested paths for the information request which are a_1's decisions on the best request route up to the agents known to a_1 by the evaluations passed to a_1. For example, if *dms* is 1, a_1 can suggest a request path of a_3 which can be either a_4 or a_5. An information message (*INFO*) contains the information accompanied by the evaluation of neighbors from the provider's point of view. The radius of the evaluations included in the *INFO* message is decided by the degree of the model sharing.

$$REQ = [\overline{Info\_id}, SUGGESTED\_PATH]$$
$$INFO = [\overline{Info}, EVAL\_NEIGHBOR]$$

$\overline{Info\_id}$ is a vector of requested information identifications, and $\overline{Info}$ is the information sent to the requester from the provider.

EVAL_NEIGHBOR is an evaluation of potential information sources among the neighbors. Depending on the degree of the model sharing (*dms*), *EVAL_NEIGHBOR* contains the models of up to *dms* hops away from the requesting agent. *EVAL_NEIGHBOR* can be represented in two forms depending on the necessity. The first is a chain of the most reliable neighbors with length *dms*, and the second is a tree of the evaluations with depth *dms*. While the chain of the most reliable neighbors is compact, the tree structure is more descriptive. In Figure 3, assuming *dms* is 1, if a_1 requests information by sending a *REQ* message, $REQ = [< r_1 >, null]$, without a suggested path since a_1 does not have any evaluations about the neighbors of a_3, a_3 responds with an *INFO* message, $INFO = [< r_1 = 1 >, < a_4 = 0.5, a_5 = 0.9 >]$, including a_3's evaluations about its neighbors; in this case, a single number for simplicity.

SUGGESTED _ PATH is determined by considering EVAL _ NEIGHBOR. The requesting agent can suggest the expected request route. The suggestion can be injected into the decision-making process of the receivers. For example, a_1 sends a REQ message, $REQ = [< r_1 >, < a_5 >]$, to a_3 where a_5 is a_1's suggestion about request path from a_3

An agent must select a set of information sources. EVAL _ NEIGHBOR contains the suggested source evaluations (e.g. trustworthiness, coverage, cost). The amount of sources and information included in the result of EVAL _ NEIGHBOR is decided by the degree of the model sharing. SUGGESTED _ PATH is a set of suggestions from the receiver of the EVAL _ NEIGHBOR, and it can also affect the decisions about the best sources depending on how much the suggestions are reflected in the decision-making.

5   Conclusion

The goals and the beliefs of an agent are not independent of each other. The agent's beliefs about itself, others and the environment are based on its derived models of perceived and communicated information. In order for an agent to be autonomous, an agent must control its beliefs as well as tasks/goals. This research asserts that the degree of belief autonomy reflects the degree of control an agent maintains over its beliefs.   In other words, belief autonomy is defined as an agent's degree of dependence on external information sources to form its beliefs. We propose that an agent should select its appropriate degree of belief autonomy for a respective goal using the following factors: the information source trustworthiness, coverage and cost.

Trustworthiness of an agent can be represented by the notion of reputation which is a probability that the information provided by the agent is true. Therefore, trustworthiness is an expression for the reliability of the information sources. Coverage is a measure for an information source's contribution to an agent's information needs. The cost of getting information from a source is defined by the timeliness of information delivery from the source.

Since a respective agent only knows about a limited amount of information sources, it must also rely on other agents to share their models describing sources they know about.   The degree of the model sharing refers to the degree to which a respective agent shares its knowledge about its "neighbors" (i.e. shares trustworthiness, coverage and cost data regarding information sources it knows about).

In order for an agent to determine its most appropriate degree of belief autonomy for a goal, it must determine (1) the potential sources that might deliver information it needs given information sources it knows about as well as sources discovered through model sharing and (2) the quality and efficiency of those sources by evaluating a source's trustworthiness, coverage and cost. A respective agent evaluates the discovered sources with regard to those metrics, and selects the most appropriate sources from those potential sources. The dependences on the selected sources are controlled by an agent itself (i.e. an agent determines from whom it will gather information to form its beliefs); thus, an agent controls its degree of belief autonomy.

Acknowledgement

This research was supported in part under a subcontract with ScenPro, Inc. funded by U.S. Department of Defense, Small Business Technology Transfer (STTR) Program, Contract Number F30602-99-C-0193.

References

1. M. J. Wooldridge and N. R. Jennings, "Intelligent Agents: Theory and Practice," *Knowledge Engineering Review*, vol. 10, pp. 115–152, 1995.
2. C. Castelfranchi, "Guarantees for Autonomy in Cognitive Agent Architecture," in *Intelligent Agents: ECAI-94 Workshop on Agents Theories, Architectures, and Languages*, N. R. Jennings, Ed. Berlin: Springer-Verlag, 1995, pp. 56–70.
3. K. S. Barber and J. Kim, "Soft Security: Isolating Unreliable Agents from Society," in *Trust, Reputation, and Security: Theories and Practice, Lecture Notes in Artificial Intelligence*, M. Singh, Ed.: Springer, 2003, pp. 224–234.
4. A. F. Dragoni and P. Giorgini, "Learning Agents' Reliability through Bayesian Conditioning: a simulation study," presented at Learning in DAI Systems, 1997.
5. R. Falcone, G. Pezzulo, and C. Castelfranchi, "A fuzzy approach to a belief-based trust computation," in *Lecture Notes on Artificial Intelligence*, vol. 2631, 2003, pp. 73–86.
6. M. Schillo, P. Funk, and M. Rovatsos, "Using Trust for Detecting Deceitful Agents in Artificial Societies," *the Applied Artificial Intelligence Journal, Special Issue on Deception, Fraud and Trust in Agent Societies*, pp. 825–848, 2000.
7. K. S. Barber and J. Park, "Autonomy Affected by Beliefs: Building Information Sharing Networks with Trustworthy Providers," presented at Workshop for Autonomy, Delegation and Control at the 2nd International Conference on Autonomous Agents and Multiagent Systems (AAMAS 2003), Melbourne, Australia, 2003.

Dimensions of Adjustable Autonomy
and Mixed-Initiative Interaction

Jeffrey M. Bradshaw, Paul J. Feltovich, Hyuckchul Jung, Shriniwas Kulkarni,
William Taysom, and Andrzej Uszok

Institute for Human and Machine Cognition (IHMC),
40 S. Alcaniz, Pensacola, FL 32502
{jbradshaw, pjfeltovich, hjung, skulkarni, wtaysom, auszok}@ihmc.us
http://www.ihmc.us

Abstract. Several research groups have grappled with the problem of characterizing and developing practical approaches for implementing adjustable
autonomy and mixed-initiative interaction in deployed systems. However, each
group takes a little different approach and uses variations of the same terminology in a somewhat different fashion. In this chapter, we will describe some
common dimensions in an effort to better understand these important but ill-
characterized topics. We are developing a formalism and implementation of
these concepts as part of the KAoS framework in the context of our research on
policy-governed autonomous systems.

1 Introduction

As computational systems with increasing autonomy interact with humans in more
complex ways, there is a natural concern that they are sufficiently predictable and
controllable as to be acceptable to people [10]. In addition to traditional concerns for
safety and robustness in such systems, there are important social aspects relating to
mutual situation awareness, intent recognition, coordination of joint tasks, and efficiency and naturalness of the interaction that must be attended to [11; 24]. Since
autonomous entities cannot always be trusted to regulate their own behavior appropriately, various approaches have been proposed to allow continuous external adjustment of the bounds of autonomous behavior, assuring their ongoing safety and
effectiveness.

Policies are declarative constraints on system behavior that provide a powerful
means for dynamically regulating the behavior of components without changing code
nor requiring the cooperation of the components being governed (http://www.policy-
workshop.org/). Moreover, they have important analogues as regulatory mechanisms
in animal societies and human cultures [24]. Elsewhere we have pointed out the many
benefits of policy-based approaches, including reusability, efficiency, extensibility,
context-sensitivity, verifiability, support for both simple and sophisticated components, protection from poorly-designed, buggy, or malicious components, and reasoning about component behavior [10].

M. Nickles, M. Rovatsos, and G. Weiss (Eds.): AUTONOMY 2003, LNAI 2969, pp. 17–39, 2004.

In this chapter, we describe how policies can be used to represent and help implement adjustable autonomy and mixed-initiative interaction. Previously, several research groups have grappled with the problem of characterizing and developing practical approaches to address these issues. However, each group takes a little different approach and uses variations of the same terminology in a somewhat different fashion. In this chapter, we will briefly characterize what we see as the most significant dimensions in order to better understand this important but ill-characterized topic.

As foundation to the remainder of the chapter, section 2 describes our view of the major dimensions of adjustable autonomy; and section 3 does likewise for mixed-initiative interaction. In section 4, we briefly describe our efforts to develop a formalism and implementation of these concepts as part of the KAoS framework in the context of our research on policy-governed autonomous systems, and follow this with some concluding observations (section 5).

2   Dimensions of Adjustable Autonomy

In this section, we informally describe our view of the dimensions of adjustable autonomy[1]. Section 2.1 briefly discusses the concept of autonomy itself. In section 2.2, we give a description of the major dimensions under consideration, and in section 2.3, we outline basic concepts relating to adjustable autonomy.

2.1   Autonomy

No description of adjustable autonomy can proceed without at least some discussion of the concept of autonomy itself. The word, which is straightforwardly derived from a combination of Greek terms signifying self-government (*auto-* (self) + *nomos* (law)) has two basic senses in everyday usage[2]. In the first sense, we use the term to denote *self-sufficiency,* the capability of an entity to take care of itself. This sense is present in the French term *autonome* when, for example, it is applied to someone who is success fully living away from home for the first time. The second sense refers to the

---

[1] A formal description of these concepts is currently being developed.

[2] Here we are only concerned with those dimensions that seem to directly relevant to adjustable autonomy as we define it. Some excellent detailed and comprehensive analyses of the concept of autonomy that go beyond what can be treated in this chapter have been collected in [31] and in the current volume. We note that subtle differences in the use of the term *autonomy* sometimes affect the slant or emphasis that different researcher put on various aspects of their respective conceptualizations. Note, for example, Brainov and Hexmoor's emphasis on degree of autonomy as a relative measure of independence between an agent and the physical environment, and within and among social groups [13]. Luck *et al.* [38], unsatisfied with defining autonomy as a wholly relative concept, argue that the self-generation of goals should be the defining characteristic of autonomy, thus allowing it to be regarded in absolute terms that more clearly reflect the priority of the aspect of self-sufficiency.

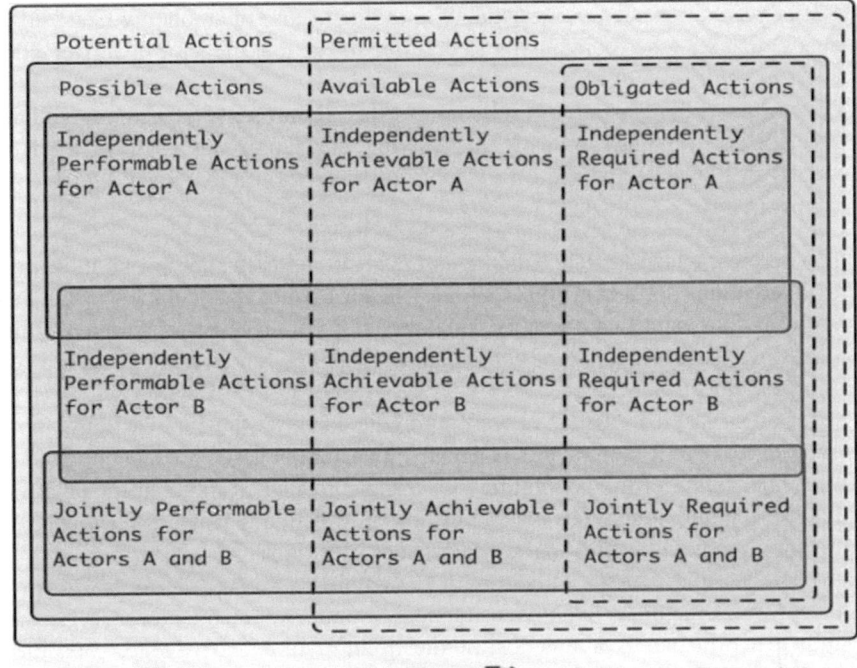

Fig. 1. Dimensions of autonomy

quality of *self-directedness,* or freedom from outside control, as we might say of a portion of a country that has been identified as an "autonomous region."[3]

2.2   Description of the Dimensions

Some important dimensions relating to autonomy can be straightforwardly characterized by reference to figure 1[4]. Note that there are two basic dimensions:

---

[3]   We note that "no man [or machine] is an island"—and in this sense of reliance and relation to others, complete autonomy is a myth.

[4]   Here we emphasize those dimensions that are most pertinent to our discussion of adjustable autonomy; see elsewhere for examples of other possible dimensions (e.g., self-impositions, norms)/ We can make a rough comparison between some of these dimensions and the aspects of autonomy described by Falcone and Castelfranchi [23]. Environmental autonomy can be expressed in terms of the possible actions available to the agent—the more the behavior is wholly deterministic in the presence of a fixed set of environmental inputs, the smaller the range of possible actions available to the agent. The aspect of self-sufficiency in social autonomy relates to the ranges of what can be achieved independently vs. in concert with others; deontic autonomy corresponds to the range of permissions and obligations that govern the agent's choice among actions.

- a *descriptive* dimension corresponding to the first sense of autonomy (self- suffi-ciency) that stretches horizontally to describe the actions an actor in a given con-text is *capable* of performing; and
- a *prescriptive* dimension corresponding to the second sense of autonomy (self-directedness) running vertically to describe the actions an actor in a given context is allowed to perform or which it must perform by virtue of *policy* con-straints in force.

The outermost rectangle, labeled *potential actions,* represents the set of all actions across all situations defined in some ontology under current consideration[5]. Note that there is no requirement that all actions that an actor may take be represented in the ontology; only those which are of consequence for policy representation and reason-ing need be included.

The rectangle labeled *possible actions* represents the set of potential actions whose performance by one or more actors is deemed plausible in a given situation [5; 21][6]. Note that the definition of possibilities is strongly related to the concept of affor-dances [28; 41], in that it relates the features of the situation to classes of actors capa-ble of exploiting these features in the performance of actions[7].

Of these possible actions, only certain ones will be deemed *performable* for a given actor[8] (e.g., Actor A) in a given situation. *Capability,* i.e., the power that makes an action performable, is a function of the *abilities* (e.g., knowledge, capacities, skills) and *conditions* (e.g., ready-to-hand resources) necessary for an actor to suc-cessfully undertake some action in a given context. Certain actions may be *independ-ently performable* by either Actor A or B; other actions can be independently per-formed by either one or the other uniquely[9]. Yet other actions are *jointly performable* by a set of actors.

Along the prescriptive dimension, declarative policies may specify various *permis-sions* and *obligations* [20]. An actor is *free* to the extent that its actions are not limited

---

[5] The term *ontology* is borrowed from the philosophical literature, where it describes a theory of what exists. Such an account would typically include terms and definitions only for the very basic and necessary categories of existence. However, the common usage of ontology in the knowledge representation community is as a vocabulary of representational terms and their definitions at any level of generality. A computational system's "ontology" defines what exists for the program—in other words, what can be represented by it.

[6] The evaluation of possibility admits varying degrees of confidence—for example, one can distinguish mere plausibility of an action from a more studied feasibility. These nuances of possibility are not discussed in this chapter.

[7] As expressed by Norman: "Affordances reflect the possible relationships among actors and objects: they are properties of the world" [43].

[8] For discussion purposes, we use the term *actor* to refer to either a biological entity (e.g., human, animal) or an artificial agent (e.g., software agent, robotic agent).

[9] Note that figure 1 does not show every possible configuration of the dimensions, but rather exemplifies a particular set of relations holding for the actions of a particular set of actors in a given situation. For example, although we show A and B sharing the same set of possible actions, this need not always be the case. Also, note that the range of jointly achievable ac-tions has overlap only with Actor B and not Actor A.

by permissions or obligations. *Authorities* may impose or remove involuntary policy constraints on the actions of actors[10]. Alternatively, actors may voluntarily enter into *agreements* that mutually bind them to some set of policies for the duration of the agreement. The *effectivity* of an individual policy specifies when it is in or out of force.

The set of *permitted actions* is determined by *authorization policies* that specify which actions an actor or set of actors is allowed (*positive authorizations* or *A+* policies) or not allowed (*negative authorizations* or *A* policies) to perform in a given context[11]. The intersection of what is possible and what is permitted delimits the set of *available actions.*

Of those actions that are available to a given actor or set of actors, some subset may be judged to be *independently achievable* in the current context. Some actions, on the other hand, would be judged to be only *jointly achievable.*

Finally, the set of *obligated actions* is determined by *obligation policies* that specify actions that an actor or set of actors is required to perform (*positive obligations* or *O+* policies) or for which such a requirement is waived (*negative obligations* or *O* policies). *Jointly obligated actions* are those that two or more actors are explicitly required to perform.

Figure 2 contrasts the general case to its extremes[12]. *Absolute freedom,* a condition representing the absence of deontic constraints governing an actor's actions, is attained when every potential action is permitted, making any action that is possible available to the actor, and any performable action achievable to it. *Absolute capability,* a condition representing the extreme of self-sufficiency, is attained when an actor is capable or performing any possible action, making any action that is available achievable to it. *Absolute autonomy* combines absolute freedom and absolute capability, meaning that only the impossible is unachievable.

2.3   Adjustable Autonomy

A major challenge in the design of intelligent systems is to ensure that the degree of autonomy is continuously and transparently adjusted in order to meet whatever performance expectations have been imposed by the system designer and the humans and agents with which the system interacts. We note that is not the case that "more" autonomy is always better[13]. as with a child left unsupervised in city streets during rush hour, an unsophisticated actor insufficiently   monitored   and   recklessly   endowed   with

---

[10] Authority relationships may be, at the one extreme, static and fixed in advance and, at the other, determined by negotiation and persuasion as the course of action unfolds.

[11] We note that some permissions (e.g., network bandwidth reservations) involve allocation of finite and/or consumable resources, whereas others do not (e.g., access control permissions). We note that obligations typically require allocation of finite abilities and resources; when obligations are no longer in effect, these abilities and resources may become free for other purposes.

[12] To simplify the diagram, the dimension of obligation is omitted. Note that absolute capability implies an absence of obligations, since any obligations in effect would typically reduce capability in some measure.

[13] In fact, the multidimensional nature of autonomy argues against even the effort of mapping the concept of "more" and "less" to a single continuum.

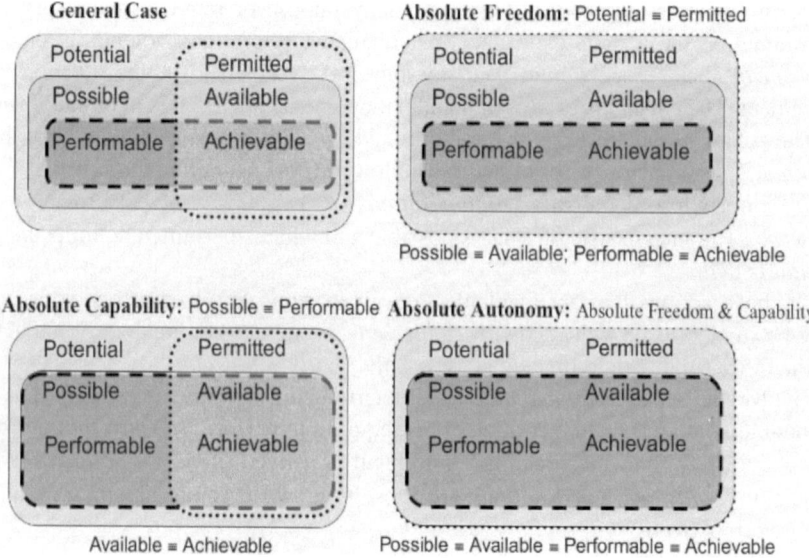

Fig. 2. The general case and its extremes

unbounded freedom may pose a danger both to others and itself. On the other hand, a capable actor shackled with too many constraints will never realize its full potential.

Thus, a primary purpose of adjustable autonomy is to maintain the system being governed at a sweet spot between convenience (i.e., being able to delegate every bit of an actor's work to the system) and comfort (i.e., the desire to not delegate to the system what it can't be trusted to perform adequately)[14]. Assurance that agents will operate safely within well-defined bounds and that they will respond in a timely manner to external control is required for them to be acceptable to people in the performance of nontrivial tasks. People need to feel that agents will handle unexpected circumstances requiring adjustment of their current state of autonomy flexibly and reliably. To the extent adjustable autonomy can be successfully implemented, agents are kept, to the degree possible, from exceeding the limits on autonomy currently in effect, while being otherwise free to act in complete autonomy within those limits. Thus, the coupling of autonomy with adequate autonomy adjustment mechanisms gives the agent maximum opportunity for local adaptation to unforeseen problems and opportunities while assuring humans that agent behavior will be kept within desired bounds.

All this, of course, only complicates the agent designer's task, a fact that has lent urgency and impetus to efforts to develop broad theories and general-purpose frameworks for adjustable autonomy that can be reused across as many agents, domains, and applications as possible. To the degree that adjustable autonomy services can be competently implemented and packaged for convenient use within popular development platforms, agent designers can focus their attention more completely on the

---

[14] We note that reluctance to delegate can also be due to other reasons. For example, some kinds of work may be enjoyable to people—such as skilled drivers who may prefer a manual to an automatic transmission.

unique capabilities of the individual agents they are developing while relying on the extant services to assist with addressing cross-cutting concerns about human-agent interaction.

We now consider some of the dimensions on which autonomy can be adjusted.

Adjusting Permissions. A first case to consider is that of adjusting permissions. Reducing permissions may be useful when it is concluded, for example, that an agent is habitually attempting actions that that it is not capable of successfully perform-ing—as when a robot continues to rely on a sensor that has been determined to be faulty (figure 3). It may also be desirable to reduce permissions when agent delibera-tion about (or execution of) certain actions might incur unacceptable costs or delays.

If, on the other hand, an agent is known to be capable of successfully performing ac-tions that go beyond what it is currently permitted to do, its permissions could be in-creased accordingly (figure 4). For example, a flying robot whose duties had previously been confined to patrolling the space station corridors for atmospheric anomalies could be given additional permissions allowing it to employ its previously idle active barcode sensing facilities to take equipment inventories while it is roaming [27] [11].

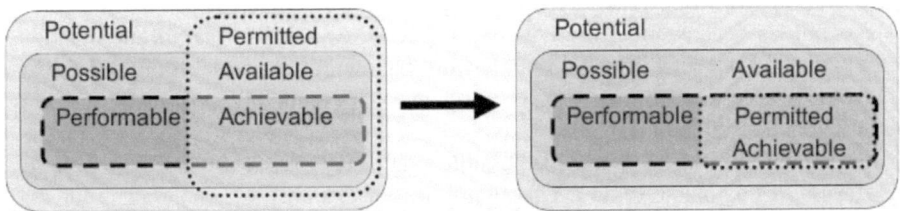

Fig. 3. Reducing permissions to prevent outstripping capabilities

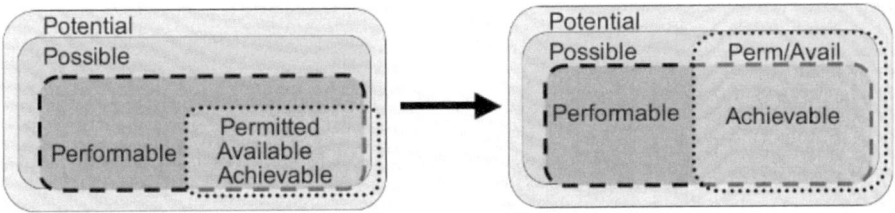

Fig. 4. Increasing permissions to take full advantage of capabilities

Adjusting Obligations. On the one hand, "underobligated" agents can have their obli-gations increased—up to the limit of what is achievable—through additional task as-signments. For example, in performing joint action with people, they may be obliged to report their status frequently or to receive explicit permission from a human before proceeding to take some action. On the other hand, an agent should not be required to perform any action that outstrips its permissions, capabilities, or possibilities[15]. An

---

[15] In some cases, rather than rejecting commitments to unachievable obligations outright, it may be preferable to increase permissions, capabilities, or possibilities (if possible), thus transforming an unachievable obligation into one that is achievable.

"overcommitted" agent can sometimes have its autonomy adjusted to manageable levels through reducing its current set of obligations (figure 5). This can be done through delegation, facilitation, or renegotiation of obligation deadlines. In some circumstances, the agent may need to renege on its obligations in order to accomplish higher priority tasks.

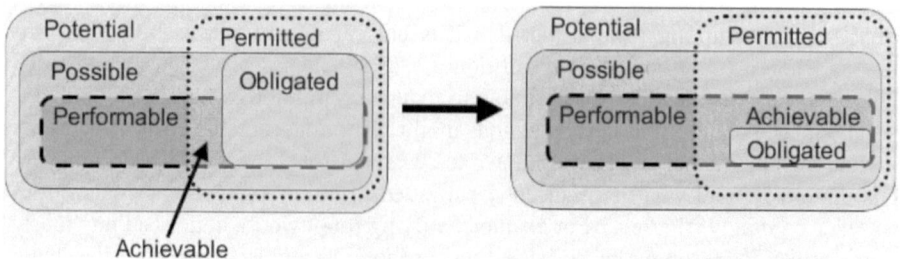

Fig. 5. Decreasing obligations to match capabilities

Adjusting Possibilities. A highly-capable agent may sometimes be performing below its capabilities because of restrictions on resources available in its current situation. For example, a physical limitation on network bandwidth available through the nearest wireless access point may restrict an agent from communicating at the rate it is permitted and capable of doing[16].

In some circumstances, it may be possible to adjust autonomy by increasing the set of possibilities available to an agent (figure 6). For example, a mobile agent may be able to make what were previously impossible faster communication rates possible by moving to a new host in a different location. Alternatively, a human could replace an inferior access point with a faster one.

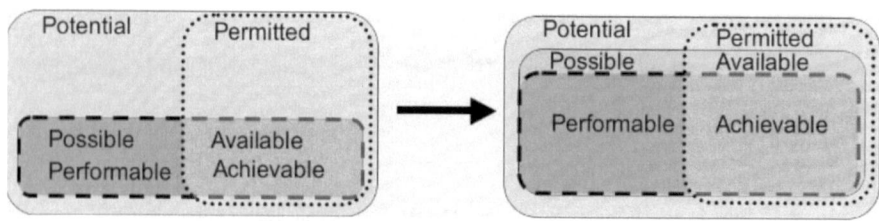

Fig. 6. Increasing possibilities to leverage unused capabilities

Sometimes reducing the set of possible actions provides a powerful means of enforcing restrictions on an agent's actions. For example, an agent that "misbehaved" on the network could be sanctioned and constrained from some possibilities for action by moving it to a host with restricted network access.

---

[16] Besides constrained resources, other features of the situation may also limit the possibility of certain actions, e.g., the darkness of nighttime may prevent me from reading.

Adjusting Capabilities. The autonomy of an agent can be augmented either by increasing its own independent capabilities or by extending its joint capabilities through access to other actors to which tasks may be delegated. An agent's capabilities can also be affected by changing current conditions (e.g., adding or reducing needed resources).

An adjustable autonomy service aimed at increasing an agent's capabilities (as in figure 7) could assist in discovering agents with which an action that could not be independently achieved could be jointly achieved. Or if the agent was hitting the ceiling on some computational resource (e.g., bandwidth, memory), resource access policies could be adjusted to allow the agent to leverage the additional assets required to perform some action. Finally, the service could assist the agent by facilitating the deferral, delegation, renegotiation, or reneging on obligations in order to free up previously committed resources (as previously mentioned in the context of adjusting obligations).

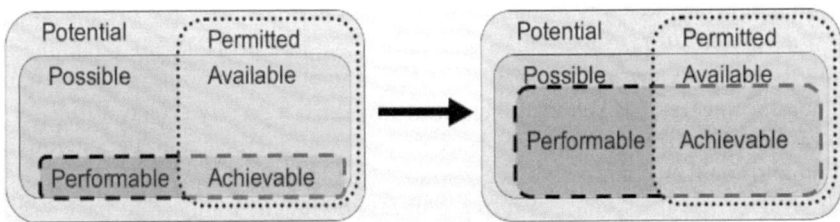

Fig. 7. Increasing an agent's capabilities to take better advantage of the set of actions available to it

Having described the principal dimensions of autonomy and the kinds of adjustments that can be made, we now analyze the concept of mixed-initiative interaction from that perspective[17].

3   Mixed-Initiative Interaction

It is enlightening to look at the place of mixed-initiative interaction in the onward march of automation generally[18]. The concept of automation—which began with the

---

[17] In this chapter we do not address the question of how to evaluate the quality of autonomy adjustment and mixed-initiative interaction. See [19; 29; 33] for a sampling of perspectives on this issue. We note that there are many criteria that can play into such an assessment, including *survivability* (ability to maintain effectiveness in the face of unforeseen software or hardware failures), *safety* (ability to prevent certain classes of dangerous actions or situations), *predictability* (assessed correlation between human judgment of predicted vs. actual behavior), *controllability*(immediacy with which an authorized human can prevent, stop, enable, or initiate agent actions), *effectiveness* (assessed correlation between human judgment of desired vs. actual behavior), *adaptability* (ability to respond to changes in context), and *task performance* (overall economic and cognitive costs and benefits expressed as utility).

[18] See [6; 16; 30; 32; 44] for insightful perspectives on these and related topics.

straightforward objective of replacing whenever feasible any task currently performed by a human with a machine that could do the same task better, faster, or cheaper— became one of the first issues to attract the notice of early human factors researchers. These researchers attempted to systematically characterize the general strengths and weaknesses of humans and machines [26]. The resulting discipline of *function alloca- tion* aimed to provide a rational means of determining which system-level functions should be carried out by humans and which by machines.

Over time it became plain to researchers that things were not as simple as they first appeared. For example, many functions in complex systems are shared by humans and machines; hence the need to consider synergies and conflicts among the various performers of joint actions. Also, the suitability of a particular human or machine to take on a particular task may vary by time and over different situations; hence the need for methods of function allocation that are dynamic and adaptive [30]. More- over, it has become clear that function allocation is not a simple process of transfer- ring responsibilities from one component to another [8]. Automated assistance of whatever kind does not simply enhance our ability to perform the task: it changes the nature of the task itself [16; 42]. Those who have had a five-year-old child help them by doing the dishes know this to be true—from the point of view of an adult, such "help" does not necessarily diminish the effort involved, it merely effects a transfor- mation of the work from the physical action of washing the dishes to the cognitive task of monitoring the progress (and regress) of the child.

With all these complications, even the pioneers of function allocation have been constrained to admit only limited success in implementing this concept in practice [48]. And so it is that any researcher proposing to design and develop systems mani- festing mixed-initiative behavior must approach the task with humility—since such systems will not only manifest all the complexities heretofore mentioned, but also attempt the ambitious aim to delegate the task of dynamic and adaptive function allo- cation to the automated elements themselves.

The concept of *mixed-initiative interaction,* involving some combination of hu- mans and/or agents has been succinctly described by Allen as follows:

> "Mixed-initiative refers to a flexible interaction strategy, where each agent can contribute to the task what it does best. Furthermore, in the most gen- eral cases, the agents' roles are not determined in advance, but opportunisti- cally negotiated between them as the problem is being solved. At any one time, one agent might have the initiative—controlling the interaction— while the other works to assist it, contributing to the interaction as required. At other times, the roles are reversed, and at other times again the agents might be working independently, assisting each other only when specifi- cally asked. The agents dynamically adapt their interaction style to best ad- dress the problem at hand" [1, p. 14].

The following subsections define the concept of mixed-initiative interaction in more detail. We first describe the essential characteristics of joint activity and joint action (3.1). Then we show how the concepts of joint activity and adjustable auton- omy come together to enable mixed-initiative interaction (3.2). Finally, we show with an example how the kinds of policies mentioned in the section on adjustable autonomy come into play within mixed-initiative interaction (3.3).

3.1 Joint Activity and Joint Action[19]

An understanding of joint activity must be at the heart of any formulation of mixed-initiative interaction. Our concept of joint activity relies on the work of Herbert Clark [17], who borrows the following definition from Levinson [37, p. 69]:

> "I take the notion of an activity type to refer to a fuzzy category whose focal members are goal-defined, socially constituted, bounded, events with constraints on participants, setting, and so on, but above all on the kinds of allowable contributions. Paradigm examples would be teaching, a job interview, a jural interrogation, a football game, a task in a workshop, a dinner party, and so on[20]".

Although there are many variations to joint activity, there are several core elements that seem to be common to them all:

- *Intention to produce a genuine, multi-agent product:* The overall joint activity should be aimed at producing something that is a genuine joint project, achieved differently (e.g., faster, better) than any one party, or two parties working alone could do.
- *Interdependency:* It follows that there must be interdependencies among the parties' actions; if the parties' actions have no interdependency, then they are not involved in joint activity (although they may be involved in something that might be thought of as "parallel" activity).
- *Coordination:* There must be coordination with regard to elements of interdependency and mixed-initiative interaction.
- *Coordination devices:* There must be devices, mutually understood by the parties, that guide coordination. Some of these devices are discussed in more detail below.
- *Common ground:* There must be shared knowledge and interpretation; although the parties' knowledge and interpretations need not be exactly alike, they should have enough commonality to enable the joint activity to move positively in the direction of its goal.
- *Repair:* When there is evidence of loss of common ground—loss of sufficient common understanding to enable joint activity to proceed—there are mechanisms engaged that aim to restore it, to increase common understanding.

---

[19] See Klein and Feltovich [36] for a more complete discussion of the implications of Clark's work for coordination of joint action. We rely heavily on their analysis in this section.

[20] The relationship between language and joint activity is described by Clark as follows: "When people use language, it is generally as part of a joint activity…. The argument is that joint activities are the basic category, and what are called discourses are simply joint activities in which conventional language plays a prominent role. If we take language use to include such communicative acts as eye gaze, iconic gestures, pointing, smiles, and head nods—and we must—then all joint activities rely on language use. Chess may appear to be nonlinguistic, but every chess move is really a communicative act, and every chess game a discourse" [17, p. 58].

Joint activity is a *process*, extended in space and time. There is a time when the parties enter into joint activity and a time when it has ended. These are not "objective" points of time that would necessarily be agreed on by any "observer-in-the-world," but most importantly are interpretations arrived at by the parties involved [17, p. 84]. In some circumstances the entry and exit points may be very clear such as when two people play a classical duet; the same would probably not be said of two musicians doing jazz improvisation.

The overall structure of joint activity is one of embedded sets of actions, some of which may also be joint and some of which may be accomplished more or less individually. All these actions likewise have entry and exit points, although as we have mentioned earlier, these points are not epistemologically "objective." Synchronizing entry and exit points of the many embedded phases involved in complex joint activity is a major challenge to coordination[21].

So, how does coordination happen? Given a structure of embedded actions—some of which may be joint actions—as well as overall joint activity, this appears to be two questions. How does coordination take place in the more local joint acts that make up an overall joint activity, and how does coordination take place at the more macro level of the overall joint activity itself. With regard to the first, the "coordination devices" [17, pp. 64-65] play a major role:

- *Agreement:* Coordinating parties are sometimes simply able to communicate their intentions and work out elements of coordination. This category includes diverse forms of signaling that have shared meaning for the participants, including language, signs, gestures, and the like.
- *Convention:* Often, prescriptions of various types apply to how parties interact. These can range from rules and regulations, to less formal codes of appropriate conduct such as norms of practice in a particular professional community, or established practices in a workplace. Coordination by convention depends on structures outside of a particular episode of joint activity.
- *Precedent:* Coordination by precedent is like coordination by convention, except that it applies to norms and expectations developed within an episode of the ongoing process of a joint activity (or across repeated episodes of such activity if the participants are a long-standing team that repeats conduct of some procedure): "That's the way we did it last time."
- *Salience:* Salience is perhaps the coordination device that is most difficult to understand and describe[22]. It has to do with how the ongoing work of the joint activity arranges the workspace so that next move becomes highlighted or otherwise apparent among the many moves that could conceivably be chosen. For example, in a surgery, exposure of a certain element of anatomy, in the course of pursuing a particular surgical goal, can make it clear to all parties involved what to do next. Coordination by salience is a sophisticated kind of coordination pro-

---

[21] Clark, in fact, defines joint actions in terms of coordination: "A joint action is one that is carried out by an ensemble of people acting in coordination with each other" [17, p. 3].

[22] Part of the complication is the relationships among these mechanisms. For example, conventions and precedents may be essential in salience "assignment."

duced by the very conduct of the joint activity itself. It requires little or no overt communication and is likely the predominant mode of coordination among long-standing, highly practiced teams.

Coordination across the entire course of an extended joint activity is in some ways similar and in some ways different from the more local coordination involved in individual joint actions (and the subactions of which they are composed). For instance, there may be "scripted" conventions for conducting an entire procedure just as there are for conducting more local components of it. That is, joint activities can be more or less open in execution, depending on the presence of applicable norms, policies and the like. In addition to regulatory coordination mechanisms, there are other kinds of macro guides that serve to coordinate across the course of an entire joint activity. Examples are plans and policies for some activity worked out in advance by the participants, or the prior extensive outline worked out by the authors involved in writing a joint academic manuscript. It has been argued that some of the reasons for "standardizing" procedures are to aid coordination and to prevent untoward interactions so that some earlier move does not clobber some necessary later move (e.g., [47]). Of course, any of these overarching coordination devices usually needs to be revisited, and very likely adjusted, as the actual work unfolds.

3.2   Toward a Better Understanding of Mixed-Initiative Interaction

With this description of joint activity, coupled with the discussion of the dimensions of adjustable autonomy in section 2, we are prepared to better understand mixed initiative interaction. To make this discussion more concrete, we will illustrate with reference to different sorts of vacuum cleaners that embody a spectrum of autonomy.

Representing the "most manual" left end of the spectrum, a plain old vacuum is directly operated as an extension of the woman's arm (figure 8)[23]. Apart from the ongoing sweeping and sucking action of the motor, every action is taken at the initiative and direction of the human.

On the "most autonomous" right end of the spectrum, we can imagine a "fully autonomous" vacuum that not only does the actual vacuuming on its own, but also decides when it is time to vacuum and turns itself on, decides when it is time to stop and turns itself off, and even retreats to a closet where it plugs itself into the wall and recharges in anticipation of its next sortie[24].

We see the process of taking initiative as a particular manifestation of autonomy. In everyday use, the term *initiative* refers to the right, power, or ability to select, begin, execute, or terminate some activity. We speak of doing something "on one's own

---

[23] Thanks to Ron Francis for permission to use this reproduction of his oil painting entitled "Vacuum Cleaner." Of this painting, Francis writes, "It is easy for someone to get a little lost in this world, and not be noticed. This person is harmlessly vacuuming her back yard. My mother once had a dressing gown just like hers."

[24] We could of course take this to a further extreme where the vacuum not only is responsible to recharge itself, but also takes responsibility for paying its share of the electric bill, hires itself out in its spare time to earn money for the bill, repairs itself, and on ad infinitum to the further reaches of unlimited autonomy.

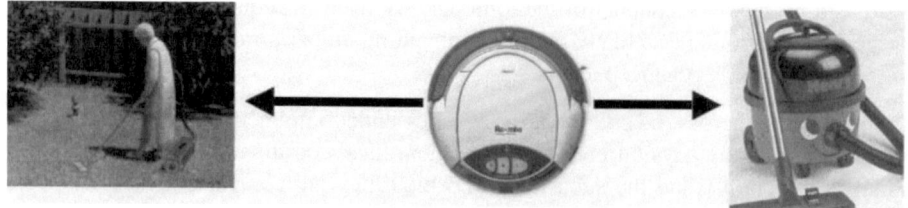

Fig. 8. Three vacuum cleaners illustrating a spectrum of autonomy

initiative" as referring to a situation where the individual has relied on his or her own discretion to act independently of outside influence or control.

Mixed-initiative interaction of necessity requires that both parties be somehow involved in directing at least some shared aspects of the joint activity. It is obvious that neither of the two extremes represented by the manual and the totally autonomous vacuum qualify as "mixed-initiative interaction"—in the one case the person is taking all the initiative, and in the other the person need take none.

Somewhere between these extremes of human and machine responsibility for the interaction is the original basic model of iRobot's Roomba. Its design is fixed such that the user must always be the one to take responsibility to switch the vacuum on, tell it how long to run, and put it away and recharge it. Once it is commissioned, the Roomba is always fully responsible for figuring out where to go and what to do until its battery runs low, it completes its work cycle, or the user manually stops it and carts it away.

Although the Roomba is arguably semi-autonomous, it is our view that its interaction with the user could not be classed as mixed-initiative. True it is that each party has a reasonable degree of autonomy from the other. It is also obvious that the action of vacuuming could have only been achieved through the participation of both parties, each party having contributed something unique to the task. What is missing, however, is the chance for either party to determine opportunistically who should perform which tasks and when they should be done. The tasks that the human does are tasks that only the human can do, the actions the vacuum takes cannot be performed by the human[25]. and these respective roles and responsibilities are fixed in advance rather than negotiated at runtime.

---

[25] One could argue that in some sense the human can clumsily take the initiative with respect to some of the actions normally performed by the vacuum (e.g., in determining where the vacuum should move by lifting it up and carrying it elsewhere or, in more advanced models, interrupting the normal pattern of movement and manually directing the vacuum's movement), but this hardly qualifies as mixed-initiative interaction in the sense we are describing it.

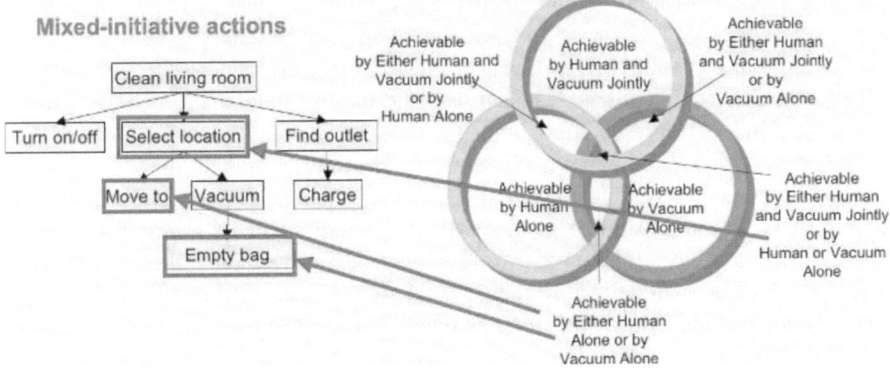

Fig. 9. Opportunities for mixed-initiative interaction with a vacuum cleaner

In short, we can say that necessary conditions for mixed initiative interaction are:

- Engagement in a joint activity of the kind described by Clark;
- At least some aspects (e.g., proposing, deciding, initiating, executing, monitoring, terminating) of some actions supporting the joint activity can be achieved individually by two or more of the participants;
- No set of policies uniquely fixes who must perform all aspects of these actions, or when they must be performed; thus any one of the participants capable of doing so is permitted to take initiative as circumstances unfold[26].

3.3   The Role of Policy in Mixed-Initiative Interaction

To understand the role that policy could play in mixed-initiative interaction, we extend the previous example to include a hypothetical mixed-initiative vacuum cleaner (figure 9). Let's assume at the outset that there is an overall activity "Clean living room," which is something that can only be achieved by the human and vacuum working jointly.

Every action that is part of the joint activity need not be a joint action. For example, the action of "Turn on/off," in our example, is something that is achievable by the human alone; and the "vacuum" action is something that is achievable by the vacuum alone.

There are three nodes in the tree of actions are candidates for mixed-initiative interaction. "Select location" is something that could either be done by both parties jointly, or by either one alone; we assume that "move to" and "empty bag" are actions that could be achieved by either the human alone or the vacuum alone but not both working together.

Recalling Levinson's definition of joint activity, we can see the role of policy in representing "focal members [who] are goal-defined, socially constituted, bounded, events with constraints on participants, setting, and so on, but above all on the kinds of allowable contributions." As the participants begin to engage in joint activity, they

---

[26] And in fact one of the participants *must* take initiative for the action to proceed.

bring with them a history of agreements, conventions, precedents, and salience-related expectations that will serve to coordinate their joint actions.

Although, thanks to lifelong experience, most humans come pre-packaged with a host of ready-made conventions, expectations, and the like that cover everyday situations, some kind of representation of these sorts of conventions and expectations needs to be explicitly "added in" to artificial systems in order to help them work well with people. This is a different concept of automation than the one that has previously been the basis for the design of generations of "strong, silent" systems that permit only two modes: fully automatic and fully manual [16]. In practice the use of such systems often leads to situations of human "underload," with the human having very little to do when things are going along as planned, followed by situations of human "overload," when extreme demands may be placed on the human in the case of agent failure.

In contrast, within systems capable of mixed-initiative interaction, policies and other relevant information needed for coordination can be explicitly represented within the artificial system in some form of "agreement" intended to govern selected aspects of joint activity among the parties. While this should not be mistaken as a requirement for full-blown agent anthropomorphism, it is clear that at least some rough analogues to human coordination mechanisms will be required in order to assure effective teamwork among people and agents.

With reference to our hypothetical mixed-initiative vacuum, we now describe examples of policies from five categories:

- Policies affecting initiative,
- Policies affecting delegation,
- Notification policies,
- Supervisory policies, and
- Policies constraining human actions.

These categories and examples are intended to be illustrative, not comprehensive.

Policies Affecting Initiative. Note that the heading of this subsection refers to "policies affecting initiative" and *not* "policies determining initiative." Of necessity, the decision about whether and when to take initiative relative to a particular joint action is determined by the agent consulting its own reasoning processes, and not by the policy-related components. Policies, however, can affect the process of initiative-taking in a number of ways, or even terminate it altogether. For example, some event may trigger an obligation policy that uniquely requires one or the other of the parties to initiate some aspect of the joint action, thus foreclosing future opportunities for mixed-initiative interaction. For example, consider the following positive obligation policy:

```
O+: If the vacuum finds a baby on the floor, then the
human must immediately tell the vacuum where to move
to next.
```

Once the obligation is in effect, the decision has been made about where the vacuum moves to next, and the possibility of further mixed-initiative interaction on that decision is gone. If, however, the human happens to possess a new advanced model of the vacuum cleaner that is baby-safe, this requirement may be waived by a negative obligation policy of the following sort:

> O-: If the vacuum in question is a super-duper model, it is not obliged to have the human tell it where to move to next when a baby is found on the floor.

Other obligations may require the termination of some joint action and the notification of interested parties by one or the other of the actors [18]:

> O+: If an actor determines that the joint action has been achieved, has become unachievable, or has become irrelevant, it is required to terminate its efforts with regard to the joint action and to notify the other parties.

There may also be policies of other sorts relating to the taking of initiative. For example, policies might be specified that not only affect what the parties can do and when they can do it, but also constrain which parties of a joint action can decide what they can do and when they can do it, or how they need to negotiate about who's going to decide these things.

Policies Affecting Delegation. Besides the examples of obligation policies mentioned above, authorization policies may be needed in some scenarios—for example, to allow actors to take the initiative in delegating to other actors in teamwork scenarios:

> A+: The vacuum is permitted to delegate vacuuming an area to any other vacuum of the same make and model (or better) in the room.

Or:

> A-: The vacuum is forbidden from delegating vacuuming tasks to the toaster.

Note that delegation can be handled as just another kind of action that may or may not be authorized or obliged. Actions that add, modify, or delete policies can be constrained in a similar way without resorting to special "meta-level" mechanisms.

Notification Policies. The fact that the actor who will take the initiative on various aspects of joint actions is not determined in advance means that there must be adequate means for each party to determine the other parties' state and intentions and coordinate its own actions accordingly[27]. Obligation policies can be used, for example, to make sure that the vacuum reports relevant aspects of its state and intentions at appropriate times:

> O+: The vacuum must notify the human about the state of its battery, its position, and the estimated time remaining to finish the room every five minutes.

---

[27] Given some requirement for notification, there is also a role for context-sensitive policies and personal preferences regarding the salience, latency, and mode of notification. How one might factor in such considerations is discussed in [14; 46].

Obligation policies can also be used to avoid conflicts or duplication of effort:

```
O+: If the human tries to empty the bag while the vac-
uum is already trying to empty the bag, then signal the
human.
```

Supervisory Policies. Humans will not expect even the most competent vacuum to be trusted to do the right thing in every situation. This fact motivates the requirement for policies that guard against certain actions taking place:

```
A-: The vacuum is not permitted to operate after mid-
night.
```

Similar considerations motivate obligations that require the vacuum to gain approval before proceeding with certain actions:

```
O+: The vacuum must obtain permission from the operator
before entering a new room.
```

An appropriate negative authorization policy tied to this policy could prevent the vacuum from performing this action (or perhaps any other action) until operator approval had been obtained. Note that this policy applies whether the "operator" is human or automated.

Policies allowing the vacuum to take initiative in executing fallback behavior when the human supervisor is not available may also be important [15]:

```
A+: The vacuum is permitted to enter a new room (just
this once) if permission from the human has been re-
quested more than ten minutes ago and the human has not
responded.
```

Policies Constraining Human Actions. Of course, there may also be situations where trust of the human operator is limited[28]. This fact motivates the requirement to be able to define policies that constrain human actions:

```
O+: The vacuum must prevent the human from deliberately
crashing it into an obstacle when its movements are un-
der manual control.
A-: Unauthorized humans are forbidden from telling the
vacuum where to move.
```

4   Work in Progress

Several groups have grappled with the problem of characterizing and developing practical approaches for implementing adjustable autonomy in deployed systems (e.g., [3; 4; 11; 22; 23; 25; 39; 40; 45]). Each one takes a little different approach and uses variations of the same terminology in a somewhat different fashion. For example, in some approaches agents explicitly reason about whether and when to transfer decision-making control to a human or other entity whereas in the work of others it is

---

[28] We note that if the vacuum is preventing the human from operating it dangerously, it is actually the authority of the (presumably human) administrator who defined the policy that is preventing the unsafe operator actions, not the vacuum itself.

presumed that the human is the one that provides guidance to the agent on such matters. Some approaches compute strategies for adjustable autonomy offline whereas others allow dynamic interactive adjustment at runtime. Some approaches integrate reasoning about adjustable autonomy with the agent's own planning mechanisms while others deliberately create redundant mechanisms independent of the planner.

In research funded by the Office of Naval Research, DARPA, the Army Research Labs, and NASA, we are currently conducting research to develop and evaluate formalisms and mechanisms for adjustable autonomy and policies that will facilitate mixed-initiative interaction in conjunction with a testbed that integrates the various capabilities of TRIPS, Brahms, and KAoS[29] [9].

KAoS policy services already enable the specification, management, conflict resolution, and enforcement of semantically-rich policies defined in OWL [52][30]. On this foundation, we are building Kaa (KAoS adjustable autonomy)[31] a component that permits KAoS to perform self-adjustments of autonomy consistent with policy.

To the extent circumstances allow Kaa to adjust agent autonomy with reasonable dynamism (ideally allowing handoffs of control among team members to occur anytime) and with a sufficiently fine-grained range of levels, teamwork mechanisms can flexibly renegotiate roles and tasks among humans and agents as needed when new opportunities arise or when breakdowns occur. Such adjustments can also be anticipatory when agents are capable of predicting the relevant events [7; 23].

5   Concluding Observations

Whereas typical approaches to adjustment of agent autonomy are concerned with generating plans for what an agent should *do*, KAoS is one of the few that aim to

---

[29] KAoS a collection of componentized policy and domain management services compatible with several popular agent frameworks, including Nomads [49], the DARPA CoABS Grid [35], the DARPA ALP /Ultra* Log Cougaar framework (http://www. cougaar.net), CORBA (http://www.omg.org), Voyager (http://www.recursionsw.com/osi.asp), Brahms (www. agentisolutions.com), TRIPS [2; 9], and SFX (http://crasar.eng.usf.edu/research/publications. htm). While initially oriented to the dynamic and complex requirements of software agent applications, KAoS services are also being adapted to general-purpose grid computing (http://www.gridforum.org) and Web Services (http://www.w3.org/2002/ws/) environments as well [34; 53]. KAoS has been deployed in a wide variety of applications, from coalition warfare [12; 50] and agile sensor feeds [51], to process monitoring and notification [14], to robustness and survivability for distributed systems [http://www.ultralog.net], to semantic web services composition [53], to human-agent teamwork in space applications [11], to cognitive prostheses for augmented cognition [10].

[30] Where expression of a policy require going beyond description logic, judicious extensions to the semantics are possible within KAoS (e.g., through the use of semantic web rule languages).

[31] In Rudyar Kipling's Jungle Book, the human boy Mowgli was educated in the ways and secrets of the jungle by Kaa the python. His hypnotic words and stare charmed the malicious monkey tribe that had captured the boy, and Kaa's encircling coils at last "bounded" their actions and put an end to their misbehavior. (A somewhat different Kaa character and story was later portrayed in the Disney movie.)

specify how agent behavior should be *constrained*. Regarding the usefulness of this perspective, Sheridan observes:

> "In democracies specification of 'shoulds' is frowned upon as an abridge-ment of freedom, and bills of basic rights such as that of the United States clearly state that 'there shall be no law against...', in other words declara-tions of unconstrained behavior. In such safety-sensitive industries as avia-tion and nuclear power, regulators are careful to make very specific con-straint specifications but then assert that those being regulated are free to comply in any manner they choose.
>
> Vicente and Pejtersen assert that constraint-based analysis accommo-dates much better to variability in human behavior and environmental cir-cumstance. They make the point that navigating with a map is much more robust to disturbance and confusion over detail than navigating with a se-quence of directions" [48, pp. 212-213].

References

1. Allen, J. F. (1999). Mixed-initiative interaction. *IEEE Intelligent Systems*, September-October, 14–16.
2. Allen, J. F., Byron, D. K., Dzikovska, M., Ferguson, G., Galescu, L., & Stent, A. (2001). Towards conversational human-computer interaction. *AI Magazine*, 22(4), 27–35.
3. Allen, J. F., & Ferguson, G. (2002). Human-machine collaborative planning. *Proceedings of the NASA Planning and Scheduling Workshop*. Houston, TX,
4. Barber, K. S., Gamba, M., & Martin, C. E. (2002). Representing and analyzing adaptive decision-making frameworks. In H. Hexmoor, C. Castelfranchi, & R. Falcone (Ed.), *Agent Autonomy*. (pp. 23–42). Dordrecht, The Netherlands: Kluwer.
5. Barwise, J., & Perry, J. (1983). *Situations and Attitudes*. Cambridge, MA: MIT Press.
6. Billings, C. E. (1996). *Aviation automation: The Search for a Human-Centered Approach*. Hillsdale, NJ: Lawrence Erlbaum Associates.
7. Boella, G. (2002). Obligations and cooperation: Two sides of social rationality. In H. Hexmoor, C. Castelfranchi, & R. Falcone (Ed.), *Agent Autonomy*. (pp. 57–78). Dordrecht, The Netherlands: Kluwer.
8. Boy, G. (1998). *Cognitive Function Analysis*. Stamford, CT: Ablex Publishing.
9. Bradshaw, J. M., Acquisti, A., Allen, J., Breedy, M. R., Bunch, L., Chambers, N., Fel-tovich, P., Galescu, L., Goodrich, M. A., Jeffers, R., Johnson, M., Jung, H., Lott, J., Olsen Jr., D. R., Sierhuis, M., Suri, N., Taysom, W., Tonti, G., & Uszok, A. (2004). Teamwork-centered autonomy for extended human-agent interaction in space applications. *AAAI 2004 Spring Symposium*. Stanford University, CA, AAAI Press,
10. Bradshaw, J. M., Beautement, P., Raj, A., Johnson, M., Kulkarni, S., & Suri, N. (2003). Making agents acceptable to people. In N. Zhong & J. Liu (Ed.), *Intelligent Technologies for Information Analysis: Advances in Agents, Data Mining, and Statistical Learning*. (pp. in press). Berlin: Springer Verlag.
11. Bradshaw, J. M., Sierhuis, M., Acquisti, A., Feltovich, P., Hoffman, R., Jeffers, R., Prescott, D., Suri, N., Uszok, A., & Van Hoof, R. (2003). Adjustable autonomy and hu-man-agent teamwork in practice: An interim report on space applications. In H. Hexmoor, R. Falcone, & C. Castelfranchi (Ed.), *Agent Autonomy*. (pp. 243–280). Kluwer.

12. Bradshaw, J. M., Uszok, A., Jeffers, R., Suri, N., Hayes, P., Burstein, M. H., Acquisti, A., Benyo, B., Breedy, M. R., Carvalho, M., Diller, D., Johnson, M., Kulkarni, S., Lott, J., Sierhuis, M., & Van Hoof, R. (2003). Representation and reasoning for DAML-based policy and domain services in KAoS and Nomads. *Proceedings of the Autonomous Agents and Multi-Agent Systems Conference (AAMAS 2003)*. Melbourne, Australia, New York, NY: ACM Press,

13. Brainov, S., & Hexmoor, H. (2002). Quantifying autonomy. In H. Hexmoor, C. Castelfranchi, & R. Falcone (Ed.), *Agent Autonomy*. (pp. 43–56). Dordrecht, The Netherlands: Kluwer.

14. Bunch, L., Breedy, M. R., & Bradshaw, J. M. (2004). Software agents for process monitoring and notification. *Proceedings of AIMS 04*.

15. Chalupsky, H., Gil, Y., Knoblock, C. A., Lerman, K., Oh, J., Pynadath, D. V., Russ, T. A., & Tambe, M. (2002). Electric Elves: Agent technology for supporting human organizations. *AI Magazine, 2*, Summer, 11–24.

16. Christofferson, K., & Woods, D. D. (2002). How to make automated systems team players. In E. Salas (Ed.), *Advances in Human Performance and Cognitive Engineering Research, Vol. 2*. JAI Press, Elsevier.

17. Clark, H. H. (1996). *Using Language*. Cambridge, UK: Cambridge University Press.

18. Cohen, P. R., & Levesque, H. J. (1991). *Teamwork*. Technote 504. Menlo Park, CA: SRI International, March.

19. Cohen, R., & Fleming, M. (2002). Adjusting the autonomy in mixed-initiative systems by reasoning about interaction. In H. Hexmoor, C. Castelfranchi, & R. Falcone (Ed.), *Agent Autonomy*. (pp. 105–122). Dordrecht, The Netherlands: Kluwer.

20. Damianou, N., Dulay, N., Lupu, E. C., & Sloman, M. S. (2000). *Ponder: A Language for Specifying Security and Management Policies for Distributed Systems, Version 2.3*. Imperial College of Science, Technology and Medicine, Department of Computing, 20 October 2000.

21. Devlin, K. (1991). *Logic and Information*. Cambridge, England: Cambridge University Press.

22. Dorais, G., Bonasso, R. P., Kortenkamp, D., Pell, B., & Schrekenghost, D. (1999). Adjustable autonomy for human-centered autonomous systems on Mars. *Proceedings of the AAAI Spring Symposium on Agents with Adjustable Autonomy. AAAI Technical Report SS-99-06*. Menlo Park, CA, Menlo Park, CA: AAAI Press,

23. Falcone, R., & Castelfranchi, C. (2002). From automaticity to autonomy: The frontier of artificial agents. In H. Hexmoor, C. Castelfranchi, & R. Falcone (Ed.), *Agent Autonomy*. (pp. 79–103). Dordrecht, The Netherlands: Kluwer.

24. Feltovich, P., Bradshaw, J. M., Jeffers, R., & Uszok, A. (2003). Social order and adaptability in animal, human, and agent communities. *Proceedings of the Fourth International Workshop on Engineering Societies in the Agents World*, (pp. 73–85). Imperial College, London,

25. Ferguson, G., Allen, J., & Miller, B. (1996). TRAINS-95: Towards a mixed-initiative planning assistant. *Proceedings of the Third Conference on Artificial Intelligence Planning Systems (AIPS-96)*, (pp. 70–77). Edinburgh, Scotland,

26. Fitts, P. M. (Ed.). (1951). *Human Engineering for an Effective Air Navigation and Traffic Control System*. Washington, D.C.: National Research Council.

27. Gawdiak, Y., Bradshaw, J. M., Williams, B., & Thomas, H. (2000). R2D2 in a softball: The Personal Satellite Assistant. H. Lieberman (Ed.), *Proceedings of the ACM Conference on Intelligent User Interfaces (IUI 2000)*, (pp. 125–128). New Orleans, LA, New York: ACM Press,

28. Gibson, J. J. (1979). *The Ecological Approach to Visual Perception.* Boston, MA: Houghton Mifflin.

29. Guinn, C. I. (1999). Evaluating mixed-initiative dialog. *IEEE Intelligent Systems,* September–October, 21–23.

30. Hancock, P. A., & Scallen, S. F. (1998). Allocating functions in human-machine systems. In R. Hoffman, M. F. Sherrick, & J. S. Warm (Ed.), *Viewing Psychology as a Whole.* (pp. 509–540). Washington, D.C.: American Psychological Association.

31. Hexmoor, H., Falcone, R., & Castelfranchi, C. (Ed.). (2003). *Agent Autonomy.* Dordrecht, The Netherlands: Kluwer.

32. Hoffman, R., Feltovich, P., Ford, K. M., Woods, D. D., Klein, G., & Feltovich, A. (2002). A rose by any other name... would probably be given an acronym. *IEEE Intelligent Systems,* July–August, 72–80.

33. Horvitz, E. (1999). Principles of mixed-initiative user interfaces. *Proceedings of the ACM SIGCHI Conference on Human Factors in Computing Systems (CHI '99).* Pittsburgh, PA, New York: ACM Press,

34. Johnson, M., Chang, P., Jeffers, R., Bradshaw, J. M., Soo, V.-W., Breedy, M. R., Bunch, L., Kulkarni, S., Lott, J., Suri, N., & Uszok, A. (2003). KAoS semantic policy and domain services: An application of DAML to Web services-based grid architectures. *Proceedings of the AAMAS 03 Workshop on Web Services and Agent-Based Engineering.* Melbourne, Australia,

35. Kahn, M., & Cicalese, C. (2001). CoABS Grid Scalability Experiments. O. F. Rana (Ed.), *Second International Workshop on Infrastructure for Scalable Multi-Agent Systems at the Fifth International Conference on Autonomous Agents.* Montreal, CA, New York: ACM Press,

36. Klein, G., & Feltovich, P. J. (in preparation). Multiple metaphors for complex coordination. In

37. Levinson, S. C. (1992). Activity types and language. In P. Drew & J. Heritage (Ed.), *Talk at Work.* (pp. 66–100). Cambridge, England: Cambridge University Press.

38. Luck, M., D'Inverno, M., & Munroe, S. (2002). Autonomy: Variable and generative. In H. Hexmoor, C. Castelfranchi, & R. Falcone (Ed.), *Agent Autonomy.* (pp. 9–22). Dordrecht, The Netherlands: Kluwer.

39. Maheswaran, R. T., Tambe, M., Varakantham, P., & Myers, K. (2003). Adjustable autonomy challenges in personal assistant agents: A position paper.

40. Myers, K., & Morley, D. (2003). Directing agents. In H. Hexmoor, C. Castelfranchi, & R. Falcone (Ed.), *Agent Autonomy.* (pp. 143–162). Dordrecht, The Netherlands: Kluwer.

41. Norman, D. A. (1988). *The Psychology of Everyday Things.* New York: Basic Books.

42. Norman, D. A. (1992). Cognitive artifacts. In J. M. Carroll (Ed.), *Designing Interaction: Psychology at the Human-Computer Interface.* (pp. 17–38). Cambridge: Cambridge University Press.

43. Norman, D. A. (1999). Affordance, conventions, and design. *Interactions,* May, 38–43.

44. Sarter, N., Woods, D. D., & Billings, C. E. (1997). Automation surprises. In G. Salvendy (Ed.), *Handbook of Human factors/Ergonomics, 2nd Edition.* New York, NY: John Wiley.

45. Scerri, P., Pynadath, D., & Tambe, M. (2002). Adjustable autonomy for the real world. In H. Hexmoor, C. Castelfranchi, & R. Falcone (Ed.), *Agent Autonomy.* (pp. 163–190). Dordrecht, The Netherlands: Kluwer.

46. Schreckenghost, D., Martin, C., & Thronesbery, C. (2003). Specifying organizational policies and individual preferences for human-software interaction. *Submitted for publication.*

47. Shalin, V. L., Geddes, N. D., Bertram, D., Szczepkowski, M. A., & DuBois, D. (1997). Expertise in dynamic, physical task domains. In P. J. Feltovich, K. M. Ford, & R. R. Hoffman (Ed.), *Expertise in Context: Human and Machine.* (pp. 195–217). Menlo Park, CA: AAAI/MIT Press.

48. Sheridan, T. B. (2000). Function allocation: algorithm, alchemy or apostasy? *International Journal of Human-Computer Studies,* 52(2), 203–216.

49. Suri, N., Bradshaw, J. M., Breedy, M. R., Groth, P. T., Hill, G. A., Jeffers, R., Mitrovich, T. R., Pouliot, B. R., & Smith, D. S. (2000). NOMADS: Toward an environment for strong and safe agent mobility. *Proceedings of Autonomous Agents 2000.* Barcelona, Spain, New York: ACM Press,

50. Suri, N., Bradshaw, J. M., Burstein, M. H., Uszok, A., Benyo, B., Breedy, M. R., Carvalho, M., Diller, D., Groth, P. T., Jeffers, R., Johnson, M., Kulkarni, S., & Lott, J. (2003). DAML-based policy enforcement for semantic data transformation and filtering in multi-agent systems. *Proceedings of the Autonomous Agents and Multi-Agent Systems Conference (AAMAS 2003).* Melbourne, Australia, New York, NY: ACM Press,

51. Suri, N., Bradshaw, J. M., Carvalho, M., Breedy, M. R., Cowin, T. B ., Saavendra, R., & Kulkarni, S. (2003). Applying agile computing to support efficient and policy-controlled sensor information feeds in the Army Future Combat Systems environment. *Proceedings of the Annual U.S. Army Collaborative Technology Alliance (CTA) Symposium.*

52. Tonti, G., Bradshaw, J. M., Jeffers, R., Montanari, R., Suri, N., & Uszok, A. (2003). Semantic Web languages for policy representation and reasoning: A comparison of KAoS, Rei, and Ponder. In D. Fensel, K. Sycara, & J. Mylopoulos (Ed.), *The Semantic Web— ISWC 2003. Proceedings of the Second International Semantic Web Conference, Sanibel Island, Florida, USA, October 2003, LNCS 2870.* (pp. 419–437). Berlin: Springer.

53. Uszok, A., Bradshaw, J. M., Jeffers, R., Johnson, M., Tate, A., Dalton, J., & Aitken, S. (2004). Policy and contract management for semantic web services. *AAAI 2004 Spring Symposium Workshop on Knowledge Representation and Ontology for Autonomous Systems.* Stanford University, CA, AAAI Press,

Founding Autonomy: The Dialectics Between (Social) Environment and Agent's Architecture and Powers[1]

Cristiano Castelfranchi and Rino Falcone

Istitute for Cognitive Sciences and Technologies – CNR – Rome - Italy
{castel, falcone}@ip.rm.cnr.it

Abstract. 'Autonomy', with 'interaction' the central issue of the new Agent-based AI paradigm, has to be recollected to the internal and external *powers* and resources of the Agent. Internal resources are specified by the Agent architecture (and by skills, knowledge, cognitive capabilities, etc.); external resources are provided (or limited) by accessibility, competition, pro-social relations, and norms. 'Autonomy' is a relational and situated notion: the Agent - as for a given needed resource and for a goal to be achieved- is autonomous *from* the environment or *from* other Agents. Otherwise it is 'dependent' on them. We present this theory of Autonomy (independence, goal autonomy, norm autonomy, autonomy in delegation, discretion, control autonomy, etc.) and we examine how acting within a group or organization reduces and limits the Agent autonomy, but also how this may provide powers and resources and even increase the Autonomy of the Agent.

Keywords: Autonomy, Agent architecture, Powers, Dependence

1 Introduction

Since the introduction of the new "interactive" paradigm of Artificial Intelligence [Bobrow, 1990], autonomy has become a very relevant notion. In fact, there has been a looseness of interest about all the centralized functions (design, control, programming, planning) and in particular, with the birth of *Social Artificial Intelligence* -where intelligencies are modeled as "agents"- computation has changed its own meaning, consisting now mainly of interaction and communication. At the same time, information is always more local and distributed. Within this new paradigm (particularly relevant for theoretical and applied domain such as Robotics, Artificial Life, Multi-Agent Systems) autonomy of adaptive or intelligent agents is a must, and the problem is how to engineer it and how design a good trustworthy interaction with those entities endowed with initiative, learning, subcontracting, delegation, and so on.

We will provide an *operationalised* notion of autonomy based on: a notion of *dependence*, a notion of *delegation*, the characterisation of the agent's *architecture* and, the joining of groups or organizations. We will consider, on the one side, how autonomy is a "relational" notion: it is in fact about the relationship between an active entity and its environment or other influencing active entities (social autonomy). On

---

[1] This paper has been partially founded by the TICCA Project: joint research venture between the Italian National Research Council CNR- and Provincia Autonoma di Trento; and by the Progetto MIUR 40% "Agenti software e commercio elettronico" and by the PAR project of the University of Siena.

M. Nickles, M. Rovatsos, and G. Weiss (Eds.): AUTONOMY 2003, LNAI 2969, pp. 40–54, 2004.

the other side, this relationship also derives from *internal, architectural properties* of the former entity. Now we in fact intend to generalise and make more clear the role of architecture in the agent's autonomy. Autonomy actually is a matter of *power* (the capability of maintaining and of satisfying adaptive functions and goals through appropriate behaviours); this requires specific "powers" [Cas90] that can be distinguished in "external" and "internal".

External powers are conditions for actions, material resources, social resources and facilitation, etc. that allow an agent to successfully act and satisfy its need and functions.

Internal powers are internal material resources, knowledge, skills, that make an agent able and in condition of pursuing a goal. Essential part of this power are its "cognitive" capabilities: the ability to perceive, to plan, to decide, to have an internal goal (purposive behaviour), to learn, etc.

Through agent's architecture we can identify certain *specific* powers of the agent. For example, if an agent has not a visual apparatus (it is blind) while needs this kind of information, it is non autonomous as for visual information; if an agent is lacking the ability to decide (for example for some brain damage) it is non autonomous as for deciding. Since, as we saw [Cas90], lack of power (combined with the power of others) produces social dependence, if another agent (and only it) can provide to the former what it lacks *for structural reasons* the former agent is non autonomous *from* the latter.

So, with respect to the agent's architecture we can say that, autonomy (and lack of autonomy) *is due to the agent's architecture* (see later) since architecture (both cognitive and physical ones) defines what *kind* of resources one needs, and it also provides specific capabilities. On the other hand, it remains that -given an architecture- one can be dependent for structural reasons or for accidental reasons (lack of acquirable resources and conditions).

Another interesting phenomenon is the conditioning of the agent's autonomy deriving from more or less structured groups (teams, organizations, and so on). In fact, the agent's autonomy in these contexts is modified because accessibility of resources and permitted/possible agent's functions are changed; not only external resources are changed, but also the internal ones.

2   Autonomy

"Autonomy" is a relational notion. One is autonomous *as for* a given action or goal (and not for another), and *from* something or somebody [Cas95, Cas03].

2.1   Non-social Autonomy: Autonomy from the Environment

It seems that the most relevant notion of autonomy for Artificial Intelligent Agents is the «social autonomy»: autonomy from other Agents or from the user. But this is only partially true. The Agents we need are entities able to act in an environment, on the basis of the perception of this environment (also during the execution of the action). Being an Agent requires a certain degree of autonomy from these stimuli and from the environment itself. The Agent's behaviour cannot be completely determined and predictable on the basis of the current input, like a billiard ball under mechanical forces. Agents (at this level) are at least goal-oriented systems, not simply causal entities. More precisely, their «autonomy from stimuli» is guaranteed:

- by the fact that their behaviour is teleonomic: it tends to certain specific results due to internal constraints or representations, produced by design, evolution, or learning, or their previous psychological *history*;
- by the fact that they do not simply receive an input but they (actively) *perceive* their environment and the effects of their actions;
- by the fact that they are not 'pushed' by external forces but invest and employ their own energy;
- by the fact that they move themselves in searching selected *stimuli*, in other words, they determine and select environmental *stimuli*;
- by the fact that they have *internal states* with their own exogenous and endogenous evolution principles, and their behaviour also depends on such internal states.

Let us more deeply analyze some of these points.

Internal States and Behaviour
One of the most important features of Agents is that they have *internal states* [Sho93] and that their reaction to a given stimulus or their processing of a given input depends on their internal states. This seems to be one of the main differences between an "Agent" and a software component, module or function. This implies that Agents react in different ways or give different results to the same input, depending on their internal states that have their own internal dynamics and evolution. So they have different reactive styles, either stable or contingent. As argued in [Cas95], this property (internal state mediation between input and output) is a very basic aspect of autonomy: autonomy relative to the stimulus, to the world. In Agent-based computing, there is a need to introduce different treatments of the same input or different processing reactions, that cannot be decided on the basis of external parameters and tests, or input conditions. These different "computations" are conditional to "internal" parameters which evolve independently of the Agent sending the input and are externally not completely predictable.

Thus the first relevant feature of autonomy in Agents is *in their relation with their environment*. And this is important especially because of their «social autonomy» and of their acting remotely i.e. far from and without our monitoring and intervention (control), on the basis of their local and contingent information, reactively and adaptively. They should manage by themselves and in a goal-oriented way (i.e. possibly achieving their tasks) their relation with the environment (physical or virtual).

Social Autonomy (see § 3) means that the agent is able and in condition to pursue and achieve a given goal not *depending* on the others' intervention.

2.2   Architecture and Autonomy

As we said the Agent autonomy is relative to its architecture that specify what it needs for acting and what is able to do. Let's start our argument by assuming as a good operational reference a BDI architecture - in particular PRS Architecture (Fig. 1)[2] which is a minimal, sufficient model of an *intentional system* or *cognitive agent*, i.e. *a goal-governed system whose actions are internally regulated by explicitly represented goals and whose goals, decisions, and plans are based on beliefs*. Both goals and beliefs are cognitive representations that can be internally generated, manipulated, and subject to inferences and reasoning. Since a cognitive agent may have more than one goal active in the same situation, it must have some form of choice/decision, based on some *reason* i.e. on some belief and evaluation.

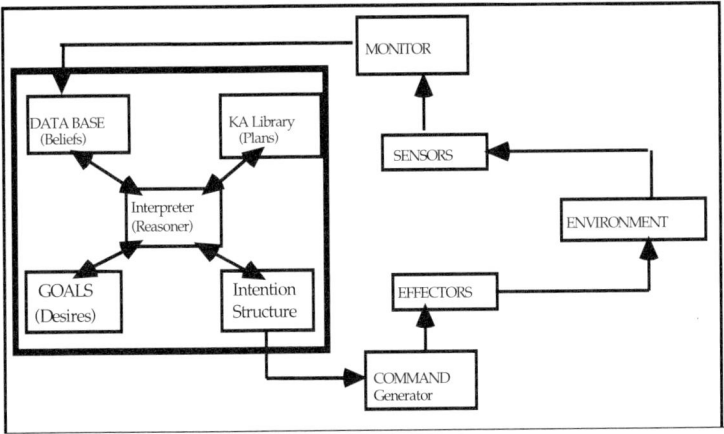

Fig. 1. PRS archictecture (from Georgeff and Ingrand,1989)

This model shows that in order to adaptively 'act' a cognitive agent X needs:

- *information* from current environment (sensors), and the ability to *interpret/understand* this input relative to its knowledge and goals (in this perspective the *monitor* or the data base modules are a bit reductive);
- a *plan library*; and the ability to retrieve from it an appropriate plan;
- a *knowledge* or beliefs base; and the ability to reason on it and to make *relevant inferences*;
- a set of *motives/goals/tasks* and the capability to *activate* them (either internally and *pro-actively* or by *reacting* to external stimuli) or to *drop* them;
- the capability to *choose* among alternative plans, and among incompatible goals;
- the ability to *select an appropriate intention* (a preferred, realistic goal endowed with its plan) and to *persist* in pursuing it (until appropriate);
- the capacity to *generate commands* to some effectors on the world; and to materially *execute* this command controlling these effectors;
- the capacity of *monitoring* the action results during and after its execution and to use this as a feedback on the process, at several levels.

---

[2] We might use any other BDI model [Had96] or any other agent architecture -for example Chaib-draa's more "psychological" model [Cha95]- provided that it is minimally complex and articulated.

We should add to these components and abilities, the *external* resources and opportunities that allow or prevent the agent from successfully executing its action. Let us distinguish at least between

- material or practical resources and conditions (including agents to be delegated), and
- deontic resources or conditions (permission, authorisation, role, etc.).

In our model this define the set of (internal and external) *powers* [Cas90] that make X able/in condition to act purposively and autonomously. If X has these *powers* it 'can' and 'may'. If some of these internal or external *powers* or *resources* is lacking, the agent is not able to (cannot/may not) pursue/achieve its goal. Given this, we have *all the basic dimensions of X autonomy or dependence* (which is the complement of autonomy [Sic94; Sic95]).

3   Social Autonomy

One can identify two notions of autonomy at the social level (i.e. relative to other Agents):

- one is autonomy as *independence*, self-sufficiency;
- the other one is autonomy in *collaboration*: how much an Agent is autonomous when it is working for another Agent: helping it, collaborating with it, obeying to it.

Actually the second type is a sub-case of the former one [Cas00], but it deserves specific attention.

3.1   Autonomy as Independence

In this sense *an Agent is completely autonomous (relative to a given goal) when does not need the help or the resources of other Agents neither to achieve its goal nor to achieve that goal in a better way.*

This notion of autonomy (it would be better to use the term «self-sufficiency» or «independence» as opposed to «depencence») is particularly relevant in MultiAgent systems [Sic94]. But it is very relevant also in interaction between user and Agents, or among independent software Agents. In fact, if the Agent is supposed to act on our behalf, if we have to delegate it some task, it must be able to satisfy this task by itself (not necessarily alone, also involving other Agents). It must posses the necessary capability and resources: either practical (to execute the task) or cognitive (to solve the problem), or social (to find out some cooperation). Thus, relatively to the delegated task it should be non-dependent on the delegating Agent, and able to deal with its dependence relations with others.

In other term we have to *trust* the Agent as for its competence and capability, which is one of the two fundamental part of trust (the other is reliability) [Cas98a].

3.2   Autonomy in Collaboration

This notion of autonomy is strictly related to the notion of *delegation* [Cas98, Fal97] (*on behalf of*) in a broad sense, which is another definitional feature of AI Agents.

This is the notion of autonomy in acting for realising an assigned task and relatively to the delegating Agent. This notion has several very important facets.

We have first to explain the notion of delegation (on behalf of) which is usually meant in a too strong and strict sense, related to institutional, formal delegation, authorisation, and so on. We need on the contrary a very basic and general notion of delegation as reliance. In [Cas98] we give the following broad definitions of delegation and adoption, the two complementary attitudes of the delegating agent and of the delegated one (we will call A the delegating-adopted Agent (the *client*) and B the delegated-adopting Agent (the *contractor*)).

In delegation an Agent A needs or likes an action of another Agent B and includes it in its own plan. In other words, *A is trying to achieve some of its goals through B's actions; thus A has the goal that B performs a given action.* A is constructing a MA plan and B has a *part*, a share in this plan: B's task (either a state-goal or an action-goal).

In adoption an Agent B has a goal since and for so long as it is the goal of another Agent A, that is, B usually has the goal of performing an action since this action is included in A's plan. So, also in this case B plays a part in A's plan (sometimes A has no plan at all but just a need, a goal).

At this basic level there may be no interaction or communication between the Agents, the delegation (adoption) being only in the mind of one of the Agents (unilateral delegation/adoption), and delegation (adoption) can be established also between a cognitive and a non cognitive Agent.

We assume that *delegating an action necessarily implies delegating some result of that action.* Conversely, *delegating a goal state always implies delegating at least one action (possibly unknown to A) that produces such a goal state as result.* Thus, we consider the action/goal pair as the real object of delegation, and we will call it ''task'.

Depending on the *levels of delegation* and of the entitlement of the delegated Agent, the are several levels of autonomy[3]; at least:

– *executive autonomy*: the Agent is not allowed to decide anything but the execution of the entire delegated plan [Con95];
– *planning autonomy*: the Agent is allowed to plan by itself, the delegated action is not completely specified in the delegation itself;
– *goal autonomy*: the Agent is allowed to have/find its own goals.

Even ignoring the possible autonomous goals of the delegated Agent (its full autonomy), we can characterise different degrees of autonomy in delegation as follows.

The autonomy of the contractor vis-à-vis the client increases along various dimensions:

– the more *open the delegation* (the less specified the task is), or
– the more *control actions* given up or delegated to B by A, or
– the more delegated decisions (*discretion*),
the more autonomous B is from A with regard to that task.

---

[3] There are also *degrees* of autonomy. One can be more or less autonomous along various dimensions: for example, on how many goals (and how much important) you are non-dependent from others; how much I relaxed the control and I trust you; how much you can decide about the means: can you only search for various, alternative solutions or can you also choose among them?

Let us see also these aspects carefully.

Given a goal and a plan to achieve it, A can delegate goals/actions (tasks) at different levels of abstraction and specification [Cas98]. We can distinguish several levels, but the most important are the following ones:

- *pure executive delegation vs. open delegation;*
- *domain task delegation vs. planning and control task delegation (meta-actions)*

The object of delegation can be minimally specified (*open delegation*), completely specified (*close delegation*) or specified at any intermediate level.

Pure Executive Delegation

The delegating (delegated) Agent believes it is delegating (adopting) a completely specified task: what A expects from B is just the execution of an elementary action (what B believes A delegated to it is simply the execution of an elementary action).

Open Delegation

The client is delegating an incompletely specified task: either A is delegating a complex or abstract action, or it is delegating just a result (state of the world). The Agent B must realize the delegated task by exerting its autonomy.

The object of the delegation can be a practical or *domain action* as well as a *meta-action* (searching, planning, choosing, problem solving, and so on). When A is open delegating to B some domain action, it is necessarily also delegating to B some meta-action: at least searching for a plan, applying it, and sometimes deciding between alternative solutions; A exploits intelligence, information, and expertise of the delegated Agent. We call B's *discretion* concerning the task the fact that it is left to B some decision about it.

The importance of *open delegation* in collaboration theory should be examined.

First, it is worth stressing that *open delegation* is not only due to A's preferences, practical ignorance or limited ability. It can also be due to A's ignorance about the world and its dynamics: *fully specifying a task is often impossible or not convenient,* because some local and updated knowledge is needed for that part of the plan to be successfully performed. Open delegation ensures the *flexibility* of distributed and MA plans.

The distributed character of the MA plans derives from *open delegation*. In fact, A can delegate either an entire plan or some part of it (*partial delegation*) to B. The combination of the *partial delegation* (where B might ignore the other parts of the plan) and of the *open delegation* (where A might ignore the sub-plan chosen and developed by B) creates the possibility that A and B (or B and C, both delegated by A) collaborate in a plan that they do not share and that *nobody* entirely knows: that is a *pure distributed plan* [Cas98]. However, for each part of the plan there will be at least one Agent that knows it.

Second, in several cases we need local and decentralized knowledge and decision in our Agents. The Agent delegated to take care of a given task has to choose between different possible recipes; or it has to build a new plan, or to adapt abstract or previous plans to new situations; it has to find additional (local and updated) information; it has to solve a problem (not just to execute a function, an action, or implement a recipe); sometimes it has to exploit its "expertise". In all these cases the Agent takes care of our interests or goals autonomously and "remotely" i.e. far from and without our monitoring and intervention (control). This requires an *open delegation*: the delegation "to bring about that *g*"; so, the Agent is supposed to use its knowledge, its intelligence, its ability, and to have some autonomy of decision. If we delegate our goal to an Agent without simply eliciting (or requiring) a specified behavior, the Agent has to reason, choose, and plan *for* this goal; therefore, we need explicitly represented mental attitudes.

On the basis of Delegation theory we can characterise several traditional aspects of autonomy.

Autonomy as Discretion

Discretion is a possible form of autonomy. An Agent autonomous in planning has a certain "discretion": it can find and choose the plan to be executed, or it can build a new plan.

The agent is autonomous in 'deciding' at certain level: the level of sub-plans or the level of goals. It can for example decide not to keep a promise or violate an obligation.

Autonomy as Initiative and Lack of Control

Agents «operate independently of its author (...) without a direct control (...) and have their own goals and motivations» [Bic98]. They have to achieve their (assigned) goals *remotely* i.e. far from and without our monitoring and intervention (control). This is in particular the notion of autonomy used in robotics. But it mixes up several of the aspect we analysed. So let us focus at this point on two features: initiative and lack of control.

Initiative in Delagation

- *motu proprio*: an autonomous Agents could not executing the required action/plan under the direct and immediate *command* of their client or user. Their behaviour is not fired or elicited by the user or the request, but by the Agent's autonomous relation with its environment and the other agents. It will execute the task where and when appropriate, also depending on its internal state (this can be considered also as another aspect of *discretion*).
- *spontaneous interaction* or *pro-active help*: Agents can act *for us* (or for the client) also without any request or beyond the request. They might spontaneously help the other or over-help it (doing more or better than requested - [Cas98]) or spontaneously starting an interaction for example by offering some help. They may anticipate the user/client request and even desires and interests.

Control

Any true action (purposive behaviour) is based on control, is driven. This means that any action (α) is a combination of three more elementary actions: the *Executive action* (Ex_α), the *Monitoring action* (M_α), and the *Intervening action* (I_α).

- Ex_α is the part of action α that produces the final expected result if the enabling conditions are satisfied at any time during the execution of it.
- M_α is the evaluation and verification of the enabling conditions of Ex_α and of its correct execution with respect each step of it.
- I_α is a set of (one or more) action(s) whose execution has the goal of repairing, supporting, partially inhibit, stopping, and so on, the execution action (Ex_α) to guarantee the final expected result of Ex_α or at least for reducing possible damages or undesidered effects. These actions could be introduced following the results of the monitoring (M_α). They could be also introduced independently from these results but just on the basis of some rules or previsions (for example, it

could be known that in some special world conditions, Ex_α it is unable to produce all its results without a specific additional action).

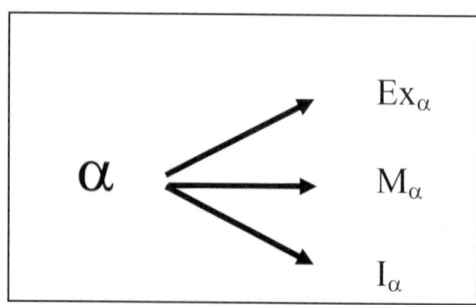

Analyzing the intrinsic nature of the actions we have to say that even if it is possible to distinguish among these three different functions (execution, monitoring, intervention/adaptation) in the same action, not always it is possible to allocate these different functions to different agents.

In particular, there are cases in which just the actor who is executing the core action (Ex_α) can be able to monitor some parts of it or of its conditions (more, it is the only agent that can decide to monitor); these monitoring actions, that in principle would be part of the M_α set, will be included, in our model, in the executive action (Ex_α). At the same way, there are intervening actions that only the actor who is executing the core action can be able to realize (and decide to realize). Also in this case, we will consider these actions part of the Ex_α. In other words, the possibility of allocating to different agents the various sub-actions becomes an additional criterion for considering these actions of different nature.

When the delegator is delegating a given object-action, what about its control actions?

Following [Cas98] there are at least four possibilities:

i)   The client delegates the control to the Agent: it does not (directly) verify the success of the delegated action;
ii) The client delegates the control to a third Agent;
iii) The client gives up the control: nobody is delegated to control the success of the task (blind trust);
iv) The client maintains the control for itself.

But there are also different degrees: for example, in maintainig the control the client can either continuously monitor and inspect, showing no trust at all and leaving very little autonomy to the Agent, or it can check only sometimes or at the end of the execution.

Autonomy as Self-motivation and Rationality

A goal-autonomous agent is endowed with its own goals. We claim [Cas95] that an Agent is fully socially autonomous if:

1) it has its own Goals: endogenous, not derived from other Agents' will;
2) it is able to make decisions concerning multiple conflicting goals (being them its own goals or also goals adopted from outside);

3) it adopts goals from outside, from other Agents; it is liable to influencing

4) it adopts other Agents' Goals as a consequence of a choice among them and other goals

5) *it adopts other Agents' Goals only if it sees the adoption as a way of enabling itself to achieve some of its own goals* (i.e. the Autonomous Agent is a Self-Interested Agent).

6) It is not possible to directly modify the Agent's Goals from outside: any modification of its Goals must be achieved by modifying its Beliefs. Thus, the control over Beliefs becomes a filter, an additional control over the adoption of Goals.

7) it is impossible to change automatically the Beliefs of an Agent. The adoption of a Belief is a special «decision» that the Agent takes on the basis of many criteria. This protects its *Cognitive Autonomy*.

Let us stress the importance of principle (5):

An autonomous and rational Agent makes someone else's goal its own (i.e. it adopts it) only if it believes adoption to be a means for achieving its own goals.

Notice that this postulate does not necessarily coincide with a "selfish" view of the Agent. To be *self-interested* or *self-motivated* is not the same of being *selfish*. The Agent's "own" goals, for the purpose of which he decides to adopt certain aims of someone else, may include *benevolence* (liking, friendship, affection, love, compassion, etc.) or impulsive (reactive) behaviours/goals of the altruistic type.

Norm Autonomy

An advanced form of (social) autonomy is norm-autonomy. In weak sense it consists of the internal capability of taking into account external norms and deciding of obeying or violating them. This Agent is not completely constrained and determined by impinging laws and prescriptions [Con95] (for a rather different interpretation see [Ver00]). In a more radical sense, it would be the case of an Agent which is the source of its own laws, norms, obligations ("auto-nomos"); it is not subject to other authorities and norms (see §4.2).

4   Socially Situated Autonomy: How Groups Affect Member's Autonomy

When an agent enters an organization, a regulated society or group, it accepts or simply encounters some restrictions relative to its previous or potential behaviour. Something what it was previously able or in condition to do now is no longer possible; it looses some of its powers, it depends on some other agents for those actions or goals. In other terms it suffers (and/or accepts) some *negative interference*. It must for example at least appropriately coordinate its actions with the others' behaviours.

However, the most interesting case of this loss of power is when this is due not simply to practical and material interference, but when it is due to the fact that the group, the organization (or the 'site' where the mobile agent should work [Cre00] is regulated by particular conventions or norms. In this case the agent looses or renounces part of its former *freedom*. Its lack of power is of deontic origin, and it becomes **deontically dependent**. We mean that the situated agent can be equally able and in condition to do its action α and to achieve its goal g as the natural result of α, but it is not allowed to do so; or better it has to ask and have some *permission*. It has in fact a new goal $G = (g\ AND\ g')$, where g' is a set of additional constraints, clearly conditioning the achievement of g (non violating, avoiding sanctions, having approval, and so on). As for g the agent is autonomous, but in order to achieve G, it has no power and is dependent on the group and on having a permission.

Any stable or functional community, group or society puts restrictions on the behaviours of its members, and members must either explicitly or implicitly, either consciously and by decision or by learning adapt to, accept those restrictions.

Let's discuss this quite relevant phenomenon.

If the Agent's autonomy happens to be seriously limited while entering in a group or organization, why should the agent chose to do so and accept those limitations?

Autonomy is not a value *per se*, a final end; it is a good condition for pursuing goals; it is a sign of empowerment and power, and *power is in fact what matters*. In this perspective the relation with the group should be reconsidered. The group (and in general the society) can both enlarge the powers and resources of the Agent, and even its autonomy, and restrict them.

Let us analyze a bit more carefully the impact of the group on the agents' internal and external powers.

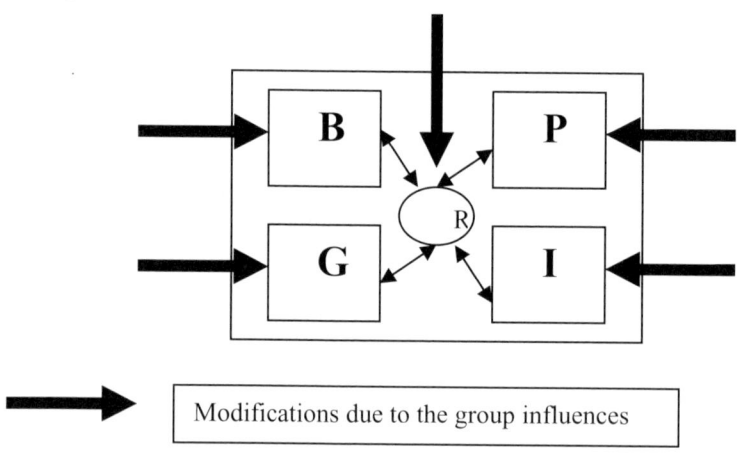

Modifications due to the group influences

As for the internal "resources" specified by the reference architecture, we can say that:

The group can reduce the numbers of **plans** that the agent can use, both by limiting the external conditions for the execution of the related actions, or by making unusual, non-permitted a given plan or obligatory another plan. Thus in plan choice the Agent is conditioned by its group.

However, on the other side, the group can also provide to the agent (by imitation, teaching, collective reasoning, data bases, etc.) new plans that it was not able to invent or to know or new plans that are intrinsically multi-agent (not executable being alone).

The group can limit the **beliefs** of the Agent, while reducing its free access to perception or data, while inducing it to a revision of previous or independent beliefs, even influencing the process of deciding to believe or not to believe. The Agent is less autonomous in believing. Nevertheless, the group also enriches very much the knowledge of the Agent by sharing it, by exchanging it, by distributed discover, by teaching, etc.

The group can limit the **goals** of the Agent by education and culture, by limiting exposure to examples. And even more it can limit the goals that the Agent will choose (intentions). But, on the other side, the group builds goals into the Agent, or provides goals through requests, jobs, tasks, obligations. In our model of BDI in fact Desires are not the only sources of Intentions [Cas98b]. We prefer a broader category of 'Goals' that includes also exogenous (adopted) goals, included 'duties'. Part of those goals reduce the autonomy of the Agent in doing what it would 'desire' to do, by introducing internal 'conflicts' (conflicting goals). But other goals help the Agent to acquire powers and independence, and to realize its motivations.

As we just said, the group can limit the goal that the Agent will choose, i.e. its **intentions**. For example, while playing a role in an organization the agent cannot decide to do whatever it would like to do. First, it 'has to' achieve its tasks. Second, it cannot realize the organizational goal as it likes and prefers: it must apply specific routines, plan operators; it cannot freely intend and solve the problems.

Finally, social relationships can also modify the **procedures** for the control of the different modules; in other words, they could change the model of interaction among beliefs, goals, plans and intentions with respect to the isolated situation (the case of the agent outside the group). For example, consider the prescription of applying a set of plans without any check on the preconditions of those plans (about their satisfaction); or the necessity to activate a specific intention just as an external order (not only ignoring the level of priority of that intention but without any check on the necessary beliefs for that intention, or with the presence of contrary beliefs with respect the probably success of that intention).

The same holds for **external resources**: *material* (something that the agent is not able to produce or to find) or *social*: relations and partners; permissions and role empowerment.

Being socially situated and related can increase accessible resources for example by 'exchange': on the one side the agent makes itself *dependent* on the producer or seller; but – on the other side – after buying it is able and in condition to do whatever it likes ('possess').

Social relations also limit very much powers (for example by competition) and autonomy, but they also provide new powers and new kinds of autonomy, like 'rights': I have no longer to ask for means or permission, I'm *entitled* to do what I like (if/when I like) and the others are 'obliged' (limited!) to let me do it.

4.1   The Trade-off Between the Advantages of Independence and of Society

In [Cas90] we said that society multiply the powers of the individuals and empower them; now we say that society limits the powers of the individual and subordinated them. Where is the truth? Both phenomena are real, and they are even linked to one the other. As Durkeim has explained, while 'entering' a society (S) we *make* ourselves dependent on each other, but we also multiply our powers and thus our welfare by 'exchange' of powers and resources, and by the creation of new powers: collective powers, institutional powers [Cas03b]. To lose (part of our) autonomy can be a very convenient choice since we are able and in condition to satisfy many more and new desires and needs of us.

On the one side, the individual accepts, undergoes such a reduction of freedom because it is convenient for it: social empowerment and realization is greater than renounced powers and freedom.[4] There is sort of social exchange between A and S: A renounces to freedom and autonomy, obeys and receives from S empowerment, as the possibility to enter social exchange and cooperation, assistance, rights, and as collective benefits, and as protection from action forbidden to the others members (A's protected rights) and protection from others external to S.

On the other side, S is able to empower A precisely thanks to the fact that A (and all the other members) follows the rules.

What A can receive from S as "reciprocation" is quite various and complex: for example

- membership per se', identity, approval, etc. (intrinsic social motives);
- institutional power; A's action count-as and thanks to those powers A can satisfy some of its goals;
- avoiding sanctions;
- access to resources, entering to a social and economic exchange network, credit for this, some power it lacks (training, knowledge, etc.); possible cooperation.

Also Norms both

- limit, restrict powers and freedom (via prohibitions, and obligations) [Hex03] [Sch01]; and
- give powers; either by consenting to them, providing entitlement and rights, permission; or by creating them (conventional powers).

4.2   Normative Autonomy and Normative Monitoring

If an agent is not *normatively autonomous* (§ 3), i.e. it is not the source of its own laws, norms, obligations ("auto-nomos"); and in general if an agent is subject to obligations (although freely and spontaneously contracted), it is subject to *monitoring* (surveillance).

This normative-monitoring is aimed at ascertaining whether there is violation or not, whether the obligation (the command, the law, the commitment, the contract, ...) has been satisfied.

The function of this monitoring can be that of bringing B's action to its realization. In this case monitoring, and more precisely letting B to believe that it is monitored, is a pressure on B and should increase the probability that it performs the task.

Sometimes, monitoring is a prevention act able to block and prevent violation.

However, this is not the most important and general function of normative-monitoring. It can be non effective for inducing *B* to do the task or to avoid violation, but is useful for *A* (the monitoring agent or its mandatary) for applying sanctions, which is a *post-hoc* activity.

This is the general function of monitoring an agent has for its obligations. This is useful for the future (learning and education) and for the MA systems (not spreading

---

[4]   In human societies -and also in designed MAS- belonging to social systems is not always a real and rational choice. Sometimes the individual "borns" in S and does not have or conceive or perceive any alternative; or what it receives from society is belonging in itself, or better the satisfaction of its important goal of 'social identity', of feeling part of S, being a member of S.

violation; example; confirming authority and the credibility of surveillance and laws; etc.) Thus, this is another interesting case of *monitoring for post-hoc intervention*, not for adjusting action to lead it to success.

This kind of monitoring (surveillance) on the one side it is usually a *sign* of a system not normatively autonomous but subject to some "authority"; on the other side, it is limiting the autonomy because B is not able of behaving as it likes while not violating or avoiding sanctions.

It is rather important to distinguish between different kinds and use of monitoring, for example between on-line monitoring for intervention oriented to adjusting (control), vs. monitoring for post-hoc intervention. For example, in the case of normative-monitoring given its nature it is clear that it is not so reasonable to delegate it to A, to the same agent subject to the obligation. It has rational reasons for deceiving about its own violation! You can monitor yourself for adjusting or avoiding damages, but you cannot be very much trusted as for denouncing yourself and expose yourself to sanctions. Trivers' hypothesis in Sociobiology [Tri71] is precisely that the feeling of guilt has been selected in *homo sapiens* in order that you yourself monitor your fairness and normative respect and punish yourself (suffering) and are leaning to pay some penalty (reparation) or search for punishment.

5 Conclusive Remarks

We have argued that autonomy has to be recollected to the internal and external *powers* and resources of the Agent. Internal resources are specified by the Agent architecture (and by skills, knowledge, cognitive capabilities, etc.); external resources are provided (or limited) by accessibility, competition, pro-social relations, and norms. 'Autonomy' is a relational and situated notion: the Agent -as for a given needed resource and for a goal to be achieved- is autonomous *from* the environment or *from* other Agents. Otherwise it is 'dependent' on them. After presenting this theory of autonomy (independence, goal autonomy, norm autonomy, autonomy in delegation, discretion, control autonomy, etc.) with special attention on autonomy in social cooperation, we examined how acting within a group or organization reduces and limits the Agent autonomy, but also how this may provide powers and resources and even increase the autonomy of the Agent.

References

[Bic98]   Bickmore, T., Cook, L., Churchill, E., and Sullivan, J., Animated Autonomous Personal Representatives. In Autonomous Agents'98, Minneapolis May 9–13, ACM Press, pp. 8–15.

[Cas90]   Castelfranchi, C. (1990), Social power: a point missed in Multi-Agent, DAI, and HCI. In Y. Demazeau and J.P. Muller (eds.), *Decentralized A.I.*, North-Holland, Amsterdam, 1990.

[Cas95]   Castelfranchi, C., Guaranties for Autonomy in Cognitive Agent Architecture. In M.J. Woolridge and N. R. Jennings (eds.) *Intelligent Agents I*, Berlin, Springer, 1995.

[Cas98]   Castelfranchi, C., Falcone, R., (1998) Towards a Theory of Delegation for Agent-based Systems, *Robotics and Autonomous Systems*, Special issue on Multi-Agent Rationality, Elsevier Editor, Vol. 24, Nos. 3–4, pp. 141–157.

[Cas98a]   Castelfranchi C. & Falcone, R., Principles of Trust for MAS: Cognitive Anatomy, Social Importance, and Quantification. ICMAS'98, Paris 2–8 July 98, AAAI-MIT Press.

[Cas98b]   Castelfranchi, C. (1998). Modelling social action for AI agents. *Artificial Intelligence, 103,* 157–182.

[Cas00]   Castelfranchi, C. (2000), Founding Agent's "Autonomy" on Dependence Theory, Proceedings of the European Conference on Artificial Intelligence (ECAI-00), Berlin, August 2000.

[Cas03]   Castelfranchi, C., Falcone R. (2003), From Automaticity to Autonomy: The Frontier of Artificial Agents, in Hexmoor H, Castelfranchi, C., and Falcone R. (Eds), Agent Autonomy, Kluwer Publisher, pp. 103–136.

[Cas03b]   Castelfranchi, C., (2003), The Micro-Macro Constitution of Power, *ProtoSociology, An International Journal of Interdisciplinary Research* In the Special Issue Understanding the Social II – Philosophy of Sociality, Edited by Raimo Tuomela, Gerhard Preyer, and Georg Peter, Double Vol. 18–19.

[Cha95]   Chaib-draa, B., (1995), Coordination between agents in routine, familiar, and unfamiliar situations. Int. Journ. of Intelligent & Cooperative Information Systems.

[Con95]   R. Conte and C. Castelfranchi. *Cognitive and Social Action*, UCL Press, London, 1995.

[Cre00]   Cremonini M., Omicini A., and Zambonelli F., (2000) Ruling agent motion in structured environments. In Marian R. Bubak, Hamideh Afsarmanesh, Roy Williams, and Bob Hertzberger, editors, *High Performance Computing and Networking — Proceedings of the 8th International Conference (HPCN Europe 2000),* volume 1823 of *LNCS,* pages 187–196, Amsterdam (NL). Springer-Verlag.

[Fal97]   R Falcone and C Castelfranchi. "On behalf of ..": levels of help, levels of delegation and their conflicts, *4th ModelAge Workshop*: "Formal Model of Agents", Certosa di Pontignano (Siena),1997.

[Fai98]   Faiz ul Haque Zeya, Subject: Re: The differences between Intelligent agent and Autonomous agent. In Agent List Date: Fri, 27 Mar 1998

[Had96]   Haddadi A. and Sundermeyer K., (1996), Belief-Desire-Intention Agent Architectures. In G.M. O'Hare and N.R. Jennings (eds.) Foundations of Distributed Artificial Intelligence, Wiley & Sons, London.

[Hex03]   Hexmoor H. A Model of Absolute Autonomy and Power: Toward Group Effects, In Journal of Connection Science, Volume 14, No. 4. Taylor & Francis Ltd, 2003.

[Sch01]   Schillo, M., Zinnikus, I., Fisher K., (2001), Towards a Theory of flexible holons: Modelling Institutions for making multi-agent systems robust. In 2nd Workshop on Norms and Institutions in MAS, Montreal, Canada.

[Sho93]   Shoham,Y., Agent-oriented programming. In *Artificial Intelligence,* 60, 1993.

[Sic94]   Sichman, J.S., Conte, R., Castelfranchi, C., Demazeau, Y. A Social Reasoning Mechanism Based On Dependence Networks. In Proceedings of the European Conference on Artificial Intelligence - ECAI'94, Amsterdam August 8–12, 188–192.)

[Sic95]   Sichman, J.S., (1995), Du Raisonement Social Chez les Agents. PhD. Thesis, Pollytechnique de Grenoble, 1995.

[Tri71]   Trivers R.L., (1971), The evolution of reciprocal altruism, in Quarterly review of Biology, XLVI.

[Ver00]   Verhagen, H. *Normative Autonomous Agents.* PhD. Thesis, University of Stockolm, May, 2000.

Agent Autonomy Through the $_3M$ Motivational Taxonomy

Steve Munroe and Michael Luck

School of Electronics and Computer Science,
University of Southampton, Southampton, UK
{sjm01r, mml}@ecs.soton.ac.uk

Abstract. The concept of autonomy applied to computational agents refers to the ability of an agent to act without the direct intervention of human users in the selection and satisfaction of goals. Actual implementations of mechanisms to enable agents to display autonomy are, however, at an early stage of development and much remains to be done to fully explicate the issues involved in the development of such mechanisms. Motivation has been used by several researchers as such an enabler of autonomy in agents. In this paper we describe a motivational taxonomy, the $_3M$ *Taxonomy*, comprising *domain, social* and *constraint* motivations, which we argue is a sufficient range of motivations to enable autonomy in all the main aspects of agent activity. Underlying this taxonomy is a motivational model that describes how motivation can be used to bias an agent's activities towards that which is important in a given context, and also how motivational influence can be dynamically altered through the use of motivational *cues*, that are features in the environment that signify important situations to an agent.

1 Introduction

Much of computing, especially Artificial Intelligence (AI), is conceptualised as taking place at the *knowledge level*, with computational activity being defined in terms of *what* to do, or *goals*. Computation can then be undertaken to achieve those goals, as is typical in planning, for example. However, the reasons for the goals arising are typically not considered, yet they may have important and substantial influence over their manner of achievement. If goals determine *what* to do, these reasons, or *motivations*, determine *why* and consequently how. The best illustration of the role of motivation in computing is perhaps in relation to autonomous agents which, in essence, possess goals that are *generated* within, rather than *adopted* from, other agents [3]. These goals are generated from motivations, higher-level non-derivative components that characterise the nature of the agent. They can be considered to be the desires or preferences that affect the outcome of a given reasoning or behavioural task. For example, *greed* is not a goal in the classical artificial intelligence sense since it does not specify a state of affairs to be achieved, nor is it describable in terms of the environment. However, it may give rise to the generation of a goal to rob a bank. The motivation of greed

M. Nickles, M. Rovatsos, and G. Weiss (Eds.): AUTONOMY 2003, LNAI 2969, pp. 55–67, 2004.

and the goal of robbing a bank are clearly distinct, with the former providing a reason to do the latter, and the latter specifying how to achieve the former. In a computational context, we can imagine a robot that normally explores its environment in an effort to construct a map, but must sometimes recharge its batteries. These motivations of 'curiosity' and 'hunger' lead to the generation of specific goals at different times, with a changing balance of importance as time passes. Similarly, when undertaking a reasoning task, the nature and degree of reasoning that is possible must be determined by the need for it: in the face of a critical medical emergency, a coarse but rapid response may be best; in experimental trials, repeatability and accuracy are needed, often regardless of the time taken. Motivation is the distinguishing factor.

This view is based on the generation and transfer of goals between autonomous agents. More specifically, agent-based computing generally operates at the knowledge-level where goals are the currency of interaction. Goals specify what must be achieved without specifying how, and in that sense, enable individual agents to choose the best means available to them in deciding how to achieve them. Although this gives a large degree of autonomy in the dynamic construction of multi-agent systems, virtual organisations, etc., it provides little by way of direction, guidance, or meta-level control that may be valuable in determining how best to achieve overarching aims. Motivations address this both by providing the reasons for the goal, and by offering constraints on how the goal might best be achieved when faced with alternative courses of action. In that sense, motivations both release and constrain an agent's autonomy.

Motivation as an area of study is thus important in two key respects. First, and most intuitively, it is critical in understanding human and animal behaviour, and computational models can aid in testing relevant hypotheses. Second, from a computer science (and altogether more pragmatic) perspective, it potentially offers a substantially higher level of control than is available at present, and which will become increasingly important for (agent) systems that need to function in an autonomous yet persistent manner while responding to changing circumstances. Autonomous agents are powerful computational tools, but without the constraints that might be provided by motivations, may potentially lack the required behavioural control. A combination of these two aspects may also permit computational agents to better understand and reason about another (possibly animate) agent's motivations, both in application to computational multi-agent systems and to the human-computer interface.

2 Autonomy and Motivation: Towards a Taxonomy

2.1 Autonomy and Agents

The concept of a computational, or autonomous agent, as used in computer science, has many definitions and, while no canonical definition exists, it is possible nevertheless to identify a group of characteristics which, to some extent, are common to most views of what constitutes agenthood. These characteristics routinely emphasise an agent's ability to be *reactive* and *pro-active* with respect to its environment, to be *sociable* and, finally, to be *autonomous*. They find expression in a well-accepted and oft-used characterisation of agenthood that is presented below, and which we adopt here [15].

- *Reactiveness*: agents are able to perceive their environments and respond in a timely fashion to changes within it in order to satisfy their objectives.
- *Pro-activeness*: agents are able to exhibit goal-directed behaviour by *taking the initiative* in order to satisfy their design objectives.
- *Social ability:* agents are capable of interacting with other agents (and possibly humans) in order to satisfy their design objectives.
- *Autonomy*: agents operate without the direct intervention of humans or others, and have some kind of control over their actions and internal state.

The first two characteristics in the above definition, *reactiveness* and *pro-activeness*, are concerned with the concept of *action*. Agents are computational entities whose existence centres around the notion of taking action to solve problems. Agents are therefore primarily entities that *do something*, be it collecting information on behalf of a user, reallocating resources in a supply chain, physically moving objects around in robotic systems, simulating the behaviour of pedestrians in an experimental setting, and so on. Action thus ensures the usefulness of agents.

2.2   Domain Motivations

In this context, being reactive and pro-active means having the ability to *respond* to the challenges that exist in the current environment, as well as having the ability to display *goal-directedness* and *pro-activeness*. Just what an agent is expected to respond to, and be pro-active towards, is strongly linked to the agent's *role* within the system it inhabits. Such roles will, of course, vary from domain to domain, and there is thus little to be said about their general characteristics. However, if we accept that an agent must have some *domain specific* roles that demand the satisfaction of a set of goals, then it is possible to place these tasks under the control of some set of *domain motivations*, which will facilitate the *generation* of these goals at *relevant* times and determine their relative *value* to the agent.

For example, an agent may have a domain role of maintaining the cleanliness of a warehouse. Part of this role may be to satisfy a number of goals related to ensuring that the various boxes in a collection of such boxes are stored at specific predefined locations. To control the generation of such goals, we can instantiate a domain motivation for tidiness, which is sensitive to situations in which boxes are not in their right location. At some point in time the agent may become motivated to generate a set of goals, the satisfaction of which would result in a tidy warehouse. Thus using the concept of domain motivations we can meet the two characteristics of reactivity and pro-activity in the above definition.

2.3 Constraint Motivations

The ability to choose the *manner* in which to satisfy a given goal is also associated with autonomy. Motivation can be used as a natural mechanism to represent the *constraints* over the ways in which a goal may be satisfied. In this way, motivation provides a natural way to conceptualise *meta-level control* processes. Thus motivation can be used to monitor aspects of an agent's situation such as risk, cost, efficiency etc. Along with domain motivations, therefore, we argue that an agent must have a set of what we call *constraint motivations*, which perform the task of imposing restrictions on the use of resources and the importance to be placed on their use.

2.4 Social Motivation

The third point in the characterisation of agents above states the requirement of some kind of social ability. Sociality in agents is arguably the key aspect of agent-based computing that differentiates it from other kinds of computing paradigms. Sociability allows agents to combine their individual capabilities together to deal with problems that are not possible, or at least much more difficult, to solve by individual agents working in isolation. The two characteristics of action and sociability are what marks out agent-based computing from other forms of computational approaches, and are what gives the technology leverage over the complex, open and dynamic domains that are the mainstay of applications for agent technology.

Whereas most existing work on motivated agents has focused on the goal generation aspects of motivation, we see motivation as also having a direct effect on an agent's *interactions*. We view motivation as offering a natural mechanism for evaluating requests for assistance. For example, an agent in a dynamic, open environment may have many forms of relationship with other agents ranging from completely cooperative to purely selfish relationships. Moreover, an orthogonal concern is how *important* a relationship is. Importance may derive from numerous sources, such as authority rank, or from the level of dependence of one agent to another and, furthermore, numerous gradations may exist between the extremes; an agent must have some way to quantify these measures of selfishness and importance.

The notion of the importance of a relationship between two agents is central to how they will interact. Generally, the more important one agent is to another the more it will gain from helping the agent satisfy its goals. Agents will vary from having relationships in which the utility gained by one agent in the relationship has no importance to the other agent, to situations in which the utility gained by one agent is worth many times that utility to the other agent. With the above considerations in mind, we add to the other two sets of motivations a third set, *social motivations*.

2.5   The $_3M$ Motivational Taxonomy

As a result of the previous discussion, we can construct a set of motivations that we claim to be sufficient enablers of autonomy in three key areas of agent operation. We call this taxonomy the $_3M$ Taxonomy, which comprises the three sets of motivations described above: *domain motivations, constraint motivations,* and *social motivations.* These three sets of motivations govern the manner in which an agent engages in goal generation and evaluation, resource monitoring, and social interaction, and are summarised below.

Domain Motivations. These types of motivations represent the concerns and tasks that make up the agent's *functional role* in the system it inhabits. Domain motivations are responsible for the generation of the goals and actions that enable an agent to fulfill the requirements placed on it, either by the designer or the organisation of which the agent forms a part.

Constraint Motivations. These types of motivations control the ways in which domain motivations are satisfied. They represent those restrictions on behaviour and resources that may derive from societal obligations such as norms, restrictions etc, or from naturally occurring limitations such as the scarcity of a resource, or a lack of time and so on.

Social Motivations. These motivations determine the manner in which an agent interacts with other agents. For every acquaintance of an agent, there is a social motivation that represents the importance of that relationship. As interactions occur, the nature of the relationship may change such that it might become more or less important. Social motivations track such relationship issues.

Figure 1 shows the $_3M$ taxonomy composed of: domain motivation, constraint motivation and social motivation and how they all impact on an agent's decision-making.

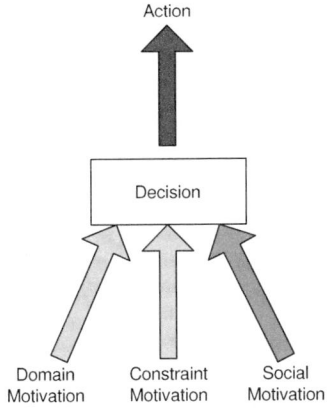

Fig. 1. The $_3M$ taxonomy

Even with the ability to take action and to engage in social interactions, the real power of agents lies in the development of methods to apply the last point in the definition regarding *autonomy*. Without autonomy, agent systems could only be applied to the most predictable problem domains, as they would be unable to cope with unexpected events and uncertainties so characteristic of the kinds of domains that we hope to use agents to address. Action without autonomy entails having to laboriously tell an agent what to do at every step, leaving us with a technology not any more advanced than most word processing packages. Allowing agents to have autonomy frees the user from having to maintain a detailed control of what, when and how an agent goes about its job. Similarly with social ability; without autonomy it would be left to the designer to explicitly state which other agents to interact with, when to do so and how. Again, autonomy here enables an agent to best decide for itself who its interaction partners should be, and under what conditions the interaction should take place.

3   A Motivational Model

As discussed in the introduction, motivations represent higher-order desires or preferences and, faced with a given decision as to how to achieve a goal, whether to respond to an external request, or to agree to the use of some resource, an agent will use motivations to determine if the decision is consistent with them or not. In this way, even in the face of large numbers of such decisions, an agent can quickly determine its priorities and proceed to organise its reasoning and actions accordingly. For example, faced with several goals to satisfy, an agent can determine which goals to achieve (and are therefore deserving of attention) in relation to its motivations. Similarly, the decision to accept or reject a request from another agent can depend on whether agreeing to help is in line with the agent's motivations at that time.

In these examples, motivation helps an agent to make decisions about what is important and therefore, which courses of action to embark upon. Effectively, motivation offers a way to bias an agent's decisions and activities so that it only engages in activities that are of value to it. Moreover, the bias or influence that a motivation may exert over an agent's activities may be *dynamically varied* by making motivation itself sensitive to an agent's external environment. These *motivational cues* provide an extra layer of control on an agent's activities.

3.1   Motivational Cues

We have discussed how motivation can be used to bias an agent's activities so that only those that are of value will be adopted. It is also possible, however, to bias a motivation's influence on the selection of activities such that at times the influence of a motivation may vary in strength. This can be achieved by defining general characteristics of situations which, when manifested in the environment, affect the motivations of an agent so that its activities begin to service the needs of whichever motivation is exerting the strongest influence. These characteristics are what we call *motivational cues* and they describe specific features of an environment that have some importance to an agent.

For example, imagine a robot agent that is interested in those classes of situations in which its environment is *untidy*. We could define a collection motivational cues that represent certain features in the environment describing situations of untidiness, such as the number of books lying around, or the length of time since a room was last dusted. In both of these cases, the environment can be classed as *untidy*, even though the features that define the situation differ. Moreover, the robot designer may only want the robot to be interested in the tidiness of the environment when it becomes *very* untidy. This suggests that the robot may need to be able to take some form of *measurement* of certain environmental features so that it can understand *to what degree* a situation falls into those classes of situations in which it is interested (in this example, untidy situations). When the robot perceive that it is within a particular situation that is important to it, it should be prompted to take some action, for example by satisfying a goal to tidy up the books. In such a way, the robot's motivations should be sensitive to a number of features in the environment that can inform it about the current situation and trigger action. Such motivational cues affect motivations by *increasing* or *decreasing* their influence on activity.

In Figure 2, we show how cues, motivations and information interact. The cues affect the strength of the motivation and that in turn affects the value that is placed on the information being considered by the agent.

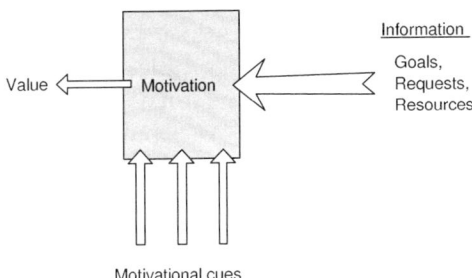

Fig. 2. Motivational cues influence the motivational the value placed on information

3.2   Goal Value

The strength of a motivation, determined by the current situation, provides a *heuristic method* for calculating the worth or *value*, and hence the importance, of a given goal. Goals generated by exceptionally strong motivations should have a proportional *value* to an agent, meaning that the state of affairs they describe assumes a high level of importance to the agent at that time. Similarly goals behind which there is only little motivational strength, should represent to the agent less valuable states of affairs. Now, as motivations can increase or decrease in strength depending on the situation, so goals may also increase or decrease in value over periods of time. For example, the value of a goal to put a box into its proper position, in the context of a clean environment would receive little value from its associated tidiness motivation. If however, the same goal were to be generated in the context of an extremely cluttered environment, then the goal should receive a larger amount of value.

3.3   Goal Relevance

An advantage of using motivation to generate goals is that it avoids generating *irrelevant* goals, that, although compatible with environmental conditions, do not have any relevance in terms of an agent's current needs. For example, imagine an agent operating without the use of motivations that can adopt a goal to tidy boxes that are out of position. The agent will adopt the goal *every time* it comes to have a belief that a box is out of position. In contrast, a motivated agent in this situation, even when it holds the belief that a box is out of position, *still* may not adopt the goal if the motivational effect of satisfying the goal is too small. A motivated agent would only adopt the goal if it is *warranted* in the current *broader situation* defined, say, as the amount of clutter existing in the agent's environment. If the environment is becoming very cluttered, then it would *make sense* to adopt the goal to put boxes in their proper positions, otherwise the agent is probably better off satisfying other, *more important* goals.

3.4   Operationalising Motivational Influence

When an agent perceives a situation that is motivationally relevant it should respond in some way, for example by generating and subsequently satisfying a goal. When such an event occurs we say that the motivation has been *matched* to cues in the environment. However, as a motivation may be matched to a number of cues, we need to be able to represent the situations in which a motivation may be matched to different amounts of cues. Generally, the more cues a motivation is matched to, the more influence that motivation should have on the agent's decision-making. For example, we may design an agent with a *tidiness* motivation that can be matched to two cues, which could be, say, paper on the floor, and a full waste-paper basket. If motivation is matched to both of these cues, it should have a greater influence over the agent's decisions than if only matched to one. We represent the strength of motivational influence by the notion of *motivational intensity*, which is a real number in the interval $[0,1]$, where 1 means maximum intensity and 0 means no intensity. Then, we can say that a motivation will increase in intensity in certain situations and decrease in others, depending on the number of cues in the environment to which it is matched.

3.5   Mitigating Motivational Intensity

So far we have discussed how, through the increase and decrease of intensity, motivation can exert a greater or lesser influence on an agent's activities. If an agent generates a goal in response to an increase in motivational intensity, then we should expect that by satisfying the goal, the agent is also satisfying or *mitigating* its motivation. Mitigating a motivation is the process of lowering intensity with the effect of lessening the motivation's influence over the activities of the agent. The operation of motivation-based agents is thus a process of becoming motivated by certain situations and then seeking ways to mitigate that motivation.

One way to link the satisfaction of goals to the mitigation of motivational intensity is to manipulate the relationships that exist between goals, cues, and motivational intensity. For example, if an agent uses the number of boxes out of position as a cue to set the strength of a *tidiness* motivation then, as more boxes move out of position, the motivational intensity should increase. Now, if the agent generates and then satisfies a goal, the goal must, in order for it to be *motivationally relevant*, affect one or more of the cues attached to the motivation. So, in satisfying a goal to place boxes into their correct positions, the agent affects the cue relating to the number of boxes out of position which, in turn, leads to the mitigation of the tidiness motivation's intensity. Figure 3 shows this relationship in which a goal affects the intensity of the motivation responsible for its generation by affecting the environment and hence the motivational cues in the environment.

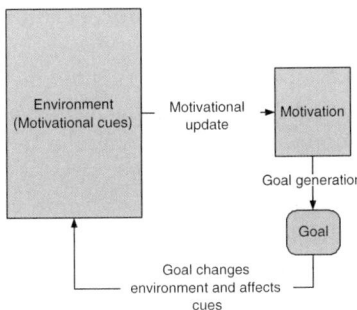

Fig. 3. The relationship between goals, cues and motivation

In this way, a goal can be said to perform a certain amount of *work* for a motivation, so that the more work a goal does (i.e. the more it mitigates a motivation), the more value it has. The amount of work done by a goal is determined by (i) the plan used to satisfy the goal, and (ii) the form of the relationship between the cue and motivational intensity. Figure 4 shows two such relationships in which the cue could represent the *number* of boxes out of position and hence might be significant for a tidiness motivation. The cue can take on any number of values and, as the number of boxes out of place increases, so the motivational intensity increases.

In each case, the relationship between cue value and motivational intensity is represented. As a cue's value increases (x axis), there is a related increase in motivational intensity (y axis). Relationship A is a linear relationship between cue values and motivational intensity, and relationship B is non-linear. In linear relationships, a change in cue value produces the same amount of motivational mitigation irrespective of the intensity of the motivation. This is shown in the figure for relationship A, where lowering the cue's value from 8 to 7 (representing

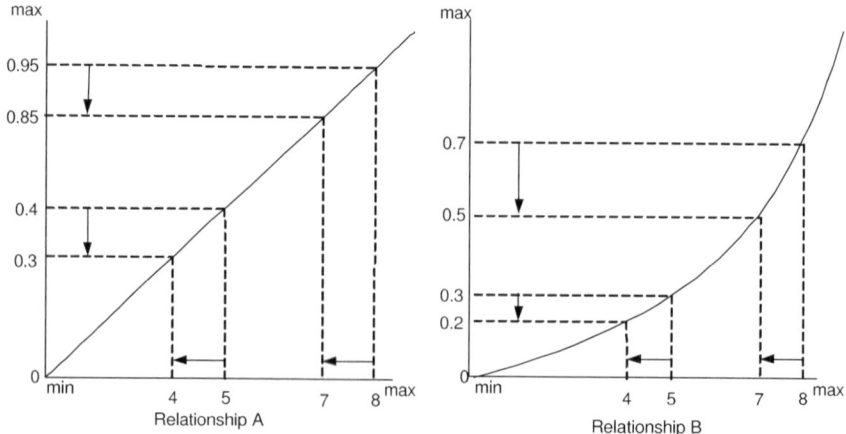

Fig. 4. Two examples of cue/intensity relationship

say, that one of the boxes out of position has now been placed back in its rightful position) produces the same amount of mitigation as changing the cue's value from 5 to 4. For type B relationships, however, changing the value of the cue by a specific amount produces different amounts of mitigation, depending on the intensity levels. Thus, changing the value of the cue from 8 to 7 now produces twice as much mitigation as changing the cue's value from 5 to 4. By defining, for each goal, the size of the effects on a cue, we can obtain different mitigation values for the same goal at different times, depending on the relationship that holds between a cue and the motivation.

3.6   Choosing Between Goals

When attempting to mitigate motivational intensity, an agent may be faced with a number of options in terms of different goals that it might have to choose between. Choosing between goals involves looking both at information in the goal and also the motivational state of an agent. Now, satisfying a goal may affect more than one motivation, possibly reducing the intensity of some while also increasing the intensity of others. For example, in satisfying a goal to eat by going to a restaurant, I may succeed in lowering my *hunger* motivation, while at the same time succeeding in raising the intensity of my motivation to *conserve money* (especially if the restaurant is expensive). In the time following my trip to the restaurant, my motivation to conserve money might make me reject the goal of taking a taxi and influence me towards choosing a goal to take a bus. Thus, in satisfying any goal, there may be effects on a number of different motivations, and these effects need to be taken into consideration when choosing between courses of action.

In our work we specify beforehand both the positive and negative effects that a goal might have on a motivation. This effect is adjusted by the intensity of the motivations that the goal affects, so that the overall motivational value,

$MValue_g$, of a goal on an agent's motivations, m, is given by

$$MValue_g = \Sigma_{i=1}^n \left(intensity_{mi} \times value_g^+\right) + \Sigma_{j=1}^k \left(intensity_{mj} \times value_g^-\right)$$

where n is the number of motivations positively affected by the goal, k is the number of motivations negatively affected by the goal, $value_g^{(+/-)}$ is the positive or negative effect of a goal and $intensity_m$ is the current intensity of motivation m.

The effect of a goal on the associated motivations is thus adjusted by the intensities of those motivations. In this way, motivation helps to place an importance value on a goal as the overall effect it has on an agent's motivations. Given a number of possible goals from which to choose, an agent is able to calculate the overall motivational effects of each goal and select the one providing the highest overall benefit.

4   Related Work

While much work has been done within psychology on motivational systems and mechanisms, the task of adapting these to *computational contexts* remains largely ignored. Some early work undertaken by Simon [10], takes motivation to be "that which controls attention at any given time", and explores the relation of motivation to *information-processing behaviour*, but from a cognitive perspective. Sloman *et al* [11,12] have elaborated on Simon's work, showing how motivations are relevant to *emotions* and the development of a *computational theory of mind*. More recently, Ferber [4] has provided a motivational taxonomy that a computational agent may possess, in which the motivational system exists as part of the agent's larger *conative system*. The taxonomy identifies four sources of motivation: (i) personal motivations, (ii) environmental motivations, (iii) social motivations and (iv) relational motivations. Unlike the $_3M$ Taxonomy, which is tied to a widely accepted definition of agency, Ferber's work targets mainly subsymbolic agents and Ferber fails to tie down his taxonomy to definite characteristics of agent activity. Ferber's taxonomy is also lacking in not having any detailed description of a motivational mechanism, nor of how motivations can be implemented in real computational systems. The importance of motivation as an enabler of autonomy in computational agents was perhaps first identified by d'Inverno and Luck [3], who discuss the importance of motivation in allowing an agent to generate its own goals, as opposed to being given them by other agents or human users. Further analysis of this view can be found in [7]. More recent efforts have extended Luck and d'Inverno's ideas to consider cooperation [5] and normative systems [6], and using motivated agents for simulating virtual emergencies [2].

Other research has concentrated on using motivation to increase an agent's ability to predict the need for future action, [9], evaluating plans [1], the merging of behaviours [13], action selection [8], and modeling organisational relationships [14]. The work described in this paper is different from these models in that we provide a detailed model of motivational cues, and a complete account of their effects on motivational intensity, the generation of activity and the mitigation of motivational intensity. Furthermore, our motivational model is grounded in the $_3M$ Taxonomy, providing a complete account of the effects of motivation over three main areas of agent activity: goal-directed behavior, social relationships and resource monitoring.

5   Conclusions and Future Work

In this paper we have described a motivational taxonomy comprising motivations to enable autonomy in three areas of agent activity, the generation of goals using *domain motivations*, the management of relationships using *social motivations* and the monitoring of resources using *constraint motivations*. The use of motivations in these areas of agent activity allows the placing of value, or importance, on goals, relationships and resources that can be dynamically altered to suit the overarching needs of an agent given its current situation. The manner in which motivations are made dynamically responsive to situations is through the use of motivational cues. Motivational cues represent features of the environment which, when manifested, cause an agent's motivations to increase or decrease in their influence over the agent's decision-making regarding the selection of goals, the response to be made to requests from other agents, and the use of resources. The paper described the possible relationships between the effects of satisfying a goal and the resulting changes that can be made to the intensity of a motivation. More specifically, the problem of how to choose between competing goals was also discussed, and a technique outlined that uses motivational intensities to weight the effects of goals on motivations, thus enabling an agent to choose a goal that has the highest positive motivational effect.

Our current efforts are directed at applying the motivational model and taxonomy in negotiation scenarios. Specifically, we are examining how an agent can autonomously discover constraints on a self-generated negotiation goal, which involves the use of constraint motivations to set the limits on the acceptable bounds for negotiation issue-value assignments. We are also developing a motivated negotiation partner-selection mechanism that evaluates prospective negotiation partners based on their past performance, and the current constraints that the agent initiating the negotiation must operate within. The overall goal of such work is to increase the autonomy of agents engaging in negotiations so that they are able to determine the kinds of negotiation outcomes that are acceptable, and which other agents might best match the those outcome requirements.

References

1. A.M. Coddington and M. Luck. Towards motivaiton-based plan evaluation. In I. Russell and S. Haller, editors, *Proceedings of Sixteenth International FLAIRS Conference*, pages 298–302, 2003.

2. S. de Lioncourt and M. Luck. Motivating intelligent agents for virtual environments. In *Proceedings of the Intelligent Virtual Agents Workshop*, Salford, 1999.
3. M. d'Inverno and M. Luck. *Understanding Agent Systems*. Springer, 2001.
4. J. Ferber. *Multi-Agent Systems: An Introduction to Distributed Artificial Intelligence*. Addison Wesley Longman, 1999.
5. N. Griffiths. *Motivated Cooperation*. PhD thesis, University of Warwick, 2000.
6. F. Lopez y Lopez, M Luck, and M. d'Inverno. Constraining autonomy through norms. In *Proceedings of the First International Conference on Autonomous Agents and Multi-Agent Systems*, 2002.
7. M. Luck, S. Munroe, and M. d'Inverno. Autonomy: Variable and generative. In C. Castelfranchi H. Hexmoor and R. Falcone, editors, *Agent Autonomy*, pages 9–22. Kluwer, 2003.
8. P. Maes. How to do the right thing. *Connection Science*, 1(3), 1989.
9. T. J. Norman and D. Long. Goal creation in motivated agents. In M. Wooldridge and N. R. Jennings, editors, *Intelligent Agents (LNAI Volume 890)*, pages 277–290. Springer Verlag, 1995.
10. H.A. Simon. *Models of Thought*. Yale University Press, 1979.
11. A. Sloman. Motives, mechanisms, and emotions. *Cognition and Emotion*, 1:217–233, 1987.
12. A. Sloman and M. Croucher. Why robots will have emotions. In *Proceedings of the Seventh International Joint Conference on Artificial Intelligence*, pages 197–202. Vancouver, B.C, 1981.
13. E. Spier and D. McFarland. A finer-grained motivational model of behaviour sequencing. In *From Animals to Animats 4: Proceedings of the fourth conference on the Simulation of Adaptive Behavior*, 1996.
14. T. Wagner and V. Lesser. Relating quantified motivations for organizationally situated agents. In *Intelligent Agents VI — Proceedings of the Sixth International Workshop on Agent Theories, Architectures, and Languages, LNAI*, pages 334–348. N. R. Jennings and Y. Lesperance (eds.), Springer-Verlag, Berlin, April 1999.
15. M. Wooldridge and N. R. Jennings. Intelligent agents: Theory and practice. *Knowledge Engineering Review*, 10(2):115–152, 1995.

A Taxonomy of Autonomy in Multiagent Organisation

Michael Schillo and Klaus Fischer

German Research Center for Artificial Intelligence,
Stuhlsatzenhausweg 3, 66123 Saarbrücken, Germany
{schillo, kuf}@dfki.de
http://www.dfki.de/~{schillo, kuf}

Abstract. Starting from a general definition of how to model the organisation of multiagent systems with the aid of holonic structures, we discuss design parameters for such structures. These design parameters can be used to model a wide range of different organisational types. The focus of this contribution is to link these design parameters with a taxonomy of different types of autonomy relevant in multiagent organisation. We also discuss the constraining effect of autonomy on the recursive nesting of multiagent organisation. As the domain for applying multiagent systems we choose a general view on multiagent task-assignment.

1 Introduction

The relationship between organisation and autonomy is of increasing importance to researchers from distributed artificial intelligence (DAI). Both concepts are of fundamental importance to the design of multiagent systems. According to Jennings, the "development of robust and scalable software systems requires autonomous agents that can complete their objectives while situated in a dynamic and uncertain environment, that can engage in rich, high-level social interactions, and that can operate within flexible organisational structures" [1]. The advantages of agents that act in organisational structures he sees are that organisations can encapsulate complexity of subsystems (simplifying representation and design) and modularise functionality (providing the basis for rapid development and incremental deployment).

We have previously presented a set of organisational forms for multiagent systems and discussed how they relate to agent autonomy [2]. In this paper, we will take the approach one step further. We will generalise from the concrete forms of organisation and describe the set of underlying design parameters from which these organisational forms (and many others) can be produced. Then we will show how these parameters match to different aspects of autonomy. For this discussion the work of Castelfranchi is important, who showed how autonomy can be founded on dependence theory [3].

Holonic multiagent systems provide the basic terminology and theory for the realisation of multiagent organisation and define the equivalent of modularity

M. Nickles, M. Rovatsos, and G. Weiss (Eds.): AUTONOMY 2003, LNAI 2969, pp. 68–82, 2004.

and recursion of traditional computer science to the agent paradigm. In a holonic multiagent system, an agent that appears as a single entity to the outside world may in fact be composed of many sub-agents and conversely, many sub-agents may decide that it is advantageous to join into the coherent structure of a super-agent and thus act as single entity. These concepts have successfully been applied to multiagent systems especially in the area of distributed scheduling. In order to ground our contribution, we consider particularly this domain, which can range from meeting scheduling, to supply web scheduling, and to service composition in the semantic web.

In the following section we present a definition of holonic organisation that both utilises recursion and allows flexibility in describing forms of holonic organisation. We will use this definition as a framework to describe design parameters for multiagent organisation in Section 3. In Section 4 we identify different types of autonomy and show how they match to these design parameters. Some issues of recursive nesting and agent autonomy are covered in Section 5.

2 Holonic Multiagent Systems – A Framework for the Definition of Multiagent Organisation

The term "holon" was originally coined by Arthur Koestler [4], according to whom a *holon* is a self-similar or *fractal* structure that is stable and coherent and that consists of several holons as sub-structures and is itself a part of a greater whole. Koestler gave biological examples. For instance, a human being consists of organs which in turn consist of cells that can be further decomposed and so on. Also, the human being is part of a family, a team or a society. None of these components can be understood without their sub-components or without the super-component they are part of.

To the outside, multiagent holons are observable by communication with their representatives. These are called the *head* of the holon, the other agents in the holon are part of the holon's *body*. In both cases, representative agents communicate to the outside of the holon in pursuit of the goals of the holon *and* coordinate the agents inside the body of the holon in pursuit of these goals. The binding force that keeps head and body in a holon together can be seen as commitments. This differentiates the approach from classical methods like object-oriented programming: the relationships are not (statically) expressed at code level, but in commitments formed during runtime. For a multiagent system consisting of the set $\mathcal{A}_t$ of agents, the set $\mathcal{H}_t$ of all holons at time t is defined in the following way.

Definition 1 (Holonic Multiagent System). *A multiagent system* $\mathcal{MAS}$ *containing* holons *is called a* holonic multiagent system. *The set* $\mathcal{H}$ *of all holons in* $\mathcal{MAS}$ *is defined recursively:*

- *for each* $a \in \mathcal{A}_t$, $h = (\{a\}, \{a\}, \emptyset) \in \mathcal{H}$, *i.e. every instantiated agent constitutes an* atomic *holon, and*

– $h = (Head, Subholons, C) \in \mathcal{H}$, where $Subholons \in 2^{\mathcal{H}} \backslash \emptyset$ is the set of holons that participate in h, $Head \subseteq Subholons$ is the non-empty set of holons that represent the holon to the environment and are responsible for coordinating the actions inside the holon. $C \subseteq Commitments$ defines the relationship inside the holon and is agreed on by all holons $h' \in Subholons$ at the time of joining the holon h.

    Given the holon $h = (Head, \{h_1, ..., h_n\}, C)$ we call $h_1, ..., h_n$ the subholons of h, and h the superholon of $h_1, ..., h_n$. The set $Body = \{h_1, ..., h_n\} \backslash Head$ (the complement of $Head$ in h) is the set of subholons that are not allowed to represent holon h. Holons are allowed to engage as subholons in several different holons at the same time, as long as this does not contradict the sets of commitments of these superholons[1].

    A holon h is observed by its environment like any other agent in $\mathcal{A}_t$. Only at closer inspection may it turn out that h is constructed from (or represents) a set of agents. The set of representatives can consist of several subholons. As any head of a holon has a unique identification, it is possible to communicate with each holon by just sending messages to their addresses. C specifies the organisational structure and is covered in detail in Section 3. As long as subholons intend to keep their commitments and as long as subholons do not make conflicting commitments, cycles in holonic membership are possible (see Example 2 below).

Example 1. Given $h = (\{h_1, h_2\}, \{h_3\}, c_1)$ and an agent k intending to request a task from h. As the head of h consists of two subholons h_1 and h_2, k has two options. It can either address h_1 or h_2. In both cases, the addressed subholons will coordinate the task performance inside of h and deliver the task result.

Example 2. Given $h_1 = (\{a\}, \{b\}, c_1)$ and an agent k addressing a. On the creation of h_1, a and b agreed on commitments c_1 that also explain in which cases a needs to act as the head (e.g. when being addressed in a certain manner or when being requested certain types of tasks). In this case, a can deduce from the way it is addressed, whether it should act as the head of h_1 or just for itself. For the same reason, cycles are not a problem in the definition. Assume we have $h_2 = (\{b\}, \{a\}, c_2)$. If k addresses a, it is clear from the definition of h_2 that a is not addressed as one of its representatives (a is not part of the head of h_2). Whether it should respond for itself or as part of the head of h_1 is decided as without the cycle in the case without h_2).

Definition 2 (Further Holon Terminology). *These notions are not required for the definition of the concept of a holonic multiagent system itself, but make it easier to discuss certain properties.*

---

[1] At this point we make the assumption that agents *act* according to their commitments. A sanctioning mechanism that punishes incorrect behaviour can build on the commitments made at runtime, as they are communicated explicitly, but is not in the scope of this work.

- *A task holon is a holon that is generated to perform only a single task. This notion is opposed to* organisational holons, *which are designed to perform a series of tasks.*
- *Delegation of tasks between two subholons h_1 and h_2 of a holon h as part of working towards the goal of h is called* intra-holon delegation. *If two holons delegate tasks and this collaboration is not part of the goal of an encompassing holon, this is called* inter-holon delegation.
- *Finally, a holon that is not atomic is called a* holarchy. *A holarchy of which all nested subholons have only a single head holon, i.e. a holarchy with a tree-like structure, is called a* hierarchical holarchy.

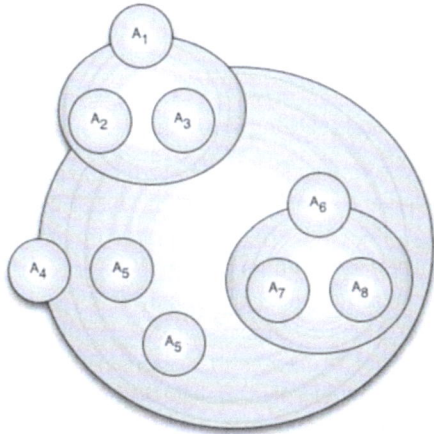

Fig. 1. An example for a holarchy, a complex nesting of holons

Figure 1 gives an example of several possible relationships. The largest entity consists of all agents depicted and has two head holons. The first head holon, named A_4 is atomic, the second one is a composed structure itself and is represented by A_1. The body of the superholon consists of another atomic holon A_5 and another composed holon which is represented by A_6. A_1 has a double function as it represents the holon that represents the outermost holon. A multimembership is demonstrated by A_7. It is member of two structures (which are both entailed by the biggest holon), in one of them it is body, in the other it is a head member. By definition, all the agents in this diagram could be replaced by further complex holonic structures, resulting in complex relationships as the ones illustrated with A_1 and A_7.

The advantages of the holonic concept are threefold. First, this technology preserves compatibility to multiagent systems by addressing every holon as an agent, whether this agent represents a set of agents or not, is encapsulated. Second, as every agent may or may not represent a larger holon, holonic multiagent systems are a way of introducing recursion to the modelling of multiagent systems, which has proven to be a powerful mechanism in software design to deal with complexity. Third, the concept does not restrict us to a specified type of association between the agents, so it leaves room to introduce organisational concepts at this point.

3   Design Parameters for Holonic Multiagent Systems

A framework that describes the different types of holon defined in Section 2 requires design parameters to span a space for design decisions. In this section we list these parameters, which not only guide the process of creating the concrete holonic structure, but which also restrict the behaviour of each agent and influence the autonomy of each agent inside the holon. They define the set of commitments C that is part of the holon's definition itself.

3.1   Mechanisms for Task Delegation and Social Delegation

Recent work on delegation, has shown that delegation is a complex concept highly relevant in multiagent systems [5,6]. The mechanism of delegation makes it possible to pass on tasks (e.g. creating a plan for a certain goal, extracting information, etc.) to other individuals and furthermore, allows specialisation of these individuals for certain tasks (functional differentiation and role performance). Representing groups or teams is also an essential mechanism in situations which deal with social processes of organisation, coordination and structuring. Following the concept of social delegation of sociologist Pierre Bourdieu [7], we distinguish two types of delegation: task delegation and social delegation. We call the procedure of appointing an agent as representative for a group of agents *social delegation*.

The activity of social delegation (representation) is in many respects different from performing tasks as described previously. For example it involves a possibly long-termed dependency between delegate and represented agent, and the fact that another agent speaks for the represented agent may incur commitments in the future, that are not under control of the represented agent. Social delegation is more concerned with performing a certain role, rather than producing a specified product. In holonic terms, representation is the job of the head, which can also be distributed according to sets of task types to different agents. Just like fat trees (multiple bypasses to critical communication channels) in massive parallel computing, distributing communication to the outside is able to resolve bottlenecks. This makes social delegation a principle action in the context of flexible holons and provides the basic functionality for self-organisation and decentralised control.

Thus, we believe it is justified to differentiate two types of delegation: task delegation, which is the delegation of (autistic, non-social) goals to be achieved and social delegation, which does not create a solution or a product but in representing a set of agents. Both types of delegation are essential for organisations, as they become independent from particular individuals through task and social delegation.

Given the two types of delegation, it remains to explain how the act of delegation is actually performed. We observe four distinct mechanisms for delegation (see also Figure 2):

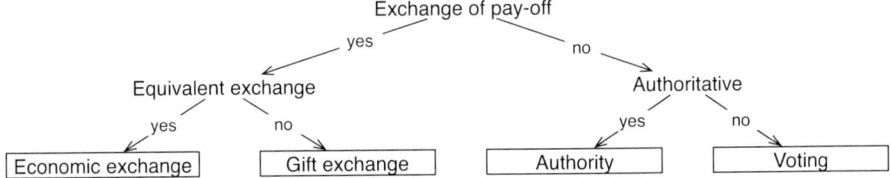

Fig. 2. Overview on four different mechanisms for delegation

- Economic exchange is a standard mode in markets: the delegate is being paid for doing the delegated task or representation. In economic exchange, some good or a task is exchanged for money, while the involved parties assume that the value of both is similar.
- Gift exchange, as an important mechanism in the sociology of Bourdieu [8, pp. 191–202], denotes the mutually deliberate deviation from the economic exchange in a market situation. The motivation for the gift exchange is the expectation of either reciprocation or the refusal of reciprocation. Both are indications to the involved parties about the state of their relationship. This kind of exchange entails risk, trust, and the possibility of conflicts (continually no reciprocation) and the need for an explicit management of relationships in the agent. The aim of this mechanism is to accumulate strength in a relationship that may pay off in the future.
- Authority is a well known mechanism, which represents the method of organisation in distributed problem solving. It implies a non-cyclic set of power relationships between agents, along which delegation is performed. However, in our framework authority relationships are not determined during design time, but at runtime when an agent decides to give up autonomy and allow another agent to exert power. This corresponds to the notion of Scott who defines authority as *legitimate* power [9].
- Another well-known mechanism is voting, whereby a set of equals determine one of its members to be the delegate by some voting mechanism (majority, two thirds, etc.). Description of the mandate (permissions and obligations) and the particular circumstances of the voting mechanism (registering of candidates, quorum) are integral parts of the operational description of this mechanism and must be accessible to all participants.

As suggested by Figure 3, all four mechanisms work for both types of delegation: for example, economic exchange can be used for social delegation as well as for task delegation. This set of mechanisms is by no means complete, however, many mechanisms occurring in human organisations that appear not to be covered here, are combinations of these four mechanisms or variations of their general principles (e.g. different voting schemes). The choice of an appropriate

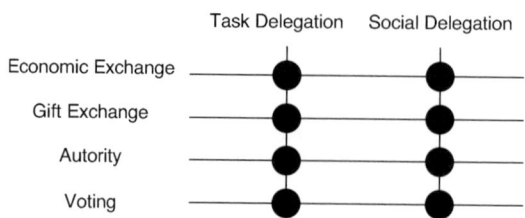

Fig. 3. The delegation matrix showing two modes of delegation and four mechanisms for performing each mode. Theoretically, every combination of mode and mechanism is possible in multiagent organisation

mechanism for the two modes of delegation represents the first two design parameters of a holon.

3.2   Membership Restrictions

A membership restriction can state that there is no limitation or that an agent is only allowed to be a member of one single non-atomic holon. It can also have the value "limitation on product", which means that the holon is free to join another holon, as long as they do not perform tasks of the same type. And, of course, other restrictions are possible.

3.3   The Set of Holon Heads

The number of permitted holon heads permitted is described by another parameter. In very egalitarian holons all subholons can receive incoming tasks and redistribute them, thus all subholons are head holons. More authoritative holons may be organised in a strictly hierarchical manner, with only a single point of access to the outside. An obvious intermediate form is the possibility to define a subset of all subholons as the holon head. The advantage of the egalitarian option is that single-points of failure or communication bottle-necks are avoided. However, the hierarchical option may ease communication with the structure by a single point of access and reduce the communicational effort to coordinate the goals of several holon heads.

3.4   Goals of a Holon

We also allow variations of the goal of a holon. A holon can be created to perform a single task, all task of a single product (i.e. a given type) or it performs all products together that it is able to achieve in collaboration (possibly only making use of a subset of the subholons). This design parameter is central for the interface of the holon to the outside: Only if the goal of a holon is defined, can the prospective subholons determine at creation of the holon if the goals of the holon coincide with any other holon they are member of. In the case of the holon head, a conflict free set of holon goals enables them to determine unambiguously for any incoming request the proper context in which they should process this request. For example, if an agent is head agent for two holons, one of which was designed to process task type t_1 and the other for t_2 and an incoming request matches t_1, then the agent knows it needs to investigate a joint processing of the request in the first holon.

3.5   Profit Distribution

Profit distribution can be done on a per task basis using economic exchange or gift exchange. Other possibilities imply that during the formation phase of the organisation agents agree on how profit is split between head and body agents ("regulation", e.g. 10:90, 20:80 etc.) or that a "fixed income" is being paid from the head to the body agents regardless of the number of tasks performed (in this case, variable costs are paid by the head plus a fixed income chosen by the designer).

3.6   Rules for Inclusion and Exclusion of Subholons

If holons are designed for handling membership flexibly during runtime, they need rules to include or exclude members. We propose two different schemes. The consensus, by which all decision-makers must agree to include or exclude a new member, or the veto scheme where one decision-maker alone veto the majority decision. Furthermore, the set of decision-makers needs to be defined as it need not be the set of all subholons. Reasonable choices are either *all subholons, a subset* thereof or *only the head holons*. In case of the exclusion, the single vote equals a veto on the membership: one subholon can then reject the membership status of another subholon.

3.7   Rules for Termination of a Holon

If holons need maximum flexibility, they need at least the possibility to leave superholons, if the membership is no longer beneficial. Already at the start of the holon, these rules can be fixed. Subholons may specify that the termination process is automatically initiated after performing a single task or by veto of a single decision-maker. Other possibilities are that the party intending to terminate the holon pays a fee to the other parties (to compensate for their loss in structure) or the holon is only terminated after a notice period. Any of the last three types requires to specify who is allowed to invoke the process. Reasonable choices are that either all subholons can do this, or only the head holons. In case of a very static system, a holon can be defined to have no option for termination.

3.8   Summary

Table 1 gives an overview over all parameters and their possible values, which shows the complexity of the possible relationships between subholons. The choice from this set of parameters defines the part C of the holon tuple that regulates the commitments among subholons at creation time of the holon. In case subholons are only included later, they are required to agree to the commitments at the time of joining the holon.

Table 1. Overview of the design parameters for multiagent organisation

Parameter	Possible values
Mechanism for task delegation	− Economic exchange − Gift exchange − Authority − Voting
Mechanism for social delegation	− Economic exchange − Gift exchange − Authority − Voting
Membership restrictions	− Exclusive membership − Restriction on product − None
Goal of the holon	− One task − One product − All products
Set of holon heads	− All subholons are head holons − Some subholons are head holons − One subholon is head holon
Profit distribution	− Case by case negotiation − Fixed share − Salary
Rules for inclusion and exclusion	− Consensus − Single vote
Decision maker for inclusion and exclusion	− All or some subholons − All head members
Rules for termination of the holon	− Automatic after task − Veto − After payment or notice period − No termination
Initiator for termination	− Consensus − Consensus among head members − Any member or any head member

4   Agent Autonomy and Multiagent Organisation

As described by Wooldridge et al. [10], autonomy is an integral part of the agent
definition. However, autonomy is not a quantifiable notion, but rather consists
of qualitatively different types. As Castelfranchi [3] showed, there are several
distinct types of autonomy that correspond to different types of *dependency*.
Therefore, autonomy of agents is a phenomenon with qualitatively different
aspects: an agent can be autonomous (independent) or dependent on others
concerning information, the interpretation of information, planning, its motivations and goals, resources, and authority ('being allowed to do X', deontic
autonomy) and these dependencies directly relate to losses of autonomy (e.g.

loss of goal autonomy, resource autonomy etc.). Therefore, an analysis of dependencies between agents in holonic organisations will also lead to an analysis of agent autonomy.

Beyond the dependencies in pure agent interaction, new dependencies are created if agents engage in long-lasting holonic structures. The following is a taxonomy for these different types of dependencies between agents, according to the choice of the aforementioned design parameters. These types of dependencies are taken and reformulated from [3], except for *representational, exit* and *processing* dependence. We do not share the distinction made by Castelfranchi between *goal-discretion* and *goal-dynamics* dependence, because goal-dynamics dependence is concerned with the timing of goals. As we are concerned with multiagent systems for task-assignment and distributed scheduling, timing is part of the goal description in our context, and (only) therefore the distinction is not relevant here. Also for reasons that lie in our application domain, we combine skill dependence and resource dependence into a single topic, as the resources define the skills in task-assignment.

4.1   Skill and Resource Autonomy

As formulated by Castelfranchi, skill dependence of agent Y on agent X means that the action repertoire of Y is not sufficient for achieving a goal G. Resource dependence of Y on X is the fact that Y depends on the resources of X to achieve its goal G (these resources include time). For our purposes, performing a skill requires the allocation of resources, these two dependencies always come together: if Y needs the skills of X it also requires some of X's resources, requiring resources also requires the skill of using them (it is not envisaged to let an agent surrender its resource to another agent). The formation of holons for the joint performance of a job that requires the resources (and skills) of different agents always includes this kind of dependence. Therefore, all subholons loose to some extent skill and resource autonomy.

4.2   Goal Autonomy

Goal dependence of Y on X is the dependence of Y to choose its own goals. Consider as an example that one would expect artificial agents to have constraints on their goals formulated by their user. As described above, joining a holon includes the definition of a holon goal. If this includes the commitment to perform certain jobs only inside the holon, a participating agent is constrained (and therefore dependent) in formulating new goals: it would be a breach of commitments to perform a job of the same type with a set of agents outside the holon. As these agents can no longer freely choose their goals, they no longer possess goal autonomy.

4.3   Representational Autonomy

Representational dependence of Y on X means that Y depends on X in order to represent it to other agents, either always (Y has no contact to other agents) or only in a special context (X represents Y in specific matters). This type of dependency is probably the most "social" type of dependency, as it does not directly relate to the performance of a task, but only to interaction with other members of the population. It implies loss of representational autonomy, which is of high importance as it deprives an agent of social contacts and may incur loss of opportunities to pursue other interests of the agent.

4.4   Deontic Autonomy

All types of autonomy or dependence state that the agent is allowed to perform some kind of action or not. In the case of representational dependence it is decided at design time of the holon that a body agent is not allowed to represent itself. However, here we are interested in a more complex and flexible mechanism, where we consider cases where the agent is permitted, obliged, or forbidden to do something by some other agent while the holon is active. The expressions, *permitted, obliged* and *forbidden* are the three canonical operators of deontic logic as described by von Wright [11]. Therefore, if an agent Y commits itself to wait for X to permit, forbid or oblige Y to work towards a goal G, we call this the creation of deontic dependence. As an observer, we can differentiate deontic dependence from other dependencies by the occurrence of messages between agents that contain one of the deontic operators during runtime (assuming that agents adhere to these kinds of messages).

Deontic dependence means that even if agent Y has appropriate skills and resources, this dependency can stop the agent from pursuing its goals (if it accepts the deontic dependence). Compared to other types of autonomy, deontic dependency translates into a more abstract loss of autonomy. Deontic autonomy has no physical relation to performing a task and is only manifested by the commitments made between agents (i.e. the lack thereof). In our case, this type of autonomy relates to two aspects. First, it relates to the task delegation mechanism "authority", which means that Y looses deontic autonomy if X can delegate tasks to Y by authority, i.e. to oblige Y to do the task, and permit or forbid Y to perform other tasks in the meantime. Second, there is a connection to the notion of the set of commitments that define a holon, as they can specify the necessity for further permissions or obligations that depend on other holon members.

4.5   Planning Autonomy

If an agent Y relies on X to devise a plan for its actions then Y is plan dependent. If subholons have only one representative, it may make sense to centralise planning and remove planning autonomy from the subholons: In tightly coupled organisational forms, reduction of communication costs can be achieved by this design. Also, if X has authority over Y it needs to be aware of Y's work-plan in order to decide which jobs to delegate to it. In this case X can (but may not) also devise a plan for Y, which again may be used to save communication. Note that these two cases are not necessarily identical, as one agent may have two superiors that have authority over different resources of a subholon. In case Y depends on the plan devised by X, it has lost planning autonomy and the relevant design parameters are the set of holon heads and the task delegation mechanism.

4.6 Income Autonomy

If Y commits itself to accept fixed payments for providing its services from X, and cannot alter these arrangements (e.g. if there is no exit option for the holon, see Section 4.7), then Y has lost influence over its income. If its income is realised by negotiation, it still depends on others, but it still has a choice and hence its autonomy. The design parameter "profit distribution" deals with this issue.

4.7 Exit Autonomy

All subholons have made certain commitments when they entered a holonic structure. As long as this structure persists, these commitments bind the members and they are not independent in this respect. Being able to exit the structure therefore corresponds to a specific type of autonomy named *exit autonomy*. The mentioned dependencies are described by the design parameters "rules for inclusion and exclusion of subholons" and "rules for termination of the holon". For example, the holon can be designed to terminate after a single task, which leaves no autonomy to the subholons to terminate the structure but instead guarantees a foreseeable end. On the other hand, holons can be designed to be terminated by each member (higher degree of autonomy), by the head members (difference in exit autonomy between head and body members) or by only a single member (small degree of autonomy).

4.8 Processing Autonomy

Processing autonomy expresses the dependency between several agents that decided to merge into a single agent. As any of the formerly different agents surrender all their abilities to process information, this autonomy lost in this process is called *processing autonomy*.

4.9 Summary

Several distinct types of dependencies between the subholons can be identified in a holonic multiagent systems. As Castelfranchi [3] argues, such dependencies are directly linked to distinct types of autonomy. Table 2 gives an overview of these types of autonomy and the corresponding holonic design parameters. Note that "processing autonomy" is the combination of several design parameters.

5 Autonomy and Holonic Nesting

Although the recursive structuring of holonic multiagent systems allows in principle the delegates of organisations to be part of other organisations (as described

Table 2. Overview of the Taxonomy of Autonomy for Holonic Organisation

Type of Autonomy	Relevant Holonic Design Parameter
Skill and resource autonomy	Not applicable in collaborative holons
Goal autonomy	Goal of holon
Representational Autonomy	Set of holon heads
Deontic Autonomy	Mechanism for task delegation
Planning Autonomy	Membership restrictions
Income Autonomy	Profit distribution
Exit Autonomy	Rules for termination of the holon, and Initiator for termination
Processing Autonomy	Goal of holon, membership restrictions, mechanism for task delegation, and rules for termination of the holon

in the previous example constellation), it is precisely the issue of autonomy that imposes restrictions on holonic nesting. Here, the concept of autonomy again demonstrates that it is not a simple scalar parameter but has qualitatively different dimensions. While some dimensions are irrelevant for the nesting of holons, others incur restrictions. Without these restrictions, body agents with a higher degree of autonomy would be introduced to an holonic form that in contradiction requires more restrictions on their autonomy.

A summary overview of the types of autonomy critical for nesting in this sense is given in Table 3. The choice of profit distribution or the set of holon heads for the substructure does not constrain the income autonomy or representational autonomy of the superstructure. All other types of autonomy need to be available on the superstructure as well. If the subholon can freely choose which jobs to perform, then so must be the superholon, otherwise super- and subholon could run into conflicting commitments. It is clear that if the subholon has deontic autonomy then the same must hold for the superholon. If the subholon has planning autonomy, then the superholon must also be master of its own schedule (the same holds for processing autonomy). Exit autonomy must be passed on to the superholon level as well. If not, the structure could face the paradox situation that it has no exit autonomy, but all subholons could use their exit autonomy and then the superholon would in fact no longer exist, without having the autonomy

Table 3. Overview of the types of autonomy that are critical for Holonic Nesting

Type of Autonomy	Critical for Nesting
Skill and resource autonomy	n/a
Goal autonomy	√
Representational Autonomy	
Deontic Autonomy	√
Planning Autonomy	√
Income Autonomy	
Exit Autonomy	√
Processing Autonomy	√

to decide to stop existence. Skill and resource autonomy is not applicable to this discussion, as the question of a superholon implies by its very nature that several subholons combine their efforts to pursue a common goal and hence are skill and resource dependent.

6   Conclusions

With this contribution we have advanced the state of the art in two ways. On the one hand we have given a classification of different dimensions of autonomy that extends the previous classification of Castelfranchi [3], for example by the notions of exit or representational autonomy, which we consider essential for the modelling of multiagent organisation. On the other hand, we have shown the connection between this classification and the design parameters for multiagent organisation.

The design parameters describe such crucial properties as membership rules, mechanism of internal delegation and representation, etc. The design parameters describe the design decisions that need to be made to create a holonic structure. Beyond this design framework we also presented a list of options for each of the design parameters. For example, we listed a number of possible mechanisms for terminating a holon. It turns out that the choices for each design parameter are interwoven with the autonomy of the involved agents (e.g. the option to exit an organisation). For an agent it is not only a question of being autonomous or not autonomous, or more or less autonomy (in the sense of a single scalar dimension). As we view the concept, the choices for the different design parameters incur dependencies on several dimensions, which correspond to dimensions of autonomy that define holonic organisation. These dimensions of organisation define the interplay between organisation and autonomy.

Holonic multiagent systems have been identified as a superior mechanism for modelling multiagent organisation. As holonic multiagent systems provide the notion of recursive structuring, it is especially important to pay attention to nested organisations. In general, the nesting of holons is restricted by the autonomy granted to the subholons by the organisational form. Although there are some types of autonomy that are irrelevant in this respect, the general rule is that nested holonic structures may not provide more autonomy than the entailing structure.

Acknowledgements

Prof. Castelfranchi's work on autonomy and dependence was an important source of inspiration. We particularly thank him for his comments on the work in this article. Thanks are also due to our colleagues in the socionics project at the sociology department of the Technical University of Hamburg-Harburg: Dr. Michael Florian, Dr. Frank Hillebrandt and Bettina Fley. This work was funded by DFG under contract Si 372/9-2.

References

1. Jennings, N.: Agent-based computing: Promise and perils. In: Proceedings of the 16th International Joint Conference on Artificial Intelligence (IJCAI-99). (1999) 1429–1436
2. Schillo, M.: Self-organization and adjustable autonomy: Two sides of the same medal? Connection Science **14** (2003) 345–359
3. Castelfranchi, C.: Founding Agent's "Autonomy" on Dependence Theory. In: Proceedings of the 14th European Conference on Artificial Intelligence (ECAI 2000), Berlin, Germany, IOS Press (2000) 353–357
4. Koestler, A.: The Ghost in the Machine. Hutchinson & Co, London (1967)
5. Castelfranchi, C., Falcone, R.: Towards a Theory of Delegation for Agent-based Systems. Robotics and Autonomous Systems **24** (1998) 141–157
6. Norman, T.J., Reed, C.A.: A Model of Delegation for Multiagent Systems. In: Foundations and Applications of Multiagent Systems, Lecture Notes in Artificial Intelligence, vol. 2403. Springer Verlag (2002) 185–204
7. Bourdieu, P.: Sozialer Raum und Klassen. Suhrkamp, Frankfurt am Main (1985)
8. Bourdieu, P.: Pascalian Meditations. Polity Press, Cambridge, UK (2000)
9. Scott, W.R.: Organizations: Rational, Natural and Open Systems. Prentice Hall Inc., Englewood Cliffs, N.J. (1992)
10. Wooldridge, M., Jennings, N.: Intelligent agents: Theory and practice. The Knowledge Engineering Review **10(2)** (1995) 115–152
11. von Wright, G.H.: Handlung, Norm und Intention. de Gruyter (1977)

Autonomy and Reasoning
for Natural and Artificial Agents

Harko Verhagen

Department of Computer and Systems Sciences
Stockholm University and The Royal Institute of Technology
Forum 100, SE-16440 Kista, Sweden
verhagen@dsv.su.se

Abstract. In this article, I will present recent thoughts and theories on autonomy for natural and artificial agents. Even though the recent work on autonomy for artificial agents has interesting aspects, it excels in being unsystematic and a lack of references to theories outside of agent research supporting one or the other. Embedding these discussions in a broader framework of discussions in philosophy and sociology will enable us to sketch a broader yet more detailed picture. It will also enable us to discuss the reasoning of artificial agents.

1 Introduction

Is there a need for theories of the artificial? Are not the thoughts of philosophers from the ancient Greeks onwards enough? Such questions spring to mind as we read the discussions of autonomy in artificial agent literature. Surely there is more than enough written on autonomy by philosophers? Yes, there is and we should not pass them by as some authors do. But on the other hand, artificial agents are special. Here, we have direct access to the internal machinery, we can tinker around in it almost at will in order to create the type of agent we like or need. This direct access to the mind creates new and challenging questions. To answer these, we need to look at the discussions on natural agents to find some answers to basic questions so we can concentrate on the special issues.

In this article, I will try to convince the reader that there are some basic questions that have (preliminary) answers. These questions and answers are echoed in artificial agent research. I will focus on autonomy and reasoning. I will present these issues for natural agents, following that these topics are presented from the artificial agent literature. Finally, in the discussion, directions for further research are pointed out. But first of all, let us have a look at a general definition of autonomy.

2 A General Definition of Autonomy

Both within and outside multi-agent research there are several definitions of autonomy. I will take as a starting point the definitions of autonomy and autonomous as found in the Merriam-Webster dictionary [20]. Autonomy:

M. Nickles, M. Rovatsos, and G. Weiss (Eds.): AUTONOMY 2003, LNAI 2969, pp. 83–94, 2004.
© Springer-Verlag Berlin Heidelberg 2004

1. the quality or state of being self-governing; especially: the right of self-government
2. self-directing freedom and especially moral independence
3. a self-governing state

Etymologically, autonomy comes from the Greek "autos" (meaning self) and "nomos" (meaning rule, or law), and was originally applied to the idea of the self-governing Greek city-state. In philosophy, and agent research, these general definitions have been sharpened in several ways. Let us take a look at the theory-building for natural agents first.

3   Autonomy for Natural Agents

Discussions of autonomy (or more precisely acting autonomously) for natural agents go back to Aristotle and Plato and were reintroduced to modern philosophy by amongst others Kant [17].

For Kant the self-imposition of universal moral law is autonomy. This moral law is created by the agent itself and based on reason, rather than imposed from the outside or issued by a higher authority (these situations are called heteronomy). Being the creator of one's own laws is the ultimate form of self-government. These moral laws are the ground of both moral obligation generally and the respect others owe to us (and we owe ourselves). Practical reason as the ability to use reasons to choose our own actions presupposes that we perceive ourselves as free. Freedom means lacking barriers to our action that are in any way external to our will.

Thus, autonomy in philosophy is closely connected to freedom of the will, and autonomous action is based on choosing one's actions because of reasons since freedom of the will expresses the capacity of rational agents to choose a course of action from among various alternatives. This freedom of the will is on its turn closely connected to moral responsibility. As Velleman [23] puts it:

> Autonomy is self-governance. The will as a desire-driven faculty of practical reason i.e., the will is a locus of autonomy since it contains the motives which can restrain, redirect, and reinforce other motives for action in accordance with the agent's conceptions of those motives (p. 173).

3.1   Types of Autonomy for Natural Agents

As we saw, a basic type of autonomy is *moral autonomy*, which refers to the capacity to impose the (objective) moral law on oneself, i.e. govern oneself. Another type of autonomy is *personal autonomy*, which is a trait that individuals can exhibit relative to any aspects of their lives, not limited to questions of moral obligation [10].

In moral autonomy, the capacity to govern oneself means one must be in a position to act competently and from desires that are in some sense one's own. Here we see two sorts of conditions often mentioned in conceptions of autonomy:

competency conditions and *authenticity conditions*. Competency includes various capacities for rational thought, self-control, and freedom from pathologies, systematic self-deception, and so on. Authenticity conditions include the capacity to reflect upon and endorse (or identify with) one's desires, values, and so on. The most influential models of authenticity in this vein claim that autonomy requires second-order identification with first order desires, we will get into more detail on this further on.

Within personal autonomy, four flavours of personal autonomy can be distinguished [3]. The main focus in personal autonomy is on the special relation an agent has to its own motives.

The first approach to personal autonomy can be called coherentist. Here, the governing self in relation to ones own motives is represented by reflexive attitudes – higher-order attitudes toward the mental states that move an agent to act. An agent is an autonomous agent, the coherentists argue, if and only if it accepts its motives, or identifies with them, or approves of them, or believes that they make sense in terms of its long term commitments or plans. The content and the origin of these higher-order attitudes are irrelevant to whether the agent is autonomous, that is, it is unimportant if the agents hold attitudes that are irrational or not well-informed, nor does the agent have to do anything to have these attitudes, they are given. Thus, coherentists are double internalist.

The other three approaches to autonomy introduce conditions that are externalist in one or both of these ways. A reasons-responsive theory of autonomous agency implies that an agent not really governs itself unless its motives, or the mental processes that produce them, are responsive to a sufficiently wide range of reasons for and against behaving as the agent does. In other words, an agent has to understand itself what the reasons are for its behaviour, and these reasons have to move the agent.

A third popular approach to autonomous agency stresses the importance of the reasoning process itself. These so-called "responsiveness to reasoning accounts" state that the essence of self-government is the capacity to evaluate one's motives on the basis of whatever else one believes and desires, and to adjust these motives in response to one's evaluations. Thus, the agent can deduce what follows from it's beliefs and desires, and act accordingly.

A fourth approach to autonomy is the incompatibilist conception of autonomy which states that autonomy is incompatible with determinism. If an agent's actions can be fully explained as the effects of causal powers that are independent of the agent, then even if the agent's beliefs and attitudes are among these effects, the agent does not govern them, and so it does not govern itself.

The above mentioned theories of autonomy mainly focused on aspects of cognitive reasoning. Other factors that tie in to reasoning are e.g. personal characteristics (values, physical properties, relations to others, etc.). These may enter the reasoning process in various ways and are not always under control of the agent. We cannot for instance change our body at will when we need to, however handy a third hand may be, it is something that we can not change at will, limiting our autonomy in this sense.

Another addition to the individualistic theories of autonomy presented above is relational autonomy [18]. Relational conceptions of autonomy underline the role that relatedness plays in both persons' self-conceptions (relative to which autonomy must be defined) and self-government itself [8].

3.2   The Reasoning of Natural Agents

Following Kant [17], philosophy distinguishes theoretical and practical reasoners. Theoretical reasoners are passive bystanders to the events on the basis of which they predict future events. Practical reasoners are not mere observers of the passing scene but try to determine their responses to what they observe, and find a reason to do one thing rather than another. Since no fact can play the role of a reason unless someone takes it to be a reason, practical reasoners necessarily have the ultimate authority over the powers that move them, thus they are autonomous. So, reasoning equals having reasons for acting.

One might be tempted to think that acting in accord with ones preferences (desires, intentions) could suffice. However, as described above in the section on personal autonomy, current theory within philosphy has it that this does not suffice. The most prominent exponent of this is Frankfurt [11], who defines autonomous agency as a matter of ones actions flowing from desires that one reflectively endorses having. What distinguishes autonomous action and choice, on this view, is the presence of second-order attitudes (desires) regarding ones first-order desires. The mere existence of second-order desires does not suffice for Frankfurt. In order for the will to be free it is not enough that an agent wants it's desires; the agent must *also* want to want it, *and* want it to be effective in leading the agent to act. Such a second-order desire for a first-order desire to be effective in action Frankfurt [11] terms a "second-order volition". He states that if the first-order desire that leads one to act is endorsed by a second-order volition, then one "identifies" with it; it is one's own, autonomous desire.

This hierarchical model of desires has since been discussed and criticised. The first of these criticisms is quite simple. If autonomy is conferred upon one's lower-order desires by one's second-order desires, what confers autonomy on these desires? A similar objection to this regress objection is the objection from authority. Why should one assume that an agent's second- order desires are more indicative of what the agent really wants than the agent's first order desires? A third objection is this. On this hierarchical account, an agent's first-order desires are autonomous if they exhibit "volitional harmony" (i.e., are in accord with and endorsed by) the agent's second-order desires. However, an agent could alter its second-order desires, rather than its first-order desires, to achieve this volitional harmony. Finally, any model of the autonomy of desires that is based solely upon an agent's motivational structure falls foul of problems of manipulation, in which an agent may have a first-order desire that is endorsed by a second-order volition, both of which have been implanted into the agent by an external agent.

The hierachical model of autonomy has developed since the 1970's, in part as an answer to the above mentioned criticism. Frankfurt [12] is moving away from his earlier claim that a desire is autonomous if and only if it is endorsed by a second-order volition. An agent's desire is now seen as autonomous if the agent decides that it belongs in its (given) motivational set and if the agent is satisfied with this second-order desire, that is, the agent identifies with it. This answers the regress objection because now the theory does not depend upon a hierarchy of desires conferring autonomy upon each other. It also solves the problem of authority for the same reason; the second-order desire has more authority because the agent decides to identify with it, with this decision being guided by the current structure of that motivational set that one recognises as one's own. Thirdly, if an agent does decide to alter his higher-order desires so that they are in accord with his lower-order desires this will not be autonomy reducing, since it will be the higher-order desires that do not fit with that motivational set that the agent recognises as its own.

What is interesting regarding the reasoning of natural agents is what elements can disturbe the autonomy og the agent, i.e., what sort of information can an agent take into account that is external to it? Most theories of autonomy simply state that certain influences on an agent's intention-forming process interfere with, or pervert this process, but fail to explain what distinguishes internal from external forces.

We can categorise these deterministic externalities that endanger the freedom of will in the following way [19]:

1. physical (including biological)
2. internal (psychological)
3. social (theological, relational)

This categorisation can be generalised using Habermas' overview of action cetegories [15] which I will present in the next section.

Habermas' Theory of Communicative Action. From a sociological point of view, something is missing. In his monumental work on the theory of communicative action [15], Habermas discusses and in a sense unifies several different sociological schools of thought (i.e., Max Weber, Emile Durkheim, George Herbert Mead and Talcot Parsons). Habermas distinguishes four action models. Each action model makes presumptions about the kind of world the agents live in which has consequences for the possible modes of rational action in that model.

The first action model is teleological action. Agents are goal directed and try to maximize their choice of means to obtain a goal. This is the rational choice model. The central issue in this action model is the choices the agent makes between different action alternatives, based on maximizing utility. Agents can thus try to influence the world, and the rationality of the behavior of the agents can be evaluated with respect to the efficiency of their behavior. Adding other agents, with respect to whom the agent acts in a strategic manner (strategical action model), to the decision making model does not change the ontological principles. The agents may need to model the desires and actions of the other agents but these are still part of the objective world of existing states of affairs. Agents act with respect to this world according to their beliefs about the existing states of affairs and their intentions to bring about desired states of affairs in that world.

The second action model is the normatively regulated action model. Social agents are assumed to belong to a group and follow the norms that members of that group are obliged to follow. Following norms is taken as to behave according to expectations. The objective reality is extended by a social reality of obliging norms (acknowledged as such by the group). The rationality of the behavior of an agent is not only related to the objective reality (teleological and strategical action model), but also to the social reality. The conformity between the norms of the group and the behavior of the agents and the relation between the norms and the generalized interests of the agents (and thus if it is wise of the agents to confirm to those norms) are part of this social rationality. Agents act with respect to an objective world and a social world, namely the normative context that defines the possible interactions and legitimate interagent relationships between the agents.

The third action model is the dramaturgical action model. In this action model the inner world of the agents is considered. Based on Goffman's [14] dramaturgical analysis of social life, this action model has as a core the presentation of the self of an agent to an audience. This representation of the self may or may not be truthful. The agent makes use of the fact that its inner self is only admissible to itself. The inner self is defined as the constellation of beliefs, desires, intentions, feelings, and needs of an agent. Habermas views this inner self as a reality in its own right. When presented in a truthful and authentic way, and at the same time connected to the shared evaluation criteria and interpretations of needs, the subjective point of view of the agent can gain an intersubjective value. Truthful is not the same as true in objective sense, opening the door for lying and manipulation or insincerity. Agents act with respect to an objective world and a subjective world formed by the totality of subjective experience to which the agent has a privileged access.

The fourth and final action model is the communicative action model. This action model unites the three functions of language specified by the three previous action models. In the strategical action model, language is used by an agent to reach its own goals possibly via influencing other agents by use of language, the normative action model uses language to actualize already existing normative agreements and the dramaturgical model uses language to allow for one to express oneself. In the communicative action model, language is used to bring about *mutual understanding* on all three previous levels. The agents use language to claim the truth of their utterances, the normative correctness of their speech acts in view of the context, and the sincerity of their intentions being formulated. Testing for rationality of actions is here no longer the privilege of the observer, but is done by the agents themselves to realize a common definition of the situation described in terms of relations between the speech act and the three worlds (i.e., the objective, social, and subjective world) this speech act has relations with. In the cooperative process of interpretation, all participating agents have to incorporate their own interpretation with that of the other agents so that the agents have a sufficiently shared view of the external (i.e., objective and social) world in order to coordinate their actions while pursuing their own goals.

4 Autonomy for Artificial Agents

In [25], the basic set of agent properties consists of autonomy (i.e., the agent has some kind of control over its internal state and actions), social ability (i.e., the ability to interact with other agents which may include humans), reactivity (i.e., the ability to act upon perceptions of the environment of the agent) and pro-activeness (i.e., the ability to display goal-directed behavior so that the agent's behavior is not only ruled by environmental changes or social interaction). In [13], an autonomous agent is defined as a system situated within and part of an environment that senses its environment and acts on it, over time, in pursuit of its own agenda and so as to effect what it senses in the future. This definition more or less implies that there also exist nonautonomous agents, but the authors fail to give a definition of these agents.

Most definitions of agency do share one main characteristic, namely that an agent is an entity that acts autonomously. This definition also turns expressions like *autonomous agents* into tautologies (although according to [9] we might reserve the term autonomous agents for agents that are both situated, i.e., are part of a society, and embodied, i.e., have a physical body that can move in the real world, e.g., a mobile robot).

Autonomy is relative to something external to the agent; an agent is not autonomous in an abstract sense but is autonomous on some level with respect to another entity, be it the environment, other agents, or even the developers of the agent. This point is also made in [5] where it is stated that an agent's autonomy is necessarily limited since it is situated. Unlimited autonomy (i.e., the solipsistic paradigm) makes an agent autistic since the agent does not allow for any input from its environment.

4.1 Types of Autonomy for Artificial Agents

The autonomy of agents can be of various kinds. One meaning of autonomy is the degree to which movements are determined by external stimuli (*autonomy from stimuli* in [5]), in other words the degree to which the behaviour of an agent is directly controlled by another entity at runtime (note that the above cited definition of an agent by [26] uses both the term proactiveness and the term autonomy to capture this definition of autonomy). Another meaning is the independence of the human mind: cognitive autonomy, the degree to which one's choices and actions based on them are governed by other agents (see also [5]). At the level of a multiagent system, autonomy is the autonomy a society has to give itself laws. In previous work [24] I have called this *norm autonomy*, i.e., the control an agent has over its set of norms. Autonomy of the agents forming a multiagent system with respect to its designer(s) or even users is related to

the self-organizing capabilities of a multiagent system. The more autonomous
the agents are, the higher the multiagent systems self-organizing capabilities
are, thereby making *emergent* behavior more feasible. Autonomy in this sense
may be equaled to the control an agent has of different factors that play a role
in its decision making process. It also implies that the agents' and multi-agent
system's behaviour is less predictable for it is less deterministic.

Recent work on autonomy for artificial agents includes among others [7], [1],
and [4]. In [7], autonomy is defined in terms of power, namely independence (thus
absence of power held over one). Autonmous is seen as non hetero-directed (the
observent reader will here recognize the dichotomy Kant described between autonomous and heteronomous), i.e. behaviour that is not totally directed or driven
from outside the agent (whatever outside may be). This refers to autonomy vis-
a-vis the environment. This autonomy is then supplemented with autonomy at
a social level, where the authors describe two different notions:

1. autonomy as indepence, self-sufficiency
2. autonomy in collaboration

The first case adresses agents that do not need any help or resources in order
to achieve its goal or execute an action. The second notion adresses the autonomy
an agent has when working on behalf of another agent, in other words autonomy
in its agency. Other types of autonomy described in [7] include executive autonomy (mainly stating the agent is autonomous as to what means or plans it will
use to reach a goal given to it by another agent), goal autonomy, and deontical
autonomy (autonomy with respect to the agent's organizational context). These
building blocks are then incorporated in the theory on delegation, dependence,
and control being worked on by this research group. For the current article this
work is rather unimportant.

In [1], the authors start with the definition of autonomy as "behaviour that
is independent and self-governing" and more specifically state that this implies
agents may actually choose to act contrary to the desires of other agents, a mys-
terical twist since the appearance of other agents and desires was not prepared
for. However, the agents are also thought to be intentional (thus excluding reactive agents from the discussion of autonomy). Distinctions between local and
non-local limitations on autonomy are made, examples of the first are contraints
on intentional and physical abilities, the second type of limitations include e.g.,
social values (such as norms, laws, etc.). When discussing [6], the authors also
include will in their definition (in the sense that dependence that can be ended
at will means an agent is autonomous with respect to the agent it depends
upon) and futhermore introduce first-order and second-order autonomy. First-
order autonomy expresses relations at the level of actions, whereas second-order
autonomy relates to the depedency relations perse. If for example the environments state influences the amount of choices available to an agent, there is said
to be a second order effect on the agent's first order autonomy. Also, the authors
wish to make a distinction between being autonomous and acting autonomous.
Finally, they also blen din personal traits that are said to influence the autonomy.
Combined with group membership, which is also taken to influence autonomy,

autonomy is seen as a combination of socially warrented freedoms and internally perceived freedoms.

In [4], agents are defined as goal-directed, meaning that the agents' behaviour is guided by internal representations of the effects of behaviour alternatives. Motivations generate these goals. The authors describe several forms of autnomy for artificial agents:

1. autonomy with respect to the user (agent as assistant, mixed initiative or adjustagble autonomy is the passing of control between the user and the agent; U-autonomy)
2. autonomy with respect to other agents (social autonomy). Social autonomy has several versions
   (a) I-autonomy (intention autonomy, to adopt a goal or not)
   (b) O-autonomy (to obey a norm or not)
3. autonomous with respect to the environment (E-autonomous)
4. autonomous from itself for one of its motivations, second-order desires called emotionsby the authors (A-autonomous)

Parunak [21] described two types of autonomy (in the sense that an agent decides by itself when it executes, in contrast to objects in object-oriented programming), viz. *dynamic autonomy* and *deterministic autonomy*. Dynamic autonomy is the ability to start execution, the pro-active side of agents, which is determined by the agent's internal structure. Deterministic autonomy is the ability to deny access to the agent's resources, influencing the predictability of the agent. Strangely, Parunak states that an external observer determines this. This autonomy is then not autonomy of the agent but an indication that the external observer has an incomplete model of the agent's internal model. The deterministic autonomy is however a useful concept if the control over access to its resource is managed by the agent.

In the context of A-life, Boden (Boden 1996) defines the following three dimensions of autonomy as control over behavior:

– the extent to which behavior is directly controlled by the environment
– the extent to which the controlling mechanisms are self-generated (emerging) rather than externally imposed (deliberatively preconfigured). These include hierarchical levels of abstraction (groupings of individuals that can be considered as a unit themselves)
– the extent to which inner directing mechanisms can be reflected upon and/or selectively modified

4.2   The Reasoning of Artificial Agents

Within artificial agent research, we are familiar with the dichitomy between reactive and deliberative agents. Research based on reactive agents is widespread. The paradigm of reactive agents is closely tied to research on robotics and artificial life, cf., e.g., the work of Brooks [2] and Steels [22]. It is not limited to these domains, however. Other applications include for example simulations of animal behavior and simple assistants currently replacing simple help functions. Proponents of using reactive agents state some advantages these agents have over cognitive (or intentional or deliberative) agents. The absence of a cognitive apparatus renders an explicit model of the world obsolete. They also lack an idea of sequentiality and are thus unable to build plans. Their simplicity makes them dependent upon each other seen from the level of the behavior of the system as a whole. Reactive agents usually operate in large groups where more than one agent can react in the same way to the environment, thus making the multiagent system robust to any loss of agents. Reactive agents are especially useful in worlds where the rate of change is faster than agents can update their world model. The absence of a world model (including a model of the other agents) renders this type of agent unsuitable for e.g. research on emerging social structures since the structure observed is dictated by physical properties of the environment (including the distribution of the agents over the environment) and has no meaning to the subjects. The structure is thus only in the eye of the observer, as is the case in the teleological and strategic action model described in the section above on Habermas. The same holds for the rationality of the agents.

Deliberative agents on the other hand do have a reasoning process. A prime example of such agents are intentional agents. The intentional type of agents is in many respects the opposite of the reactive type of agents. Intentional agents have an explicit model of their intentions (what the agent wants to achieve in the current situation), desires (what the agent would like to achieve in general) and beliefs (a model of the environment, which may or may not include models of other agents and the agent itself). BDI agents form a subset of this type of agents. These agents act following the strategic action model. As a consequence of this, these agents are not to be considered social agents (in the norm autonomous sense), even though an agent architecture like e.g. GRATE* [16] makes use of social concepts such as joint intentions and joint commitments. The BDI model can be extended with other social concepts such as norms, obligations, etc.

5   Discussion

Combining Habermas' three basic action models with Vellemans model of practical reasoning, we can categorise agent actions and autonomy using the same three concepts: the outer, objective (physical) world, the inner (emotional) world, and the interagent (social) world. Different models of reasoning have different views on how these three concepts have an impact on agent reasoning. For artificial agents, we can distinguish social relations within the agent system and outside the agent system (developer and user), since they are by nature part of a hybrid social system, consisting of natural and artificial agents.

Future work will build upon the above presented arguments in order to develop a general model for the design and analysis of agent systems (where the latter can be used for natural and artificial agents alike). An extension of the current work is a description of models of agency, which will form the base of such a framework. A further analysis of social concepts such as norms, commitments, and obligations is another goal for future research. Finally, I plan to incorporate psychological theories on reasoning, reconciling (social) philosophy, psychology, and sociology in an analytical framework.

Acknowledgments

This work was in part sponsored by Vinnova through the project Agenttechnology for Intelligent Embedded Systems. Parts of this article are based on previous work and discussions with Magnus Boman, Ruud Smit, and Leon van der Torre.

References

1. G. Beavers and H. Hexmoor. Types and limits of agent autonomy. In *Proceedings of AAMAS2003 Workshop "Computational Autonomy, Potential, Risks, Solutions"*, 2003.
2. R. Brooks. A Robust Layered Control System for a Mobile Robot. *IEEE Journal of Robotics and Automation*, 2(1):14–23, 1986.
3. S. Buss. Personal autonomy. In E. N. Zalta, editor, *The Stanford Encyclopedia of Philosophy (Winter 2002 Edition)*. Stanford University, plato.stanford.edu/archives/win2002/entries/personal-autonomy/, 2002.
4. C. Carabelea, O. Boissier, and A. Florea. Autonomy in multi-agent systems: A classification attempt. In *Proceedings of AAMAS2003 Workshop "Computational Autonomy, Potential, Risks, Solutions"*, 2003.
5. C. Castelfranchi. Multi-Agent Reasoning with Belief Contexts: The Approach and a Case Study. In M. J. Woolridge and N.R. Jennings, editors, *Intelligent Agents*. Springer-Verlag, 1995.
6. C. Castelfranchi. Founding an agent's autonomy on dependence theory. In *Proceedings of ECAI'00*, pages 353–357. Berlin, 2000.
7. C. Castelfranchi and R. Falcone. In R. Falcone H. Hexmoor, C. Castelfranchi, editor, *Agent Autonomy*, chapter From Automaticity to Autonomy: The Frontier of Artificial Agents, pages 79–103. Kluwer Publisher, 2002.
8. J. Christman. Autonomy in moral and political philosophy. In E. N. Zalta, editor, *The Stanford Encyclopedia of Philosophy (Fall 2003 Edition)*. Stanford University, plato.stanford.edu/archives/fall2003/entries/autonomy-moral/, 2003.
9. P. Davidsson. *Autonomous Agents and the Concept of Concepts*. PhD thesis, Lund University, 1996.
10. G. Dworkin. *The Theory and Practice of Autonomy*. Cambridge University Press, 1988.
11. H. Frankfurt. Freedom of the Will and the Concept of a Person. *Journal of Philosophy*, 1971.
12. H. Frankfurt. *Necessity, Volition, and Love*. Cambridge University Press, 1999.
13. S. Franklin and A. Graesser. Is It an Agent or Just a Program?: A Taxonomy for Autonomous Agents. In *Proceedings of the Third International Workshop on Agent Theories, Architectures and Languages*. Springer-Verlag, 1996.
14. E. Goffman. *The Presentation of Self in Everyday Life*. Doubleday, 1959.

15. J. Habermas. *The Theory of Communicative Action, Volume One, Reason and the Rationalization of Society*. Beacon Press, Boston, 1984. transl McCarthy, orig publ as Theorie des Kommunikativen Handels, 1981.
16. N.R. Jennings. Specification and Implementation of a Beliefe-Desire-Joint-Intention Architecture for Collaborative Problem Solving. *International Journal of Intelligent Cooperative Information Systems*, 2(3):289–318, 1993.
17. I. Kant. *Critique of Practical Reason (Kritik der praktischen Vernunft)*. Prentice-Hall, 1993 (1788).
18. C. Mackenzie and N. Stoljar, editors. *Relational Autonomy: Feminist Perspectives on Autonomy, Agency, and the Social Self*. Oxford University Press, 2000.
19. T. O'Connor. Free Will. In E.N. Zalta, editor, *The Stanford Encyclopedia of Philosophy (Spring 2002 Edition)*. http://plato.stanford.edu/archives/spr2002/entries/freewill/, 2002.
20. Merriam-Webster OnLine. The Language Center. www.m-w.com/.
21. H. V. D. Parunak. Engineering Artifacts for Multi-Agent Systems. www.erim.org/van/Presentations/presentations.htm, 1999.
22. L. Steels. Cooperation Between Distributed Agents Through Self-Organisation. In Y. Demazeau and J.P. Müller, editors, *Decentralized AI*. Elsevier, 1990.
23. J. D. Velleman. *Practical Reflection*. Princeton University Press, 1989.
24. H. Verhagen. *Norm Autonomous Agents*. PhD thesis, Department of System and Computer Sciences, The Royal Institute of Technology and Stockholm University, Sweden, 2000.
25. M. Wooldridge and N.R. Jennings. Intelligent Agents: Theory and Practice. *Knowledge Engineering Review*, 10(2):115–152, 1995.
26. M. J. Wooldridge and N.R. Jennings. Agent Theories, Architectures and Languages: A Survey. In M. J. Woolridge and N.R. Jennings, editors, *Intelligent Agents:, ECAI Workshop on Agent Theories, Architectures and Languages*. Springer-Verlag, 1995.

Types and Limits of Agent Autonomy

Gordon Beavers and Henry Hexmoor

Computer Science & Computer Engineering Department
Engineering Hall, Room 313, Fayetteville, AR 72701
{gordonb, hexmoor}@uark.edu

Abstract. This paper contends that to fully understand interactions between agents, one must understand the dependence and autonomy relations that hold between the interacting agents. Individual variations, interpersonal dependencies, and environmental factors are determinants of autonomy that will be discussed in this paper. The paper concludes with a discussion of situations when autonomy is harmful.

1   Introduction

Autonomy has been a primary concept for characterizing agents. This paper calls attention to several facets of understanding autonomy from the point of view of an agent. We are primarily interested in enabling agents to explicitly reason about autonomy while interacting with other agents. Autonomy is conceived as an internal and integral component of reasoning about interactions. This paper is not concerned with negotiation about autonomy among agents who may share a similar conception of autonomy, but rather we seek to expose the origins of this concept within a given individual as a result of the relationships in which that individual participates. This paper will investigate various ways in which autonomy can be interpreted and implemented in artificial agents, the utility of these implementations, and how these interpretations are linked to usage in a human context.

The simplest notion of autonomy is that of local determination. An agent that determines its actions for itself based only on its internal state is generally considered autonomous, that is, if the determination of the agent's behavior is local and without input from other agents, the agent is autonomous. Thus, a reactive agent running on a deterministic program would be considered autonomous, but since such an agent has no intentions, and is incapable of introspection, autonomy seems to be a concept of limited usefulness in the context of strictly reactive agents. In contrast, other investigators in the area of autonomy claim that the concept of autonomy is appropriate only for intentional agents that have introspective abilities. An intermediate position is that autonomy is essentially a social notion, which can be understood in terms of social dependencies. The continuum of positions with respect to autonomy will be surveyed in section 2.

One aspect of autonomy we will explore is individual variation among agents such as agent personality [6]. Autonomy is partly a psychological characteristic, and thus varies across agents according to personality. There are individuals who lead and others who prefer to follow. Another aspect of autonomy to be considered is interpersonal dependencies such as dependencies in the physical world and interpersonal social dependencies. Earlier papers, e.g. [2] have claimed that autonomy can be fully

M. Nickles, M. Rovatsos, and G. Weiss (Eds.): AUTONOMY 2003, LNAI 2969, pp. 95–102, 2004.

understood in terms of social independence. While this notion of autonomy is satisfactory for simple agents, it becomes inadequate as agents become more richly endowed with psychological characteristics, in particular, when agents can be considered to have a personality. A third area of exploration is social regulation and an agent's attitude toward laws, norms, and values in the agent's environment such as found in agent organizations, institutions, and general agent society.

2   Interpretations of Autonomy

It is common to refer to agents that determine their behavior without the influence of other agents as autonomous. This usage reflects the notion that an autonomous agent is independent and self-governing. These, however, are vague notions. What is generally meant is that an autonomous agent is empowered to choose to act contrary to the desires of other agents. Consider two simple agents each of which must make a choice between two options A and B. The first makes the decision based on input from a sensor monitoring some physical condition of the agent's environment. The second makes the decision based on input from another agent. The first agent acts independently of any other agent, but the second agent's action is dependent on the action of another agent. The first is autonomous with respect to the decision, but the second is not autonomous according to common notions. Is there any meaningful distinction between these agents? Both are reacting to the condition of an input. Neither has any knowledge of other agents, social constructs, or social structures. Is autonomy a fruitful concept in considering such simple agents? Probably not. We, therefore, confine our attention to intentional agents, and note the positive correlation of usefulness of the concept of autonomy with social abilities of the agent.

An intentional agent can exhibit various degrees of autonomy as determined by how much influence local and non-local there is in its decision making process. Local limitations on autonomy include constraints on intentional and physical abilities, while non-local limitations can be exerted through social values, e.g., avoid damage to the concerns of other agents, norms, e.g. keep to the right, legal restrictions, e.g., do not exceed the speed limit, and the need to cooperate.

An advocate of a more sophisticated position with respect to agent autonomy might claim that autonomy is a social notion, and as such, can only be a property of agents endowed with social properties interacting with other such agents in a social situation. Within this camp, there can be many variations with one of the simplest being that expressed by Castelfranchi in [2]. Castelfranchi asserts that "all forms of social autonomy should be defined in terms of different forms of social independence" and further that "each and any (sic) component of the architecture or necessary condition for a successful action can define a dimension/parameter of autonomy, since it can define an abstract 'resource' or 'power' necessary for the (sic) goal achievement, i.e. it can characterize a specific 'lack of power' and than (sic) a possible dependence and social non-autonomy." The question of what agent characteristics are required for an agent to be considered social must be addressed. Must an agent be intentional to be social? The Castelfranchi quote above implies that the mere existence of a dependence relation in a social setting is enough to establish an autonomy value. Is more required? Must an agent be intentional? Must an agent be capable of introspection and self-evaluation to be considered social?

Castelfranchi implies that dependency and autonomy are inverses of each other, but it is possible that an agent could be dependent on another agent and yet still remain fully autonomous, if the agent has the capability of ending the dependency relation at will. This suggests a distinction between first and second order autonomy. An agent who is dependent on another agent, but who retains control over that relation has given up first order autonomy, but retained second order autonomy. Dependencies likewise can be either first order or second order. When agent A depends on input from agent B to establish A's dependency relations, A has a second order dependency on B.

A stronger position on autonomy requires that an agent not only be social, but also be capable of reliably assessing its own capabilities. The claim is that an agent is not fully autonomous unless it has justification for believing that it is autonomous. Capability must be part of autonomy in the sense that an agent is autonomous with respect to a particular task or action only if that agent has the ability to complete the task or action. The greater the agent's warranted belief in its own capability the greater the agent's autonomy with respect to the relevant tasks.

In some environments, agent intentions and decisions are highly constrained by the environment. For example, an agent driving on an icy road has fewer options than an agent driving on a clear road. These constraints limit the agent's autonomy in the sense that a prudent agent will refrain from activities that have a sufficiently high probability of resulting in undesirable consequences. On the other hand, if the environment is safe for the agent, then choices are unconstrained, and autonomy is expanded. This example illustrates a common sense idea about the relationship between autonomy and the extent of an agent's choice, namely, the larger the number of choices that an agent has available the greater the agent's autonomy. While it is certainly the case that an agent with no choice has no autonomy, the converse does not hold. Consider an agent presented with a choice between an action that will cause its hard drive to be damaged and an action that will drain its batteries. How much autonomy does such an agent have? Autonomy evaluation must consider quality of choices as well as number of choices. Indeed, having too many choices can be detrimental, if the agent does not have the resources to properly evaluate those choices.

We now provide a critique of two claims found in [2]. First, Castelfranchi makes the assertion that "all forms of social autonomy should be defined in terms of different forms of social independence." While it is the case that dependency tends to diminish autonomy, lack of dependency does not necessarily yield autonomy, and further the effect of a dependency on an individual agent depends upon that particular agent's personality as discussed in the next section. The effect of a dependency is not uniform across all agents. A richer understanding of autonomy requires the consideration of agent mental properties that are not under the control of social relationships. Individual assertiveness and self-confidence are examples of such properties. That is, we are claiming that personality affects subjective autonomy estimates, which are an aspect of social autonomy, but subjective autonomy estimates are not definable in terms of social independence.

Second, Castlefranchi also claims that every component of the agent's architecture and every condition necessary for a successful action define a dimension of autonomy. Castelfranchi's linking of autonomy to resources and powers necessary for goal achievement would seem to make autonomy assessment objective, since, in any given situation it is an objective matter of fact which resources and powers are necessary for goal achievement. We argue that consideration of objective resources and powers is inadequate to capture a full understanding of autonomy, that autonomy has an essential subjective aspect that is overlooked by objective measures. In addition, autonomy can be affected by non-necessary, contingent conditions. Castelfranchi's concept of autonomy is objectively determined and lacks aspects of intentionality and personality.

Castelfranchi is correct to assert that autonomy is a relational notion between agents, actions, and goals. However, he incorrectly assumes that autonomy is completely determined by these relationships and the agent's "powers". Castelfranchi also correctly asserts that autonomy considerations must extend to the agent's relationship to its environment, since the ability to perceive and react in a reasonable manner to the environment is necessary for reliable goal achievement.

With respect to the first claim, Castelfranchi provides the following unproblematic assertion :(1) "If an agent Y depends on an agent X for its internal or external power/resource p relative to its goal G then Y is not autonomous from X relative to its goal G and resource p." It is, however, an unjustified leap from this assertion to (2) "all forms of social autonomy should be defined in terms of different forms of social independence." (1) is an implication, while (2) is essentially a bi-conditional. The converse of (1) is not true, that is, there is an agent who is not autonomous, to some degree, with respect to some goal because of its internal mental state, e.g., diffidence or self-doubt, but who is not dependent on any other agent or the environment. Learning and personality characteristics such as self-confidence and aggressiveness increase independence. Castelfranchi claims that autonomy is independence, nothing more. We need to distinguish "being autonomous" from "acting autonomous". Both internal and external conditions affect autonomy.

With respect to the second claim, here is an example where a contingent matter prevents goal achievement. An agent may have adequate resources and power to help a friend agent but may reason about lack of recent reciprocity from the friend and refrain from helping. This is an example where contingency of balance of reciprocity between two agents contributes to determining autonomy. Although objectively, resources and power were present for goal achievement, a subjective matter entered in autonomy decision-making. Dependence has both local and global aspects.

Consider how the gravity of a decision affects autonomy, that is, how consequences interact with autonomy. A casual treatment would say that autonomy is independent of the consideration of consequences. If an agent drives too fast on icy roads, unfortunate consequences are likely to ensue, but the agent's autonomy to select a speed is not affected by the possibility of disaster. A prudent agent will exercise its autonomy by choosing a lower speed, but its autonomy in the choice of speed is not affected by the possible consequences. This analysis is superficial. The consideration of consequences has affected the choices open to the agent, and thus has a second order affect on the agent's first order autonomy. In other momentous decisions, the connection between autonomy and consequences is harder to place. Consider an agent who is disarming a bomb by removing a timer. Cut the wrong wire and the bomb detonates. The gravity of the situation affects the manner in which the agent exercises its autonomy, namely very carefully, that is, it has a second order effect on the agent's autonomy but does not affect the agent's first order autonomy with respect to disarming the bomb. The gravity of the situation has a second order affect on autonomy in this case, since the gravity affects how the agent acts, but does not affect whether the agent acts. Autonomy is not just a matter of whether an agent chooses to act, but also how it chooses to act.

3   Individual Variations

Individuals vary in their conception of autonomy. These variations are due, in part, to differences in personality [6]. In the case of agents, we take personality to be a collection of persistent patterns of behavior such as trusting and agreeableness. The agent community has begun developing agents with synthetic personality, see [5] and [8]. deRosis and Castelfranchi mention laziness, helpfulness, dominance, and conflict-checking as agent personality traits [3]. Obviously, personality contributes to the individual's conception of autonomy. For a rough understanding of individual variations of autonomy we suggest two dimensions or scales. The first dimension we term a *rigidity/flexibility* scale. Rigidity constrains an individual's interactions by demanding (1) certainty, (2) independence, (3) norms and values, (4) boundary precision, and (5) control. Highly rigid individuals demand precision and are less flexible. Personality disorders such as perfectionism and obsessive\compulsive behavior commonly accompany high rigidity. One aspect of rigidity is the need for certainty of effects of actions (by self and others) and certainty of information available to them. With higher certainty, individuals are more confident in their actions and rely on their choices. When there is a lot of on uncertainty, an individual agent may not feel in control of its environment and may feel less confident about its decisions. Another aspect of personality is the need for independence. A highly dependent personality requires the assistance of others. A dependent agent may require constant direction and reinforcement. In the opposite extreme, an independent individual will avoid assistance from others, and be fiercely boundary conscious, which can affect its ability to be a team player. A third aspect of rigidity is mindless adherence to laws, conventions, norms, and values. A legalistic individual will be unyielding in following rules. On the other hand, an individual may seek to avoid rules and regulations. The fourth aspect of rigidity is boundary maintenance. Individuals differ in their tolerance of limits and some tolerate flexible values while others demand precise boundaries and limits. A fifth aspect of rigidity is the need for control and dominance. This is related to the need for a particular level of independence. Some individuals have higher need to be in control while others are content with less control. The relationship between rigidity and independence is in need of further investigation.

Our second dimension we will call *capacities* which is the individual's capacity to objectively perceive an agent's attributes that constrain interaction, which will be measured on a *competence* scale. Individuals that do not accurately perceive uncertainties can believe that they are in deterministic environments and may conceive of their autonomy being circumscribed. Those that perceive more uncertainties are better able to form a more complex sense of autonomy. Individuals who see themselves to be highly dependent, will sense relatively lower levels of autonomy as opposed to those who are oblivious to these dependencies. Sensing the extent to which an individual's decisions are guided by norms and values can give an individual a sense of security and lack of freedom. The ability to perceive control is another aspect of an individual's capacity. Sensing higher levels of control is generally reassuring for an individual and will lead to a sense of freedom and experience of higher autonomy. The exact relationship between control and autonomy appears to be very complex. At one end of the capacities dimension are individuals with the highest capacities and in the other end there are individuals with the lowest capacities.

Individuals fall in a unique place in this two dimensional personality space shown in Figure 1. This space is useful in suggesting the agent type to construct. For a domain that requires stable functions with moderate timing coordination precision, we suggest agents with low capacity and low rigidity. Functions are stable if they are context independent. Coordination precision is moderate if coordination constraints among functions are not strict. Automation of routine tasks in a stable environment is an example of such an environment. In environments that require stable functions but with high coordination precision, we suggest agents with low capacity and high rigidity. Coordination precision is high if coordination constraints among functions are strict. Air traffic control is an example of such an environment. For a domain that requires context sensitive functions with moderate precision, we suggest agents with high capacity and low rigidity. An environment that requires negotiation and collaboration is an example of such an environment. For a domain that requires context sensitive functions with high precision, we suggest agents with high capacity and high rigidity. Unpredictable planetary surface exploration is an example of such an environment.

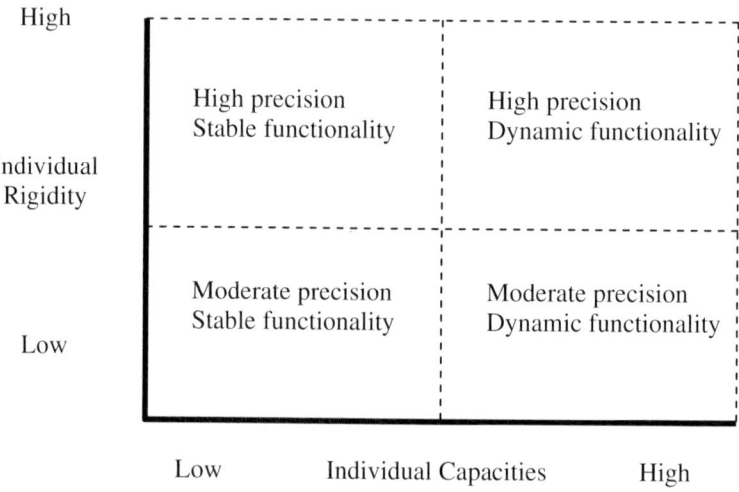

Fig. 1. Space of individual variations

4   Environmental Variations

Whereas freedom is implied from autonomy, freedom to act or decide does not imply autonomy. Freedom is the social deontic component of autonomy. Autonomy is also more than this exogenous sense of freedom. An individual must satisfy endogenous conditions, beyond socially extended autonomy, in order to be autonomous.

Autonomy can interpreted as a combination of socially warranted freedoms and internally perceived freedoms. When individuals are in groups, organizations, and institutions, their individual autonomy is altered by the properties of the group. This effect may be caused by their membership, their representation of group to others, or their participation in collective behavior. In settings where social obligations are rigid and matter a great deal, autonomy tends to be how an individual relates to those social rules. In contrast, there are many environments where few or no taboos or strict codes of conduct constrain agent behavior. In such environments, individuals are free to be unaffiliated.

Individuals, as members of groups, are subject to the social climate within which they are embedded. Their memberships and allegiances constrain their decision making and provide them with a set of behaviors, rights, duties, and obligations governed by a set of norms and values. This may augment or detract from their individual autonomies. Autonomies are affected by the individual's degree of commitment and loyalty to the groups to which they belong. As we explored above in our consideration of individual variation, individuals do vary in their tendency toward commitment and loyalty. However, here we are suggesting that the existence of groups to which an individual may belong produces a set of attributes for an agent's autonomy deliberation. We point to a need to develop reasoning methods for autonomy that account for an agent's membership in groups.

Given that an individual is a member of a particular group, when it represents the group in its interaction with non-group members, the individual assumes the group's persona and autonomy and augments it to its own. A representative must embody the essential components of the group it is representing and use it in reasoning about the group's autonomy.

In collective action or collective decision-making, the group as a whole owns the action or decision. An individual's autonomy toward the collective action or decision can only be conceived of as a contribution. This can be in terms of voting or vetoing power in the case of decision-making. In the case of physical or social action, an agent's autonomy is represented by the extent to which the individual facilitates the group's intent.

5   Limits of Autonomy: When is it too much?

Agents that interact with humans must be designed with particular attention to human safety and to not endanger human goals as well. Problems can arise when agents are fully autonomous and their actions are un-interruptible. Conversely, problems can also arise when an agent pays too much attention to human input when the human has insufficient knowledge of how the agent operates. Both agents that are excessively passive and agents that are excessively active can be harmful. Humans in the loop of agents need to be able to adjust the activity level of agents dynamically, that is, the autonomy level of agents needs to be adjustable. This seems to suggest a multi-tiered approach to agent autonomy. Agents may operate with certain nominal autonomies under normal circumstances but under heightened safety or concern conditions, agents should be switched to other modes when authorized human users manipulate their autonomy. Changing agent roles in interaction has been reported in mixed-initiative work [1]. Another promising area is empowering agents to reason about shared norms and values [7]. Since accounting for all contingencies is not realistic, agents need to have the ability to reason for themselves about whether their human supervisor would approve of their choices and assumed autonomies.

An agent's assumed level of autonomy may have unintended effects on other agents in a multi-agent setting. When tasks among agents are coupled, cooperating agents need to use a shared notion of autonomy to take into account one another's actions. Sharing autonomy is useful for harmonizing agent autonomies in order to account for one another's influence and to avoid negative influences [4]. Collective autonomy requires mutual trust among agents. Groups of agents may develop a notion of autonomy that belongs to the entire group. Individuals will conceive of auto nomy of their group. However, the group's autonomy can only be altered by collective actions such as negotiation. There can be problems when an individual in such a group misunderstands the group's autonomy when it represents the group. This can cause harm to the group or others.

5   Conclusions

We have argued that autonomy is not uniformly conceptualized, and that an account of autonomy must also account for variations between individuals and environments. This variability is shown to counter arguments by Castelfranchi and others to the effect that autonomy is determined strictly by interpersonal dependencies. This work has promise in the selection and design of complex agents that must match the requirements of their environment for autonomous behavior. We briefly highlighted situations when autonomy is harmful.

Acknowledgements

This work is supported by AFOSR grant F49620-00-1-0302.

References

1. J. Bradshaw, G. Boy, E. Durfee, M. Gruninger, H. Hexmoor, N. Suri, M. Tambe (Eds), 2002. Software Agents for the Warfighter, ITAC Consortium Report, AAAI Press/The MIT Press.
2. C. Castelfranchi, 2000. Founding Agent's Autonomy on Dependence Theory, In proceedings of ECAI'01, pp. 353–357, Berlin.
3. C. Castelfranchi and F de Rosi, 1999. Which User Model Do We Need to Relax the Sincerity Assertion in HCI? UM'99 Workshop on *Attitude, Personality and Emotions in User-Adapted Interaction*.
4. H. Hexmoor, 2001. Stages of Autonomy Determination, IEEE Transactions on Man, Machine, and Cybernetics- Part C (SMC-C), Vol. 31, No. 4, Pages 509–517, November 2001.
5. D. Moffat. 1997. Personality parameters and programs. I. R. Trappl and P. Petta, editors. Creating personalities for Synthetic Actors, Pages 120–165, Springer.
6. L.A. Pervin and O.P. John, (Editors) 2001. Handbook of Personality Theory and Research, Guilford publications.
7. D. Shapiro, 2001. Value-driven agents, Ph.D. thesis, Stanford University, Department of Management Science and Engineering.
8. R. Trappl, and P. Petta, (eds.) 1999. *Creating Personalities for Synthetic Actors*. Springer-Verlag, Berlin.

Autonomy in Multi-agent Systems:
A Classification Attempt

Cosmin Carabelea[1,2], Olivier Boissier[1], and Adina Florea[2]

[1] Centre SIMMO/SMA, ENS des Mines de Saint-Etienne,
158 Cours Fauriel, Saint-Etienne Cedex 2, F-42023, France
{carabelea, boissier}@emse.fr
[2] Computer Science Department, "Politehnica" University of Bucharest
Spl. Independentei, nr. 313, Sect. 6, 77206, Bucharest, Romania
adina@cs.pub.ro

Abstract. Despite the fact that autonomy is a central notion in most agents' definitions and it is by itself a subject of research in the multi-agent field, there is not yet a commonly agreed definition of it. This paper attempts to classify the different forms of autonomy proposed in the literature by using the *Vowels* approach. After underlining some of the most relevant properties of autonomy, the classification is used to propose a comprehensive definition of autonomy. Moreover, we identify two different perspectives on autonomy, namely external and internal, and we show that the later can be used to build an integrated analysis grid of agent architectures for multi-agent systems.

1 Introduction

In the last twenty years, multi-agent systems have proposed a new paradigm of computation based on cooperation and autonomy. An intelligent agent is "a computer system, situated in some environment, that is capable of flexible and autonomous action in order to meet its design objectives" [18]. Here, flexible means reactive, proactive and social. The agents we analyze are *goal-directed* ones, meaning that they are endowed with goals and that their behaviour is guided by an internal (mental) representation of the effects [8], however, the definitions proposed in this paper apply to any other type of agent, including reactive ones. A goal is defined in [14] as a state of the world represented in an agent's mind, which the agent wants to become true.

In the social view of agents, an agent is not an entity that exists and performs alone in a system, it is part of a multi-agent system, where it has to interact with other agents. From this point of view, two important notions in multi-agent systems are the *delegation* and the *adoption* of goals or actions [8]. Put simply, we say that the agent A *delegates* a goal to agent B if it has a plan including that B will satisfy the goal for it. If B takes one of the A's goals and make it its own, then we say that B *adopted* the goal from A. More formal definitions and discussions about these two notions can be found in [9].

M. Nickles, M. Rovatsos, and G. Weiss (Eds.): AUTONOMY 2003, LNAI 2969, pp. 103–113, 2004.
© Springer-Verlag Berlin Heidelberg 2004

Given this global picture, one of the most difficult problems in multi-agent systems is to ensure a coherent behaviour of the system while allowing each agent to choose its own behaviour, i.e., to be autonomous. Although the notion of autonomy appears in most of the existing agent definitions up to date there is no commonly agreed definition of what autonomy in multi-agent systems really is. There is a tremendous amount of work done by the multi-agent community in an attempt to clearly define and formalize what seems to be the defining characteristic of an agent, namely its autonomy. Unfortunately, not only is there not a single definition of autonomy, but the proposed definitions seem to refer to completely different concepts. This paper tries to classify the different approaches to autonomy and to propose a clear, comprehensive view of them.

This paper is organized as follows. In the next section we use the *Vowels* approach [13] to structure and present an overview of related work on autonomy. Section 3 presents two different perspectives on autonomy: external and internal. For the former we propose a comprehensive definition, while the latter is discussed from the point of view of related work and agents' architectures in the following section. Finally, in Section 5, we draw some concluding remarks and trace directions for future work.

2   Guided Tour on Related Work on Autonomy

This section presents some of the related work on autonomy in multi-agent systems. We propose to use the *Vowels* view on multi-agent systems [13] to classify the different approaches to autonomy. According to the Vowels approach, a MAS consists of four different kinds of models: the (*A*)gents, the (*E*)nvironment, the (*I*)nteraction and the (*O*)rganization. Although the (*U*)ser is not explicitly part of it, we add this fifth dimension to the other four in order to better classify the definitions of autonomy.

Each of the five vowels (dimensions or families of models) defines a source of constraints bearing on the agent, against which the agent has to position in order to behave properly in the system. Thus, for each of them, we will identify a type of autonomy existing in related work and we will propose a definition of it. For a better understanding, we will use examples dealing with some robots gathering ore on Mars. The robots (agents) have to take the ore from the mines (sources of ore) and to take it back to a base where one or several users are located. Different possible situations that may occur will be used to illustrate different types of autonomy.

2.1   U-Autonomy (User-Autonomy)

One of the most frequent forms of autonomy in related work is the agent's autonomy with respect to the user for making a decision and it is called in the sequel *U-autonomy*. In most of the cases the agent is a personal assistant of the user and it has to choose its behaviour, i.e., decide on what goal, plan or action to do next. It can make this choice on its own or it can ask the user what to do. This continuous passing of control to and from the user is called *adjustable autonomy* in [27]. An interesting related question is raised in [25]: should the agent be allowed to refuse obeying the user or not?

The most common definition for this type of autonomy is that *an agent is autonomous with respect to the user for choosing what action to perform if it can make the choice without the user's intervention.* This is why we say an agent has an adjustable U-autonomy as it *sometimes* passes the control of its actions to the user. The keyword here is 'sometimes': how to decide when to act autonomously (and make the decision by itself) and when to ask the user to choose the next action to do. We can say that the U-autonomy of an agent may vary from a complete *user-independence* to a complete *user-dependence*, passing through the intermediate position of adjustable U-autonomy.

Using our example, if a robot has found a new source of ore, it might ask the user if it should go and mine the new source or use only the known mines. A robot completely user-independent will behave U-autonomously and if the mine is better situated, more profitable, etc., it will go to mine there without asking the user. This type of autonomy may seem a desirable feature for the robot, but if there are areas labeled as dangerous in the surroundings, the user may want to restrain the U-autonomy of her robots to prevent them from making decisions that might prove wrong. Thus the need of a robot (agent) with adjustable U-autonomy that *knows* when to ask the user and when not to.

2.2 I-Autonomy (Social-Autonomy)

Another frequent usage of the term autonomy is related to the *social autonomy*. We believe this is one of the most important notions underlying the interactions between the agents in multi-agent systems. In most of the cases, when one talks about social autonomy, one refers to the adoption of goals, although the object of social autonomy can vary. An attempt to unify the different objects on which an agent can be socially autonomous is given in [10], where the authors use the dependence theory as a basis of autonomy among agents. We will not enter in details here, but we consider goal adoption as the most relevant form of autonomy for agent interactions. As it mainly deals with agent interactions, we will call this form of autonomy I-autonomy.

In [24], the autonomy of an agent is viewed as its capacity to generate its own goals based on its motivations. The autonomy is then viewed as a central notion for the multi-agent interactions, because an autonomous agent can refuse the adoption of goals from other agents. In other words, when an agent tries to delegate one of its goals to another agent, there are two possible situations: if the second agent is not autonomous, it accepts the delegated goal; if it is autonomous, the agent will decide whether to refuse it or not [22][23].

A slightly different view is taken in [1], where the goal the agent should adopt is a social one. A voting mechanism is used for taking the decision on what goal to adopt; an agent varies its degree of autonomy by modifying the weight of votes of each agent.

The most used definition for the I-autonomy with respect to goal adoption is that *an agent X is autonomous with respect to another agent Y for the adoption of a goal G if X can refuse the adoption of the goal G from Y.* The decision of accepting or not the adoption of a goal cannot be imposed by Y; it is an *internal* feature of the agent X. An agent with an adjustable I-autonomy is an agent that adjusts its behaviour to the current context: sometimes it accepts the goal adoption, sometimes it may refuse it. We can say that an agent may vary from an *autistic* one (which refuses everything) to a *benevolent* one (which accepts everything), passing by the intermediate stage of the *adjustable I-autonomy*.

Using our example, imagine that one of the robots, Y, has a broken leg and cannot go back to the base to have it fixed, so it asks another robot, X, to adopt the goal of carrying Y to the base. If the robot X is I-autonomous, it can refuse to carry Y if, for example, it believes that Y functions perfectly and asks for help just to save its energy. This illustrates the need of social autonomy in multi-agent systems: it eliminates the risk of an agent using a non-autonomous agent as a slave.

2.3   O-Autonomy (Norm-Autonomy)

It is commonly agreed that the use of autonomy introduces a degree of non-determinism in the behaviour of the system, since the answer of an autonomous agent is not known a priori [15]. One solution to this problem is the utilization of social laws, conventions or organizational structures to restrain the autonomy of the agents. In the following, we will call these organizational constructs norms [3]. In such a case, the agents should be able to recognize the norms and to reason about them [19][20]. Moreover, if we introduce norms in multi-agent systems, another problem is raised: what if, for instance, there are norm-autonomous agents, like the ones proposed in [28][29] or [12]? Such an agent will be able to decide whether to obey norms or not. The authors of [5] have proposed the use of norm-autonomous agents as a means to achieve social intelligence in the context of distributed decision-making. As it deals mainly with organization, we will call O-autonomy this form of autonomy.

The definition proposed for this form of autonomy is that *an agent is autonomous with respect to a norm if it can violate that norm*. This form of autonomy raises more difficult problems than the previous ones. First of all, it is difficult to formalize norms in a way that will allow the agents to reason upon them. Second, which are the factors that influence the behaviour of an O-autonomous agent? An agent with adjustable O-autonomy may vary from an *obeying* one (which obeys all the norms) to a *rebellious* one (which obeys none).

Assume now that there is a norm in the system of our example that says the robots cannot leave a specified area. X is a norm-obeying robot: it doesn't leave the area, but it notices another robot that left the area, broke its leg (and its antenna, so it cannot ask for help) and cannot go back. If X has an adjustable O-autonomy, it can violate the norm, go there and help its teammate.

2.4   E-Autonomy

The autonomy with respect to the environment refers to the fact that *the environment can only influence the behaviour of an agent, it cannot impose it*. This comes back to what it is called the 'Descartes problem': responses of many living systems to the environment are 'neither caused by, nor independent of the external stimuli'. All the agents that we know of have E-autonomy. However, they might choose their answers to the environmental stimuli in different ways: an agent can be strongly influenced by the environment, while an other is barely influenced by it. We say the former has only a *weak E-autonomy*, while the latter has strong *E-autonomy*.

The reactive agents are agents with weak E-autonomy, because their behaviour is strongly influenced by the environment. They do not have a symbolic representation of the environment, nor do they have a symbolic reasoning about it. Deliberative agents, on the other hand, are loosely coupled with the environment, so they have strong E-autonomy.

2.5   A-Autonomy (Self-Autonomy)

The autonomy is generally viewed as the property of the agent that allows it to exert local control over its behaviour, i.e., its behaviour cannot be imposed by an external source (environment, user, other agent, a norm). Then, what is the meaning of talking about A-autonomy (self-autonomy)? This form of autonomy can be interpreted as *the property that allows an agent to have and choose between several possible behaviours*. All agents are implicitly endowed with it, because it gives the agents the possibility to *adapt* their behaviour to different situations. In fact, it is this type of autonomy that ensures an agent can have other types of autonomy: if an agent can have only one behaviour, it makes no sense to talk about its capacity to refuse the imposing of a behaviour from an external source.

Table 1. Different types of autonomy present in related work

U-autonomy	I-autonomy	O-autonomy	E-autonomy	A-autonomy
Tambe et al. [27] (others, not mentioned)	Castelfranchi [10] Luck, d'Inverno [24] Barber et al. [1]	Dignum et al. [12] Verhagen et al. [28]	present (in different forms) in all agent architectures	present implicitly in all agent architectures

To conclude this overview we have grouped in Table 1 the different types of autonomy that are considered in the MAS field. As our discussion showed, autonomy is a very rich notion covering different aspects of agent behaviour and decision process that our grid succeeds in making explicit.

3   Autonomy – Towards a Definition

With so many forms of autonomy, is it possible to find a comprehensive definition, i.e. a definition that holds for all the above-mentioned forms of autonomy? We believe so, but before proposing such a definition, we will first draw attention on a property of autonomy. We will then present two different perspectives on autonomy.

3.1   A Definition?

In all of the above-mentioned definitions, we can notice a common property attached to autonomy : its *relational* nature. As summarized in [7], *X is autonomous with respect to Y for p*: one cannot talk about an autonomous agent without mentioning for what the agent is autonomous, i.e. *p*, the *autonomy object*, and with respect to whom it is autonomous, i.e. *Y*, the *autonomy influencer*. Going back to our previous classification, the autonomy influencer is usually another agent (I-autonomy), the user (U-autonomy), a norm (O-autonomy), but it may also be the environment (E-autonomy), or the agent itself (A-autonomy), while the autonomy object can be virtually anything [10].

But there is still another aspect of the autonomy that we should consider every time we talk about the autonomous character of an agent: the *context*. An agent can be autonomous in a situation and non-autonomous in the other. This aspect is also considered by Hexmoor [17][6], who thinks that the autonomy should not be studied without taking into consideration the context of an agent. In other words, an agent may adjust its degree of autonomy by deciding to accept or not the adoption of a goal from another agent, and this decision is made by taking into account the current context. Thus, we can further refine our relational property of autonomy: *X is autonomous with respect to Y for p in the context C.*

It is easy to notice that all forms of autonomy presented in Section 2 have some similarities. Based on these similarities, we propose the following comprehensive definition of an autonomous agent: *an agent X is autonomous with respect to Y for p in the context C, if, in C, its behaviour regarding p is not imposed by Y.* According to the nature of Y, we obtain the above-mentioned types of autonomy. If Y is dealing with an autonomous agent X, Y will not be able to impose a behaviour to X, the best it can do is to ask X to behave that way. Here, the behaviour of X doesn't mean only its external behaviour (its actions), but also its internal, 'mental' processes. In other words, Y cannot condition the beliefs, desires, decision making modules or any other *power* of an agent, if the agent is autonomous with respect to Y for that power in this context.

3.2   External and Internal Perspectives on Autonomy

According to our definition, an agent Y, for example, cannot impose an I-autonomous agent X to adopt one of its goals, it is up to X to accept or refuse that adoption. In this paper we are not interested in what is the behaviour of an agent regarding p, but only if it is imposed by an external source or not. This is the reason why we call the proposed definition *the external perspective* on the autonomy, as it is not concerned with what happens inside the agent (how the agent chooses its behaviour).

This definition and perspective on autonomy is equivalent with the one based on social dependence proposed by Castelfranchi in [10]: if an agent X is dependent on an agent Y for one of its *powers* relative to the goal G (the adoption of it), then X is not autonomous with respect to Y for the adoption of G. In other words, if X depends on Y for some power p, then the behaviour of X regarding G can be imposed by Y; it is no longer independent of Y. The problem with the definition of autonomy based on social-dependence is that it is very difficult to identify and compute the existing dependencies between agents.

The *internal perspective*, doesn't consider anymore the dependence or independence of agents but is mainly interested in how the agent will make the decision to choose its behaviour, i.e., how the agent will make use of its autonomy (its capacity to have its own behaviour, and not having it imposed from outside). We call this view on autonomy an *internal* one because it is concerned with what happens inside the agent, how it behaves in autonomous fashion.

In other words, the external view is the one of an observer (the user, another agent), while the internal view is the one of the developer (or the designer of the agent). The difference between these two views is also underlined in [2], where the authors propose to use the expression 'an agent is autonomous with respect to...' to designate the external view and the expression 'an agent has autonomy with respect to...' when talking about internal autonomy.

4   Internal View of Autonomy

4.1   Related Work

In most of related work, the above-mentioned internal perspective on autonomy is considered in the sense that the authors are interested in *how* the agents are able to exhibit autonomous behavior in various situations. They mainly deal with the definition of agents' control architectures. For example, in the case of personal assistant agents with an adjustable U-autonomy, the authors of [26] have proposed the utilization of Markov Decision Processes, so their agents are able to learn when it is appropriate to act in an autonomous manner and make the decision by themselves or when they should ask the user for that decision.

In the case of I-autonomy, where an agent should be able to refuse the adoption of a goal from another agent, the authors of [21] have proposed the agent's motivations as a basis for its decision. The autonomy is then viewed as a central notion for the multi-agent interactions, because an autonomous agent can refuse the adoption of goals from other agents if these goals don't meet its motivations [23].

In the field of I-autonomy, a slightly different approach is given in [1], where the goal the agent should adopt is a social one. A voting mechanism is used for taking the decision on what goal to adopt; an agent varies its degree of autonomy by modifying the weight of votes of each agent. Thus an agent may pass from a completely autonomous one,which assigns zero weight to the other's vote, to a non-autonomous one, which assigns zero weight to its vote.

The problem of an autonomous agent behaviour is also taken into consideration in the area of norm-autonomy. In this case, however, an O-autonomous agent has to decide whether or not to obey a norm. The way this decision is formed depends on the norm representation, but usually the agent should also take into account the punishment for violating the norm, the reward for obeying it, its current goals and motivations, the probability of being caught and so on.

4.2   Example of O-Autonomy in Agents' Architectures

This brief overview on agent autonomy opens a tremendous field of research in the MAS domain, research related to the definition of agents' architectures. The complete evolution of these architectures can be explained as a long way to define architectures able to exhibit the different kinds of A,E,I,O, and U-autonomy [4]. Presenting this entire evolution is beyond the scope of this paper. In order to illustrate our point of view we will just focus on one architecture.

We take as an example the B-DOING architecture proposed in [16]. The authors have extended the BDI model with obligations, norms and goals, as depicted in Figure 1. Thus, the agent is generating new intentions using its goals and beliefs, while the goals are generated by taking into account the current beliefs and goals, but also the obligations the agent has (the other agents' interests), its desires (its own interests) and the existing norms (the society's interests).

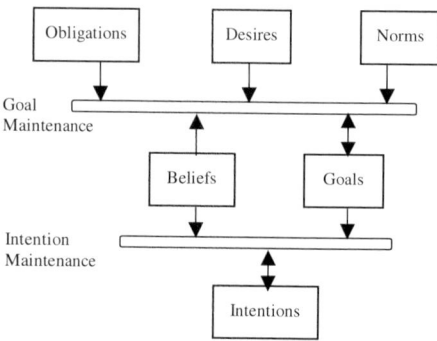

Fig. 1. The decision process of a B-DOING agent

As Figure 2 highlights, the presence of the *goal maintenance* decision module makes a B-DOING agent A-, I- and O-autonomous. For example, the obligations, what the other agents want, generate new goals for the agent. The *goal maintainence* module selects only some of these goals to be part of the agent goals. In other words, the agent can refuse the adoption of goals from other agents, it is not obliged to adopt them, so it is I-autonomous. The same holds for A- and O-autonomy.

5   Conclusions

Although the concept of autonomy appears in almost all agent definitions, currently there is not a commonly agreed definition of what autonomy in multi-agent systems really is. There is a lot of related work done in an attempt to clearly define it, but the proposed definitions seem to refer to completely different concepts. In this paper we have used the *Vowels* approach for the classification of these different types of autonomy.

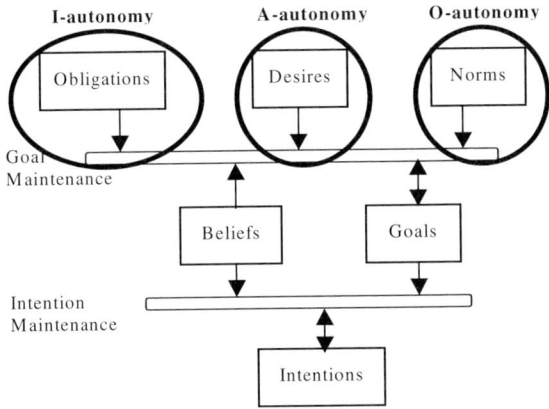

Fig. 2. A-, I- and O-autonomy in a B-DOING agent

We have thus obtained an analysis grid that allows us to identify the different categories of autonomy: A-,E-,I-,O- and U-autonomy. We have then used this result to propose a comprehensive definition of autonomy. Besides the fact that this definition holds for all the identified types of autonomy, it also proposes the context as a factor that can influence the autonomous character of an agent.

We have also identified two different existing perspectives on the autonomy: an external and an internal one. The former takes an observer (agent or user) perspective and refers to the capacity of the observed agent to behave autonomously (be autonomous), while the second concerns how such an agent is constructed. After identifying some related work in the field of internal autonomy, we have shown using an example that the concept of autonomy is also captured in agent's architectures, even if in an implicit form.

We believe autonomy to be a crucial notion in multi-agent systems and it is important to have an architecture which explicitly identifies what are the agent's parts that give it an autonomous character. Unfortunately, with one notable exception (the modification of the DESIRE architecture proposed in [11]), no such architecture has been proposed yet. As future work, we would like to use the analysis grid obtained to classify existing architectures and propose a new architecture that we would like to call the *3A architecture* – Agent with Adjustable Autonomy architecture. Such an architecture will allow the designer of an agent to clearly specify and identify what are the types of autonomy the agent is endowed with.

Acknowledgements

The Ph.D. Study of the first author is partially funded by the French Embassy in Romania. The authors would like to thank Prof. Cristiano Castelfranchi for the precious comments regarding the ideas proposed in this paper.

References

1. K. Barber, A. Goel, C. Martin: *Dynamic Adaptive Autonomy in MAS.* JETAI 12(2), 2000, p. 129–147.
2. G. Beavers, H. Hexmoor: *Types and Limits of Agent Autonomy.* In Proc. of the Workshop on Computational Autonomy, Autonomy 2003, Melbourne, Australia, 2003.
3. G. Boella, L. Lesmo: *Norms and Cooperation: Two Sides of Social Rationality.* In H. Hexmoor and R. Falcone (eds.) Agent Autonomy, Kluwer, 2002.
4. O. Boissier: *Contrôle et Coordination orientées multi-agent.* Memoire HdR, ENS des Mines de Saint-Etienne, France, 2003.
5. M. Boman, H. Verhagen: *Social Intelligence as Norm Adaptation.* In Proc. of the Workshop on Socially Situated Intelligence, 5th Int. Conf. on Simulation of Adaptive Behavior, 1998.
6. S. Brainov, H. Hexmoor: *Quantifying Relative Autonomy in Multiagent Interaction.* In Proc. of the IJCAI-01 workshop on Autonomy, Delegation, and Control: Interacting with Autonomous Agents, Seattle, WA, 2001, p. 27–35.
7. C. Castefranchi: *Guarantees for Autonomy in Cognitive Agent Architecture.* In N. Jennings and M. Wooldridge (eds.) Agent Theories, Architectures, and Languages, Heidelberg, Spinger-Verlag, 1995.
8. C. Castefranchi: *Towards an Agent Ontology: Autonomy, Delegation, Adaptivity.* AIIA Notizie, XI(3), 1998, p. 45–50.
9. C. Castefranchi: *Modelling social action for AI agents.* Artificial Intelligence, 103, 1998, p. 157–182.
10. C. Castefranchi: *Founding Agent's 'Autonomy' On Dependence Theory.* In Proc. of 14th European Conf. on Artificial Intelligence, IOS Press, 2000, p. 353–357.
11. C. Castelfranchi, F. Dignum, C. Jonker and J. Treur: *Deliberate Normative Agents: Principles and Architectures,* In N. Jennings and Y. Lesperance (eds.) Intelligent Agents VI, LNAI-1757, Springer-Verlag, 2000, p. 364–378.
12. R. Conte, C. Castelfranchi, F. Dignum: *Autonomous Norm-Acceptance.* In J. Muller, M. Singh and A. Rao (eds.), Intelligent Agents V, LNAI 1555, Springer-Verlag, 1999, p. 319–334.
13. Y. Demazeau: *From interactions to collective behaviour in agent-based systems.* In Proc. of the 1st European Conf. on Cognitive Science, Saint Malo, France, April, 1995, p. 117–132.
14. F. Dignum, R. Conte: *Intentional Agents and Goal Formation.* In M. Singh et al. (eds), Intelligent Agents IV, LNAI 1365, Springer-Verlag, 1998, p. 231–244.
15. F. Dignum: *Autonomous Agents with Norms.* In Artificial Intelligence and Law, Volume 7, 1999, p. 69–79.
16. F. Dignum, D. Kinny and E. Sonenberg: *Motivational Attitudes of Agents: On Desires Obligations and Norms.* In Proc. of the 2nd Int. Workshop of Central and Eastern Europe on MAS, 2001, p. 61–70.
17. H. Hexmoor: *Case Studies of Autonomy.* In Proc. of FLAIRS 2000, AAAI Press, 2000, p. 246–249.
18. N. R. Jennings, K. Sycara and M. Wooldridge: *A Roadmap of Agent Research and Development.* International Journal of Autonomous Agents and Multi-Agent Systems 1 (1), 1998, p. 7–38.
19. F. Lopez y Lopez, M. Luck: *Towards a Model of the Dynamics of Normative Multi-Agent Systems.* In Proc. of the Int. Workshop on Regulated Agent-Based Social Systems: Theories and Applications (RASTA'02), 2002, p. 175–193.

20. F. Lopez y Lopez, M. Luck, M. d'Inverno: *Constraining Autonomy through Norms.* In Proc. of the 1st Int. Joint Conf. on Autonomous Agents and Multi-Agent Systems, ACM Press, 2002, p. 674–681.

21. M. Luck, M. d'Inverno: *A Formal Framework for Agency and Autonomy.* In Proc. of the 1st Int. Conf. on Multi-Agent Systems, AAAI Press/MIT Press, 1995, p. 254–260.

22. M. Luck, M. d'Inverno: *Engagement and Cooperation in Motivated Agent Modelling.* In Distributed Artificial Intelligence Architecture and Modelling: Proc. of the 1st Australian Workshop on Distributed Artificial Intelligence, Zhang and Lukose (eds.), LNAI 1087, Springer-Verlag, 1996, p. 70–84,.

23. M. Luck, M. d'Inverno: *Motivated Behaviour for Goal Adoption.* In Multi-Agent Systems: Theories, Languages and Applications – Proc. of the 4th Australian Workshop on Distributed Artificial Intelligence, Zhang and Lukose (eds.), LNAI 1544, Springer-Verlag, 1998, p. 58–73.

24. M. Luck, M. d'Inverno: *Autonomy: A Nice Idea in Theory.* In Intelligent Agents VII: Proc. of the 7th Int. Workshop on Agent Theories, Architectures and Languages, Castelfranchi and Lesperance (eds.), LNAI 1986, Springer-Verlag, 2001.

25. D. Pynadath, M. Tambe: *Revisiting Asimov's First Law: A Response to the Call to Arms.* Intelligent Agents VIII Proc. of the Int. Workshop on Agents, theories, architectures and languages (ATAL'01), 2001.

26. M. Tambe, P. Scerri, D. Pynadath: *Adjustable Autonomy in Real-world Multi-Agent Environments.* In Proc. of the Int. Conf. on Autonomous Agents (Agents'01), 2001.

27. M. Tambe, P. Scerri, D. Pynadath: *Why the elf acted autonomously: Towards a theory of adjustable autonomy.* In Proc. of the 1st Autonomous Agents and Multiagent Systems Conf. (AAMAS), 2002.

28. H. Verhagen, M. Boman: *Adjustable Autonomy, Norms and Pronouncers.* AAAI Spring Symposium on Agents with Adjustable Autonomy, 1999.

29. H. Verhagen, J. Kummeneje: *Adjustable Autonomy, Delegation and Distribution of Decision Making.* In Proc. of the 1st Int. Workshop of Central and Eastern Europe on MAS (CEEMAS'99), 1999.

Autonomy and Agent Deliberation

Mehdi Dastani, Frank Dignum, and John-Jules Meyer

Utrecht University
Institute of Information and Computing Sciences
{mehdi, dignum, jj}@cs.uu.nl

Abstract. An important aspect of agent autonomy is the decision making capability of the agents. We discuss several issues that agents need to deliberate about in order to decide which action to perform. We assume that there is no unique (rational or universal) deliberation process and that the deliberation process can be specified in various ways. The deliberation process is investigated from two perspectives. From the agent specification point of view the deliberation process can be specified by dynamic properties such as commitment strategies, and from the agent programming point of view the deliberation process should be implemented through the deliberation cycle of the agent, which can be either fixed or determined by a deliberation programming language.

1 Introduction

The autonomy of software agents is directly related to their capacity to make decisions without intervention of the human users. At the lowest level these decisions concern the next action that the agent is going to perform. However, the actions are selected based on many high level decisions such as whether it will stick to its current plan to reach a goal, whether it wants to change its goal, whether to adhere to a norm or not, etc. It seems logical to start with classical decision theory for help in modelling these decisions. On the other hand intelligent agents are modelled using BDI theory. So we will first briefly explore the relation between classical decision theory, a qualitative extension of it called qualitative decision Theory, and the BDI theory. Formal details of this relation are explained in [6].

Classical decision theory [9,16] assumes that the decision making agent is able to weigh all possible alternative courses of actions before choosing one of them. Given a set of actions and a set of outcomes (states), actions are weighed by two basic attributes that are assumed to be accessible by the decision making agent. The first attribute is the probability distribution that indicates the probability of reaching a state by performing an action. The second attribute is the utility function that indicates the utility of states. Based on these attributes a certain decision rule, such as maximizing expected utility, is used to determine which action should be selected and performed. Although useful in closed systems, the problem with classical decision theory is that in open or complex agent environments, the probability distributions are not fixed or cannot be calculated due to

M. Nickles, M. Rovatsos, and G. Weiss (Eds.): AUTONOMY 2003, LNAI 2969, pp. 114–127, 2004.
© Springer-Verlag Berlin Heidelberg 2004

resource or time constraints. Therefore the agents only have partial information and cannot compare the utility of all possible states.

Qualitative decision theory [1,2] relaxes the assumption of classical decision theory that the probability distribution and the utility function should be provided, and develop representation and reasoning schemes for partial information and generic preferences to represent probabilities of states and generic preferences over those states [7]. Typically qualitative orderings are introduced that represent both the likelihood or normality (probability) and desirability (utility) of states. The maximum expected utility decision rule is replaced by qualitative rules such as Wald's criterion. A rational agent is then assumed to attempt to achieve the best possible world consistent with its likely (default) knowledge. Qualitative decision theory provides the state to be reached without indicating which actions to perform. The decision making agent is then assumed to be a planning agent who can generate actions to reach the given state.

A criticism to both classical and qualitative decision theory is that they are inadequate for real time applications that decide actions in highly dynamic environments. In such applications, the decision making agent, which is assumed to be capable of performing a set of actions, should select and execute actions at each moment in time. As the highly dynamic environment changes either during the selection or the execution of actions, the decision making agent needs to deliberate to either reconsider the previous decisions, also called intentions, or continue the commitment to those decisions (intentions). This deliberation process leads to either a reconsideration of the intentions or the continuation of the commitment to the intentions. This decision aspect is argued by Bratman [2] to be crucial to realize stable behavior of decision making agents with bounded resources. This approach has led to what is called BDI theory [14].

In BDI theory, the (partial) information on the state of the environment is reduced to dichotomous values (0-1); the propositions are believed or not. This abstraction of the (partial) information about the state of the environment is called the beliefs of the decision making agent. Through this abstraction, the information on likelihood ordering among states is lost. Similarly, the (partial) information about the objectives of the decision making agent is reduced to dichotomous values (0-1); the propositions are desired or not. This abstraction of the objectives of the decision making agent is called agent desires; the information on desirability ordering among states is lost. Finally, the information about the selected goal states is represented by dichotomous values (0-1); an agent can intend to reach a state or not. Note that the BDI systems can be extended with either qualitative orderings for likelihood and desirability of beliefs [3], or even with quantitative probabilities and utilities [13]. A decision-theoretic view of BDI theory extends the dichotomous abstraction of qualitative decision theory with the additional deliberative component, i.e. the intentions, to stabilize decision making behavior of the agent.

Unlike decision theories that propose (quantitative or qualitative) decision rules to determine which actions to select and perform, BDI theory assumes that the agent decides about which actions to perform next by reasoning on a plan to reach the selected goal state. Some constraints are proposed to identify the possible goal states and plans. Examples of static properties are realism, weak realism, and strong realism, which state in various degree that (only) states of which the agent believes that they are possible can be selected as goal states. Examples of dynamic properties are various kinds of commitment strategies that determine when the commitments to previous decisions can be dropped, i.e. when intentions can be reconsidered [13].

Whereas decision theory tries to generate the best agent decision at any point in time, the BDI theory tries to limit agent choices to the correct ones according to the constraints that are given beforehand. In order to create agents that are autonomous they have to be able to make decisions on different levels while adhering to the BDI theory. In order for autonomous agents to be programmable they probably need decision theory (which gives some procedural handles). Ideally an agent is autonomous if it would arrive at the best possible decision using some (parts of) decision theory while keeping within the constraints given by the BDI theory.

The rest of this paper is organized as follows. In the next section, various deliberation issues are explained and it is explained how they affect the decision to select actions. In section 3 it is argued that the decision-making ability of agents should be programmable in order to create agent autonomy at different levels. We therefore introduce a language to program possible deliberation processes. In section 4 we consider the specification of some static and dynamic properties of agent behavior, known from BDI literature, and indicate how these properties can be realized by programming certain deliberation processes. Finally, in section 5 we conclude the paper.

2   Deliberation Issues

Classical and qualitative decision theories propose decision rules since they have orderings on possible states. However, BDI theory cannot propose a decision rule since there is no ordering given on possible states. Note that the suggested constraints by the BDI theory cannot be considered as the counterpart of decision rules since they do not specify one single goal state that, when planned, generates a sequence of actions. For example, the realism constraint states that a goal state can be selected if it is believed to be achievable. However, given the beliefs and goals of an agent, many states can be believed as being achievable such that it should still be specified which state should be specified as the goal state.

In this section, we consider the decision making ability of agents from the implementation point of view, i.e. how the decision making ability of agents can be implemented. For the classical decision theory, the implementation of decision rules is straight-forward since it is an algorithmic function defined on the probability distribution and the utility function. Also, for the qualitative version of decision theory the decision rules can be easily implemented based on the likelihood and the preference function. Our problem is that we cannot assume an agent to have such orderings on possible states and thus have these functions available. However, starting from the higher level declarative goals plus the axioms of BDI theory we end up with a set of possible goal states rather than a sequence of actions. We should then select one goal state and propose a planning mechanism to determine the sequence of actions to reach the chosen goal state. Therefore, the implementation of the decision making ability of agents based on the BDI theory is much more complex since no ordering is provided. Even in the extended BDI framework KARO [11] which also contains capabilities, actions and results of actions the decisions on which goals and intentions are selected are deliberately left open to allow for a variety of agent types.

We argue that these choices should be made explicit by the so-called deliberation process rather than fixing them in some internal structure of the agent. In particular, the deliberation process is meant to construct and substitute the ordering information, which was lost through the dichotomous abstraction of decision theory, in a dynamic way and based on many (deliberation) issues. Because in the context of BDI agents we may assume that all relevant aspects of the environment of the agent are modelled as mental attitudes of the agent, we may assume that agents determine the course of actions based on these mental attitudes. The mental attitudes contain at least things such as beliefs and desires (or potential goals), capabilities such as actions and plans, and (defeasible) reasoning rules that can be used to reason about the mental attitudes. When other entities or mental attitudes play a role in agent decision, the corresponding deliberation aspects should be taken into consideration. For example, if we want to consider communication or sensing to play a role in agent decisions, then we should take into account the deliberation issues such as when to perform communication (receive or send) or sense actions and when to revise agent beliefs based on the acquired information. See also [13] for a good description of all possible decision points involving external influences of the agent. Given the above mentioned decision related entities, the deliberation process can be considered as consisting of reasoning about mental attitudes, selecting goals, planning goals, selecting and executing plans.

First of all an agent has to make choices about how to reason about its mental attitudes at each moment in time. For example, an agent can reason about its goals or the agent can reason about its goals only when they are not reachable using any possible plan. Some more moderate alternatives are also possible. E.g. the agent can create a plan for a goal and perform the plan. If this leads to a stage where the plan cannot be performed any further, then the agent can start reasoning about the plan and revise the plan if necessary. If the goal can still not be reached, then the agent can revise the goal. So, this leads to a strategy where one plan is tried completely and if it fails the goal is revised or even abandoned.

The deliberation process should also control the relation between plans and goals. For example, the deliberation process should control whether a goal still exists during the execution of the plan to reach that goal. If the corresponding goal of a plan is reached (or dropped), the deliberation process can avoid or allow continuing with the plan. If it is decided to halt the execution of a plan, then the deliberation process may decide to perform a kind of "garbage collection" and remove a left-over plan for the goal that no longer exists. If this would not be done the left-over plan would become active again as soon as the goal would be established at any later time. This type of deliberation choices can be exploited to distinguish and implement the so-called maintenance from achievement goals.

Another issue that should be considered by the deliberation process is related to multiple (parallel) goals and/or plans. First, it should be decided whether only one or more plans can be derived for the same goal at any time. If we allow only one current plan for each goal, the plans will all be for different goals. In this case it has to be determined whether the plans will be executed interleaved or consecutively. Interleaving might be beneficial, but can also lead to resource contention between plans in a way that no plan executes successfully anymore. E.g. a robot needs to go to two different rooms that lay in opposite directions. If it has a plan to arrive in each room and interleaves those two plans it will keep oscillating around its starting position indefinitely. Many of the existing work on concurrent planning can, however, be applied straight away in this setting to avoid most problems in this area.

3   A Programming Language for the Deliberation Process

In this section, we assume that autonomous agents have mental attitudes such as beliefs, goals, intentions and plans, as well as a reasoning capability to reason and revise their mental attitudes. From the programming point of view, mental attitudes are implemented as data structures called belief base, goal base, intention base, and plan base. These bases are initially implemented by expressions of specific languages and can change during the deliberation process. An example of languages to implement various mental attitudes is 3APL [8] The reasoning capability of agents is implemented by means of various rule bases. In particular, we assume a goal rule base, an intention rule base, and a plan rule base. These rule bases contain rules that can revise or reconsider goal, intentions, and plans, respectively. In addition, we assume a planning rule base consisting of rules that can be used to plan intentions, replan an existing plan, or backtrack from an existing plan. Finally, we assume proof engines that can check if a certain formula is derivable from the belief base, goal base, and intention base, respectively. So, we make an explicit distinction between proof engines that provide logical consequence relations for the different modalities and rule bases which regulate the updates of the different modalities.

The programming language to implement the deliberation process is considered as a meta-langauge that can express the selection and application mechanisms and the order in which decisions should be taken [5]. The selection mechanism involves decisions such as which goals or plans should be selected and executed, which rules should be selected and applied, or which mental bases should be updated by a formula. The order of decisions determine whether goals should be updated before getting executed or should the plans be executed before selecting a new goal. The meta-language is imperative and set-based. It is imperative because it is used to program the flow of control and it is set-based because it is used to select goals and rules from the set of goals and rules.

The proposed deliberation language consists of programming constructs to program directly the selection of goals, plans, and reasoning rules, to execute goals and plans, to apply rules, and to update a mental base. These programming constructs can be conditionalized on agent beliefs in order to contextualize the deliberation process. For example, a robot that has a set of goals including a transport goal (to transport boxes from position A to position B) and a goal to clean the space it operates may come in an emergency situation where it should only select and execute the transport goal.

3.1 Deliberation Terms

In order to implement the selection of goals, intentions, and plans, to apply reasoning rules to them, to execute a plan, or to update a mental base, we introduce terms that denote belief, goal, and intention formulas as well as terms that denote reasoning rules and plans. Here we use sorted terms because we need terms to denote different entities such as sets of goals, sets of intentions, sets of plans, individual goals, individual intentions, individual plans, sets of rules, individual rules, numbers, etc. These terms can be used as arguments of programming constructs that can be used to implement the deliberation cycle.

In the following we use Σ, Γ, Δ, and Π, to denote the belief base, the goal base, the intention base, and the plan base, respectively. The first three bases consists of formulae that characterize states, while the plan base consists of a set of plans (e.g. action expressions). Moreover, we use Λ_{sgrr} to denote the goal reasoning rule base consisting of rules that can be used to revise the goal base (e.g. if agent i wants x, but x is not achievable, then i wants y), Λ_{sirr} to denote the intention reasoning rule base, and Λ_{sprr} to denote the plan reasoning rule base. Generally, these (practical) reasoning rules consist of a head and a body (a goal and an intention formulae, or a plan expression), and a condition which is a belief formula. Finally, we use Ω to denote the plan rule base which can be use to plan goals, i.e. to relate plans to goals. The plan base consists of rules that have a goal formula as the head, a plan expression as the body, and a belief formula as the condition.

Definition 1. *Let S be a set of sorts including sorts for sets of beliefs (sb), sets of goal formulas (sg), sets of intention formulas (si), sets of plan expressions (sp), sets of planning rules (spr), sets of goal revision rules (sgrr), sets of intention reconsideration rules (sirr), sets of plan revision rules (sprr), individual belief formulas (ib), individual goal formulas (ig), individual intention formulas (ii), individual plan expression (ip), individual planning rule (ipr), individual goal revision rules (igrr), individual intention reconsideration rules (iirr), individual plan revision rules (iprr), numbers (N), and booleans (B).*

Let also Var_s be a set of countably infinite variables for sort s, and F be a set of sorted functions. The deliberation sorted terms T_s are then defined as follows:

- $\Sigma \in T_{sb}, \Gamma \in T_{sg}, \Delta \in T_{si}, \Pi \in T_{sp}, \Lambda_x \in T_x$ for $x \in \{spr, sgrr, sirr, sprr\}$.
- $Var_s \subseteq T_s$ for all sort $s \in S$
- if $s_1, \ldots, s_n \in S, t_i \in T_{s_i}$ and $f : s_1 \times \ldots \times s_n \to s \in F$, then $f(t_1, \ldots, t_n) \in T_s$

Some typical functions that are defined on goals are $max : sg \to sg$ that selects the subset of maximal preferred goals from the set of goals[1], $empty : sg \to B$ that determines whether the set of goals is empty, $head : ixrr \to ix$ and $body : ixrr \to ix$ for $x \in \{g, i, p\}$[2], $sel\_trans : sg \to ig$ that selects an individual goal (in this case the transport goal) from a set of goals, or $gain : ig \to N$ that determines the utility (a number) of an individual goal. Note that nested applications of functions can be used as well. For example, $gain(max(\Gamma))$ indicates the utility of the maximal preferred goal. Similar functions can be defined for practical reasoning rules and plans.

3.2   Deliberation Formulas

Given the terms denoting belief, goal, and plan formulas as well as reasoning rules, the set of formulas of the deliberation language can be defined. Moreover, we have assumed three proof engines that verify if a certain formula is derivable from the belief base, goal base, or intention base. In order to express that a certain formula is derivable from belief, goal, and intention bases, we introduce the *provable* predicate.

Definition 2. *Let $s \in S$ be a sort, $t_s, t_s' \in T_s$ be terms of sort s, $x \in \{b, g, i\}$, and provable be a predicate. We define the set of deliberation formulas DF as follows:*

- $t_s = t_s', t_s \geq t_s', provable(t_{sx}, t_{ix}) \in DF$
- *if $\phi, \psi \in DF$, then $\neg \phi, \phi \wedge \psi \in DF$*

The deliberation formula $provable(t_{sx}, t_{ix})$ for $x \in \{b, g, i\}$ expresses that the belief, goal, or intention formula denoted by the terms t_{ix} are derivable from the belief base, the goal base, and the intention base, denoted by t_{sx}, respectively. For example, let Σ be the term that denotes the belief base $\{\phi, \phi \to \psi\}$ and α be a term denoting the formula ψ, i.e. $\|[\alpha]\| = \psi$. Then, the deliberation formula $provable(\Sigma, \alpha)$ is a valid formula. Note that the predicate *provable* is only defined over deliberation terms, not over the formulas of DF itself! One important difference is that the deliberation terms do not contain negation.

3.3   Deliberation Statements

Having deliberation terms and formulas, the programming constructs or statements of the deliberation language to program the deliberation cycle of cognitive agents can now be defined.

---

[1] For this function an ordering among goals is assumed.
[2] A reasoning rule is assumed to be defined in terms of a head and a body.

Definition 3. *Let $s \in S, t_s \in T_s$ and $V_s \in Var_s$. The set of basic statements of the deliberation language is defined as follows:*

- $V_s := t_s$
- $selgoal(t_{sg}, f_c, V_{ig})$
  $selint(t_{si}, f_c, V_{ii})$
  $selplan(t_{sp}, f_c, V_{ip})$
  $selrule(t_{sxrr}, t_{sx}, V_{ixrr})$ *for* $x \in \{g, i, p\}$
- $update(t_{sx}, t_{iy})$ *for* $x, y \in \{b, g, i\}$
  $reviseplan(t_{ip}, t'_{ip})$
- $plan(t_{ii}, t_N)$
  $replan(t_{ip}, t_{spr}, f_c, t_N)$
  $btplan(t_{ip}, t_{spr}, t_N)$
  $explan(t_{ip})$

The set of deliberation statements is defined as follows:

- *Basic statements are deliberation statements*
- *If $\phi \in DF$ is a deliberation formula, and α and β are deliberation statements, then the following are deliberation statements:*
  $\alpha \; ; \; \beta$,
  IF ϕ THEN α ELSE β ,
  WHILE ϕ DO α

The first statement $V_s := t_s$ is designed to assign a sorted term t_s to a variable V_s of the same sort. The following statements are all selecting some item from a particular set of those items. The statement $selgoal(t_{sg}, f_c, V_{ig})$ selects an individual goal from the set of goals denoted by the term t_{sg}. The term denoting the selected individual goal is assigned to variable V_{ig}. The function f_c maps goals to boolean values indicating whether the goal formula satisfies the criterium c. The statement $selint(t_{si}, f_c, V_{ii})$ selects an individual intention from the set of intentions denoted by the term t_{si}. The term denoting the selected individual intention is assigned to variable V_{ii}. The function f_c indicates whether the intention satisfies the criterium c. The statement $selplan(t_{sp}, f_c, V_{ip})$ selects an individual plan from the set of plans denoted by the term t_{sp}. The term denoting the selected plan should satisfy criterion f_c and is assigned to the variable V_{ip}. The statement $selrule(t_{sxrr}, t_{sx}, V_{ixrr})$ selects a rule from the set of (goal, intention, or plan) reasoning rules denoted by the terms t_{sxrr} and assigns the term that denotes the rule to the variable V_{ii}. The selected rule should be applicable to a formula from the set denoted by the term t_{sx}.

The criterium c used in the selection functions can be used to define a preference ordering between the goals, intentions, plans and rules. So, in fact this is the place where a relation with qualitative decision theory can be made. The same argument can be made for the other selection functions. The main advantage over the classical decision theoretic approach is that the deliberation uses several independent preference orderings over different concepts. The combination of all these orderings leads to a decision on which action will be performed next.

However, unlike decision theory where all factors have to be combined into one function that determines the best action, we explicitly program this combination. Besides this advantage of having all factors explicitly available (and thus easily adjustable) the combination of these factors into a decision can be made situation dependent. therefore allowing for an adjustable preference ordering of the agent.

The statement $update(t_{sx}, t_{iy})$ updates a mental base (belief, goal, or intention base) denoted by the term t_{sx} with the formula denoted by the term t_{iy}. The statement $reviseplan(t_{ip}, t'_{ip})$ removes the plan that is denoted by the term t_{ip} from the plan base, and adds the plan that is denoted by the term t'_{ip} to it.

The final set of basic statements are all related to updating the plans of the agent in some way. The statement $plan(t_{ii}, t_n)$ generates a plan expression with maximum length t_n to achieve intention t_{ii}. The generated plan expression is assigned to the plan base of the agent. The statement $replan(t_{ip}, t_{spr}, f_c, t_N)$ uses the set of planning rules t_{spr} and generates a new plan expression to replace the plan expression t_{ip}. The new plan expression satisfies the criteria f_c and has maximum length t_N. The statement $btplan(t_{ip}, t_{spr}, t_N)$ does the same as replan except that $btplan$ uses an order among planning rules and generates the next plan expression according to that order. Finally, $explan(t_{ip})$ executes the individual plan expression denoted by the term t_{ip}. We assume that the execution of a plan has some external effects. The internal mental effects should be realized using the update statement explicitly.

In this paper, we do not consider the semantics of this deliberation language since we are only interested in the implementation of the deliberation process and how autonomous agent properties related to the agent's decision making ability can be implemented. The semantics for this language is an extension of the semantics of the meta-language already presented in [5].

3.4 Examples of Deliberation Processes

A program that implements a deliberation cycle may be a simple fixed loop consisting of the following steps:

1. Select a goal
2. Update the intention base with the selected goal
3. Select an intention
4. Generate plans to achieve the intention
5. Select a plan to execute
6. Execute the plan
7. Select a goal reasoning rule
8. Update the goal base with the body of the selected rule
9. Select an intention reasoning rule
10. Update the intention base with the body of the selected rule
11. Select a plan reasoning rule
12. Update the plan base with the body of the selected rule

This loop illustrates very clearly that the deliberation process consists of two parts. One part is the selection of goals, intentions, and plans finished with the execution of plans (steps 1 to 6). The second part deals with the reconsideration of goals, intentions, and plans (steps 7 to 12). The first part is closely related to planning. According to this example of the deliberation process reasoning rules are not used to update the goal base, the intention base, or the plan base before planning or executing them. This means that we only generate a plan, which is immediately executed. Only after this execution, reasoning rules are selected and mental bases are updated. Let Γ denote the goal base, Δ the intention base, Π the plan base, and Λ_g the goal rule base, Λ_i the intention rule base, and Λ_p the plan rule base. The above deliberation cycle can be implemented using the deliberation language as follows:

```
WHILE (¬empty(Γ)) DO
BEGIN
    selgoal(Γ, f_c, γ_ig);
    update(Δ, γ_ig);
    selint(Δ, f'_c, δ_ii);
    plan(δ_ii, t_N);
    selplan(Π, f''_c, π_ip);
    explan(π_ip);
    selrule(Λ_g, Γ, λ_g);
    update(Γ, body(λ_g));
    selrule(Λ_i, Δ, λ_i);
    update(Δ, body(λ_i));
    selrule(Λ_p, Π, λ_p);
    reviseplan(head(λ_p), body(λ_p));
END
```

Using the deliberation language, various types of domain dependent deliberation processes can be specified as well. For example, let Π be the term that denotes the plan base, Ω be the set of planning rules relating goals with plans (different rules than plan reasoning rules), $trans$ be the criterium to select a plan that is suitable to perform the transport task, $cost(\pi)$ be the term that determines how expensive is the plan that is assigned to plan π, and $gain(\pi))$ be the term that determines the utility of plan π. A part of a deliberation cycle can be specified by the following expressions of the deliberation language.

```
selplan(Π, trans, π)
WHILE (cost(π) > gain(π)) DO
BEGIN
    btplan(π, Ω, max_length);
END
IF gain(π) > cost(π) THEN explan(π)
```

This part of a deliberation cycle initially selects a plan π to perform the transport task. While the cost of this plan is higher than the utility of the transport task, it attempts to find a new plan by backtracking in the space of possible plans. When it finds a cost effective plan it will execute the plan.

4   Deliberation Properties

The properties of agent decision behavior are determined by the way the deliberation process is implemented. In this section, we consider these properties from the programming point of view and examine how to implement a deliberation process that satisfies certain properties, i.e. how to implement a certain specification of agent decision behavior. In order to answer this question, we consider static and dynamic constraints on the agent decision behavior formulated in the BDI formalism [4,15]. In the BDI formalism, the operators B, G, and I are used to express agents' beliefs, goals, and intentions, respectively. Moreover, temporal operators from CTL* (Computational Tree Logic) are used to express the dynamics of agents' mental attitudes in terms of operators A to express 'for all possible futures', E 'for some possible futures', U 'until', X 'next', F 'sometimes in future', and G to express 'always in future' [15].

Due to space limitations we will not consider a complete set of constraints but rather exemplify the implementation issues using two of the most commonly used constraints. Those are the static property called realism and the dynamic property called open-minded commitment strategy. Using the BDI formalism, these two properties are expressed by the following axioms:

$$G_i(\varphi) \rightarrow B_i(\varphi)$$
$$I_i(AF\varphi) \rightarrow A(I_i(AF\varphi) \bigcup B_i(\varphi) \vee \neg G_i(EF\varphi))$$

The first (realism) axiom characterizes the compatibility between agents beliefs and goals and states that the goal of agents should be compatible with their beliefs. For example, if the agent i wants p to be true in all possible futures, written as $G_i(AGp)$, then agent i should believe it is possible that p is true always and for all possible futures. If agent i does not have this belief, then he wants something which he does not believe is achievable, which is clearly an unrealistic behavior. In order to implement a deliberation cycle such that agents' decision making behavior satisfies this property, we should guarantee that the selection of a goal and committing to it as an intention can only take place when the goal is derivable from agent's belief.

We assume that the initial agent's belief and goal bases are compatible and illustrate how this compatibility relation can be maintained through the deliberation process. In the previous section, the statement $update(t_{sg}, t_{ig})$ is proposed which, when executed, updates the set of goals denoted by t_{sg} (goal base) with the individual goal denoted by t_{ig}. For example, let the term t_{sg} denote the goal base $\{AFp, \neg EGq\}$. Then, the execution of $update(t_{sg}, EGq)$ modifies the goal base resulting in the new goal base $\{AFp, EGq\}$. In order to implement an agent in such a way that its decision making behavior satisfies the realism axiom, we should guarantee that every occurrence of the statement $update(t_{sg}, t_{ig})$ is conditionalized with corresponding beliefs. Let Σ be the belief base of the agent and Γ be the goal base of the agent. Then, every occurrence of the statement $update(\Gamma, \alpha)$ should be replaced with the following conditional statement:

$$\text{IF } provable(\Sigma, \alpha) \text{ THEN } update(\Gamma, \alpha)$$

The second (open-minded commitment strategy) axiom characterizes the decision making behavior of agents regarding their commitments to their intentions and the conditions under which the intentions can be dropped. In particular, it states that agents remain committed to their future directed intentions (i.e. intentions of the form $AF\phi$) until they believe their intentions are reached or they do not desire to reach them in the future anymore. Following the programming constructs a future directed intention can be dropped by the instruction $update(t_{si}, \neg t_{ii})$ where t_{ii} is a term denoting a formula of the form $AF\phi$. Note that updating the intention base by the negation of a formula is equivalent with dropping the intention expressed by the formula. In order to program the open-minded commitment strategy, we should guarantee that every occurrence of $update(t_{si}, \neg t_{ii})$, where t_{ii} is a term denoting a future directed formula, is conditionalized with corresponding beliefs and goals. Let Σ, Γ, and Δ be the belief base, the goal base, and the intention base of the agent. Then, every occurrence of the statement $update(\Delta, \neg AF\alpha)$ should be replaced with the following conditional statement:

$$\text{IF } provable(\Sigma, \alpha) \lor \neg provable(\Gamma, EF\alpha) \text{ THEN } update(\Delta, \neg AF\alpha)$$

The above examples show us a kind of general strategy on implementing the BDI constraints in the deliberation cycle. Each update of a goal or intention is made conditional upon some beliefs or intentions. One could, of course, integrate these conditions in the semantics of the update function itself. However, this would mean that the agent has a fixed commitment strategy ingrained in it. By incorporating it as conditions in the deliberation program we can still (dynamically) change the agent.

5   Conclusion

In this paper we have shown how some characteristic constraints that determine the autonomy of the agent can be programmed on the deliberation level of the agent. In our opinion this shows that autonomy is closely related to the deliberation level and its properties of the agent. By programming this level explicitly it becomes clear that many choices concerning the autonomy are still open (not restricted by the BDI constraints). Most of the choices have to do with the order in which different mental attitudes are updated and with which frequency they are updated.

Although we did not mention this point above it has also shown us the difficulty of implementing some of the constraints. The second constraint we implemented suggests that an intention can be dropped whenever the agent does not have the goal that φ will be true in some possible future anymore. Although we just stated that this can be tested by using the *provable* predicate, it will be very difficult in practice to prove $EF\varphi$ for any formula φ. This is due to the fact that the future is in principle infinite and it is impossible to check whether in one possible future at some point (possibly infinitely far in the future) φ can be true. Once the *provable* predicate returns false, the condition for dropping the intention becomes true. Thus in practice there will hardly ever be a restriction on dropping an intention.

Besides this computational criticism on the BDI constraints one might also wonder whether they specify exactly the right intuition. Especially the commitment strategies quantify over possible futures. It is very hard to check all possible futures, especially in systems that contain multiple agents and where the environment is uncertain. In future research we hope to explore some more realistic commitment specifications which tie the commitment to an intention to the possible futures that the agent has an influence on itself (through its own plans). Also, we may consider commitment strategies in terms of bounded temporal operators such as bounded eventually. These restrictions of the general commitment strategy as given in BDI theory also render these strategies computationally feasible. I.e. they restrict the possible future traces to a finite number of traces of finite length.

References

1. C. Boutilier. Towards a logic for qualitative decision theory. In *Proceedings of the Fourth International Conference on Knowledge Representation and Reasoning (KR94)*, pages 75–86. Morgan Kaufmann, 1994.
2. M. Bratman. *Intention, plans, and practical reason.* Harvard University Press, Cambridge Mass, 1987.
3. J. Broersen, M. Dastani, Z. Huang, and L. van der Torre. Trust and commitment in dynamic logic. In *Proceedings of The First Eurasian Conference on Advances in Information and Communication Technology (EurAsia ICT 2002)*, volume 2510 of LNCS, pages 677–684. Springer, 2002.
4. P. Cohen and H. Levesque. Intention is choice with commitment. *Artificial Intelligence Journal*, 42(3):213–261, 1990.
5. M. Dastani, F. de Boer, F. Dignum, and J.-J. Meyer. Programming agent deliberation: An approach illustrated using the 3apl language. In *Proceedings of The Second Conference on Autonomous Agents and Multi-agent Systems (AAMAS03)*, Melbourne, Australia, 2003.
6. M. Dastani, Z. Huang, J. Hulstijn, and L. van der Torre. BDI and QDT. In *Proceedings of the Workshop on Game Theoretic and Decision Theoretic Agents (GTDT2001)*, Stanford, 2001.
7. J. Doyle and R. Thomason. Background to qualitative decision theory. *AI magazine*, 20:2:55–68, 1999.

8. K. V. Hindriks, F. S. D. Boer, W. V. der Hoek, and J.-J. C. Meyer. Agent pro-
    gramming in 3apl. *Autonomous Agents and Multi-Agent Systems*, 2(4):357–401,
    1999.
9. R. C. Jeffrey. *The Logic of Decision*. McGraw-Hill, New York, 1965.
10. D. Kinny. *Fundamentals of Agent Computation: Theory and Semantics*. Australia,
    2001.
11. J.-J. Meyer, W. van der Hoek, and B. van Linder. A logical approach to the dy-
    namics of commitments. *Artificial Intelligence*, 113(1–2):1–41, 1999.
12. J. Pearl. From conditional ought to qualitative decision theory. In *Proceedings
    of the Ninth Conference on Uncertainty in Artificial Intelligence (UAI93)*, pages
    12–20. John Wiley and Sons, 1993.
13. A. Rao and M. Georgeff. Deliberation and its role in the formation of intentions.
    In *Proceedings of the Seventh Conference on Uncertainty in Artificial Intelligence
    (UAI-91)*, pages 300–307, San Mateo, CA, 1991. Morgan Kaufmann Publishers.
14. A. Rao and M. Georgeff. Modeling rational agents within a bdi architecture. In
    *Proceedings of Second International Conference on Knowledge Representation and
    Reasoning (KR91)*, pages 473–484. Morgan Kaufmann, 1991.
15. A. Rao and M. Georgeff. Decision procedures for BDI logics. *Journal of Logic and
    Computation*, 8:293–342, 1998.
16. L. Savage. *The foundations of statistics*. Wiley, New York, 1954.

Requirements for Achieving Software Agents Autonomy and Defining Their Responsibility

Abdelkader Gouaich

Laboratoire Informatique, Robotique et Micro Electronic- UMR 5506 - 161, rue Ada - 34090 Montpellier cedex 5 - France
gouaich@lirmm.fr

Abstract. This paper addresses the problem of implementing agent-based software systems with respect to agent framework fundamental concepts such as autonomy and interaction without specifying any particular agent internal architecture. The autonomy and interaction axioms imply that a deployment environment has to be defined in order to achieve interaction among agents. This deployment environment may also encode environmental rules and norms of the agent society. The responsibility of an agent is then defined as being in adequacy with its environmental rules. Finally, a formal deployment environment, named MIC*, is presented with a simple application showing how interaction protocols are guaranteed by the deployment environment, which protects agents from non-conform actions and preserve their autonomy.

1 Introduction

Agent-based software engineering seems to be an interesting approach for the design and development of open and distributed software systems [Fer95,JW97]. According to Demazeau [Dem95] a multi-agent system can be described with four main concepts: Agent, Interaction, Organisation and Environment. Agents are autonomous goal-directed entities that populate the multi-agent system. They achieve their design goals by interacting with other agents and using resources available in their environment; interaction is defined by Ferber [Fer95] as the set of mechanisms used by agents to share knowledge or to coordinate activities. Organisation in multi-agent systems can be defined as the set of mechanisms that reduces the entropy of the system and makes it ordered, behaving as a coherent whole. The environment represents a container where agents live. According to Russell and Norvig [RN95] the environment is one of the fundamental concepts in programming and designing agent-based systems. However, there is still some confusion on the definition of this concept. In fact, situated agents works such as agent-based simulation, artificial life or robotics consider the environment as the space where agents are situated; perceive their vicinity; and access to the available resources. On the other hand, agent-based software engineering considers the agent environment as the software components surrounding agents and offering them some computing facilities. To avoid confusion, the later definition will be referenced as agent deployment environment. The presented work

M. Nickles, M. Rovatsos, and G. Weiss (Eds.): AUTONOMY 2003, LNAI 2969, pp. 128–139, 2004.

is concerned with the study of a generic formal deployment environment and its properties to implement autonomous agent-based software system.

2 Requirements to Implement Autonomous Agents

This section uses Wooldridge and Jennings agent's definition stating that [WJ95]: *"Perhaps the most general way in which the term agent is used is to denote a hardware or (more usually) software-based computer system that enjoys the following properties:*

- *Autonomy: agents operate without the direct intervention of humans or others, and have some kind of control over their actions and internal state [Cas95];*
- *Social ability: agents interact with other agents (and possibly humans) via some kind of agent-communication language [GK94];*
- *Reactivity: agents perceive their environment, (which may be the physical world, a user via a graphical user interface, a collection of other agents, the INTERNET, or perhaps all of these combined), and respond in a timely fashion to changes that occur in it;*
- *Pro-activeness: agents do not simply act in response to their environment, they are able to exhibit goal-directed behaviour by taking the initiative."*

This definition specifies some features that a physical or software entity must fulfill to be considered as an agent. Still, it does not specify how to build software agents. Consequently, developers that are in charge of implementing agent-based systems might have their own interpretation of these agent's features. For instance, autonomy is usually interpreted as concurrency. So, a simple concurrent object implements autonomous agents. These mis-interpretations of agent fundamental features have led several software engineering communities to consider agent-based systems just as a renaming of concurrent object systems. This section studies carefully agent autonomy and social ability features in order to derive the requirements on the software system implementing the agent-based system. This axiomatic approach proves that a non-agent entity is needed in order to achieve interaction among autonomous agents.

2.1 What is an Autonomous Agent?

Autonomy is given several interpretations. Two main interpretations were selected for the purpose of this paper: autonomy as self-governance and as independence.

Autonomy as Self-governance: Steels in [Ste95] tries to understand the difference between a standard software program and an autonomous agent. His paper refers to a personal communication with Tim Smithers (1992) on autonomy:

"The central idea in the concept of autonomy is identified in the etymology of the term: autos (self) and nomos (rule or law). It was first applied to the Greek

city-state whose citizens made their own laws, as opposed to living according to those of an external governing power. It is useful to contrast autonomy with the concept of automatic systems. The meaning of automatic comes from the etymology of the term cybernetic, which derives from the Greek for self-steering. In other words, automatic systems are self-regulating, but they do not make the laws that their regulatory activities seek to satisfy. These are given to them, or built into them. They steer themselves along a given path, correcting and compensating for the effects of external perturbation and disturbances as they go. Autonomous systems, on the other hand are systems that develop, for themselves, the laws and strategies according to which they regulate their behaviour: they are self-governing as well as self-regulating. They determine the paths they follow as well as steer along them"

This description defines autonomy of a system (or entity) as its ability to produce its own laws and to follow them. In contrast, an automatic entity is given what to do and follows these instructions without making new laws or changing them. This notion of autonomy is shared by Castelfranchi [Cas95] who defines an autonomous agent as a software program able to excise a choice that is relevant in the context of goals-directed behaviour. Luck and D'iverno in several works [Ld95,LD01,dL96] share also the same notion of autonomy and specify formally using Z notations [Spi87] what is an autonomous agent and what distinguish it from an object or a simple agent. From Luck and D'iverno perspective, an object is a simple software entity with some features and actions. An agent is an object increased with a set of goals to be achieved. Autonomous agent is then defined as an agent with a set of motivations steering the agent in selecting what goals have to be achieved. Guessoum and Briot [GB99] addresses also this issue and enrich a pure object communicating architecture, Actalk, with some mechanisms such as controlling message reception and selecting what behaviour to adopt, in order to build an agent platform named DIMA.

Autonomy as Independence: Social dependence network (SDN) has been introduced by Sichman et al. in [SCCD94] to allow social agents to reason about and understand their artificial society in order to achieve their goals. Within this framework, agents have external descriptions representing models about their neighbours. An external description of an agent is composed by its goals, its actions, its resources and its plans. Goals represent the state of affair that the agent want to reach; actions are operations that an agent is able to perform; resources represents the resources that an agent has control on; and finally plans are sequences of actions and resources. Notice that the authors adopt hypothesis of external description compatibility implying that all agents have the same models about others. This is unachievable in an open context where the set of agents is not static and the interaction is asynchronous. SDN framework distinguishes three forms of autonomy. An agent is a-autonomous for a given goal according to a set of plans, if there is a plan in this set that achieves the goal, and every action in each plan belongs to its capabilities. An agent is considered as r-autonomous for a given goal according to a set of plans, if there is a plan in this set that achieves the goal, and every resource in each plan belongs to its resources. Finally, an agent is s-autonomous when it is both a-autonomous and r-autonomous. According to this definition, an agent is autonomous for a particular goal if it does not depend for resources or actions on another agent. These definitions are then used to define dependency relation among agents. Following the SDN definitions, autonomy is easily proved as equivalent to the independence. Hence, agents are autonomous if and only if they are independent from all other agents.

Discussion: Sichman et al. define autonomy of agents as being independent on actions and resources from other agents. On the other hand, [Cas95,Ste95,Ld95] define agents autonomy as its ability to reason about, generate and execute its goals by its own. Our interpretation of agent autonomy is closer to autonomy as self-governance than autonomy as independence. In fact, as mentioned by [Ld95], a pocket calculator that has the resources and actions to perform to calculate some arithmetical operation is seen as autonomous agent according to SDN definitions. Our interpretation on autonomy relates more on the decisional process than on dependencies on resources or knowledge. In fact, a software agent is an autonomous entity if it behaves differently depending on its internal context. In other words, if agents have to be modelled as mathematical functions, the observer of the agent has to admit that he/she is not able to know all the parameters of this function. Consequently, this observer will never have the complete knowledge to predict the autonomous agent reaction and behaviour.

Once the agent autonomy definition has been clarified, the problem now is to know how to correctly implement autonomous agents. In other words are usual techniques such as concurrent object sufficient to implement autonomous agents? This question has been answered partially by [Ld95] with the concept of motivation. Hence, motivated-agents generate their goals by considering their current motivations. Nevertheless, introducing motivations in an agent shifts the problem from how to manage agent's goals, to how to manage agent's motivations. In fact, any external entity that has a full control of the software structure representing the agent's motivations controls the behaviour of this agent. Guessoum and Briot's approach seems also interesting to implement autonomous agents. However, as it is the case for motivated-agent, it specifies precisely what should be the agent internal architecture in order to guarantee its autonomy. Our goal is to study the minimal necessary requirements of agent-based software architecture without considering any particular internal architecture of agents.

Internal Integrity to Guarantee Agent's Autonomy: Integrity of the agent software structure is a sine qua none condition to implement autonomous agents without considering any particular agent internal architecture. The term agent software structure refers to the set of data and instructions that encode the agent and define its computational behaviours. In fact, if this structure is accessible and modifiable by another agent, the decisional process and behaviours may be altered. For instance, if the set of motivations of motivated-agent is accessible by another entity, the agent loses its autonomy. Similarly, if the decisional process is accessible by an external entity, DIMA autonomous agents lose their autonomy. Consequently, an autonomous agent should not allow any external agent to change its software structure either by setting a feature to certain value or by calling a side-effect method. On the other hand, agents are interacting entities. So, agents change the perceptions of other agents. Since, the perceptions of agents are part of their software structure, this contradicts the autonomy statement. To avoid this contradiction, the first solution is to consider that the software structure representing the agent's perception (or its inbox) does not belong to its software structure. This is not achievable in practice. In fact, software engineering requirements on modularity include the agent's perception in the agent's software structure. The other approach is to define a non-agent entity that is in charge of achieving the interaction among agents. This entity is what was identified as agent deployment environment. The deployment environment is not an autonomous agent and is considered as an automatic software system. Hence, its software structure and rules are defined once and should be followed by agents. This does not mean that dynamic deployment environment cannot be constructed. However, a dynamic deployment environment is not as single entity, but a sequence of different entities. This result is particularly interesting to establish agents' responsibility in an open and untrusted software environment. In fact, without this feature the responsibility of agents when they violate the multi-agent society norms cannot be established. This point is discussed in more details in the next section.

3   Agent Responsibility in Open Software Systems

Biological and physical environments as examples of multi-agents systems that are separated in two dimensions: the agents and the physical environment. Biological agents are subject to environmental laws and principles. Agents do not modify the environmental laws; they have to deal with them in order to achieve their goals. For instance, if a human agent's goal is to fly, he/she will never modify the physics law on gravity, but use other physics laws on aerodynamics in order to achieve its goals. This analogy can be used in order to build secured agent-based systems. In fact, the design requirements of a software system are the environmental laws. Any deployed known or unknown agent has to deal with these environmental laws. The term unknown agent refers to software agent whose software structure is completely unknown. This may be the case of agents that are designed and implemented by external organisations. In contrast, a known agent is a software agent whose software structure is accessible (open source or developed locally). In large open software systems, agents have to be assumed as unknown. Thus, the deployment environment has to identify agents that challenge its internal laws. This defines agent's responsibility as being coherent to the deployment environment rules and laws. To guarantee these points, agents are not allowed to change directly their environment. Their actions have to be discrete and explicitly represented as attempts of actions. The deployment environment considers how to react to these attempts. Attempts that violate the environmental laws are simply ignored. The emitting agent is then considered as responsible for violating the environmental laws. Furthermore, agents cannot deny this responsibility since the deployment environment is an automatic system that does not modify its internal laws. In fact, if this was not the case, agents may claim that they cannot conform rules since they are arbitrarily changed. For instance, interaction protocols are examples of agent society rules that have to be followed by agents in an open agent-based system. It would be interesting to generate a deployment environment that encodes and guarantees these interaction norms without specifying any particular agent internal architecture. This is considered as a case study and studied in section 5.

4 A Formal Agent Deployment Environment

This section presents an example of a formal deployment environment, named MIC* for {Movement, Interaction, Computation}* [GGM03], fulfilling the requirements presented before on agent autonomy and responsibility. MIC* is an abstract structure where autonomous, mobile and interacting entities are deployed. Within this framework, all interactions are conducted by explicitly exchanging interaction objects through interaction spaces. Hence, agents do not alter directly the deployment environment or the perceptions of other agents, but send their attempts as interaction objects. Interaction objects are structured: in fact, a formal addition law can compose them commutatively + to represent simultaneous interactions. Furthermore, abstract empty interaction object 0 can be defined to represent no interaction. The less intuitive part of the structure of the interaction objects concerns negative interaction objects. Negative interaction objects are constructed formally and may have no interpretation in the real world. However, they are useful for the internal model definitions and implementation of the deployment environment. For instance, the deployment environment can cancel any action, x, of the agent simply by performing an algebraic operation, $x + (-x) = 0$, that is expressed within the model notations. Finally, interaction objects defines a structure of a commutative group $(\mathcal{O}, +)$, where $\mathcal{O}$ represents the set of interaction objects and + the composition law. Interaction spaces, represented by $\mathcal{S}$, are defined as abstract locations where interaction between agents holds. They are active entities that control their local and specific interaction rules. For instance, interaction object that are sent inside an interaction space may be altered if they violate the interaction norm. Agents, represented by $\mathcal{A}$, are autonomous entities that perceive interaction objects and react to them by sending other interaction objects. As said before, agents' actions are always considered as attempts to influence the deployment environment structure. These attempts are committed only when they are coherent with the deployment environmental rules of evolution.

Having these elementary definitions, each MIC* term is represented by the following matrices:

- Outboxes Matrix: The rows of this matrix represent agents $A_i \in \mathcal{A}$ and the columns represent the interaction spaces $S_j \in \mathcal{S}$. Each element of the matrix $o_{(i,j)} \in \mathcal{O}$ is the representation of the agent A_i in the interaction space S_j.
- Inboxes Matrix: The rows of this matrix represent agents $A_i \in \mathcal{A}$ and the columns represent the interaction spaces $S_j \in \mathcal{S}$. Each element of the matrix $o_{(i,j)} \in \mathcal{O}$ defines how the agent A_i perceives the universe in the interaction space S_j.
- Memories Vector: Agents $A_i \in \mathcal{A}$ represent the rows of the vector. Each element m_i is an abstraction of the internal memory of the agent A_i. Except the existence of such element that is proved using the Turing machine model, no further assumptions are made in MIC* about the internal architecture of the agent.

4.1   Structure Dynamic

The previous part has presented the static objects to fully describe environmental situations or states. In this section, three main evolutions of this static description are characterised:

- Movement μ: A movement is a transformation μ, of the environment where both inboxes and memories matrices are unchanged, and where outboxes matrix interaction objects are changed but globally invariant. This means that the interaction objects of an agent can change positions in the outboxes matrix and no interaction object is created or lost.
- Interaction ϕ: The interaction is characterised by a transformation ϕ that leaves both outboxes and memories matrices unchanged and transforms a row of the inboxes matrix. Thus, interaction is defined as modifying the perceptions of the entities in a particular interaction space.
- Computation γ: An observable computation of an entity transforms its representations in the outboxes matrix and the memories vector. For practical reasons, the inboxes of the calculating entity are reset to 0 after the computation.[1]

These three elementary evolutions are orthogonal. For a particular agent-based system, the structure of MIC* is fully defined by having the interaction objects group; the sets of processes and interaction spaces; the sets of transformations μ, ϕ, γ and a combination order of these transformations. $\{\mu|\phi|\gamma\}*$ (MIC*) is the most general combination order, where | expresses an exclusive choice between elements. The main idea of this approach is that the dynamics of the deployment environment can be modelled using the (μ, ϕ, γ)-base. Consequently, as presented in figure 1, any state of the deployment environment can be expressed as the composition of elementry evolutions that are movement, interaction or calculus.

---

[1] The fact that the inboxes of an agent are reset also defines a local logical time for each process.

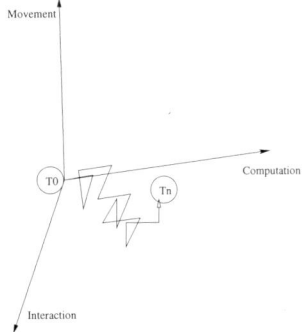

Fig. 1. Evolution of a MAS deployment environment starting from an initial term T_0 until T_n by following elementary transformations

4.2   Comparison Between MIC* and FIPA Agent Platform

The FIPA [FIP96](Federation of Intelligent Physical Agents) Agent Platform is an abstract architecture where FIPA-compliant agents are deployed. The FIPA agent platform is composed by the following components:

Agent Management System (AMS) is an agent that manages access to the agent platform. The AMS maintains a directory of logical agent identifiers and their associated transport addresses for an agent platform. The AMS is also responsible for managing the lifecycle of the agents on the platform and actions such as authentication, registration, de-registration, search, and mobility requests.

Agent Platform Communication Channel (ACC) provides communication services between agents. Agents can communicate with other agents on any FIPA platform through the Agent Communication Channel.

Directory Facilitator (DF) is an agent that manages a service directory where agents can register their description and search for other agents.

The FIPA agent platform is an example of an explicit deployment environment where the software agents are deployed [OPFB02]. The FIPA deployment environment is conceived as a middleware offering low-level computing services to agents. Both FIPA and MIC* approaches separate the agent-based system in two orthogonal dimensions: agents and the deployment environment. FIPA deployment environment meets the necessary requirements on autonomy since the interactions are conducted through the ACC (deployment environment). FIPA deployment environment is an automatic software system that follows identifiable specifications and norms (the specifications version for instance). This point was considered as helping in establishing agents responsibility. Only one interaction space is identified in a FIPA agent platform. In fact, all agents may communicate with other agents if they have their agent address. Agents code mobility is defined between different platforms, which raise some critical security problems in FIPA. In fact, FIPA deployment environment lack of action/reaction principle. Hence, agents modify the deployment environment directly by calling side effects methods. Consequently, the deployment environment may not reject illegal non-explicit actions or at least to explain what leads to an incoherent situation. This makes executing an unknown agent on a FIPA deployment environment a hazardous operation. The other fundamental difference between MIC* and FIPA deployment environment is the fact that MIC* generates a particular deployment environment for a particular agent-based system. Consequently, ad hoc application-dependent rules and norms on interaction are encoded directly in the deployment environment. This prevents agents from being disturbed by illegal interactions, which raises their autonomy. In contrast, FIPA approach delegates the control of interaction norms and protocols to agents. Consequently, responsibility of autonomous agents is defined at the FIPA specifications level and cannot be established at the application level. For instance, an agent can be identified as responsible when sending an incorrect FIPA ACL message, but the environment cannot establish its responsibility when violating an ad hoc application-dependent interaction protocol when the ACL FIPA message is grammatically correct.

5   Experiments

The experiment scenario simulates a virtual city where autonomous agents: user agent, bogus agent and service agent move and may interact. The mobility were used to show another feature of MIC* in modelling ubiquitous software systems. This is not addressed in this paper. The user agent goal is to buy a ticket from a ticket selling service. The goal of the bogus agent is to challenge the established norm on interaction. Hence, this agent follows the interaction protocol until a step n, where n is a random number. After this, the following communicative acts and actions are randomly sent.

Ticket Buying Interaction Protocol: In order to validate the presented approach, a simple[2] interaction protocol has been defined in order to buy a ticket from the ticket selling service. The notations that are used to describe this interaction protocol are the following: $(a); (b)$ expresses a sequence between two interactions a and b. This interaction is valid if and only if a happens first followed by b. $(a) \vee (b)$ expresses an exclusive choice between a and b. This interaction is valid if and only if a or b hold. $(a) + (b)$ expresses parallel interactions that hold in an unordered manner. Hence, this interaction is valid if and only if a and b happen no matter which first. Having these notations the 'ticket-buying' interaction protocol is described in figure 2. The user agent (C) initiates this protocol by sending a request to the service (S). After this, the service agent may agree to respond to the request or not. When agreeing, the service agent asks the user agent some information about the starting point; the destination of his travel and the traveller's age. After gathering these data, the service agent

---

[2] 17 messages are exchanged in order to buy a ticket

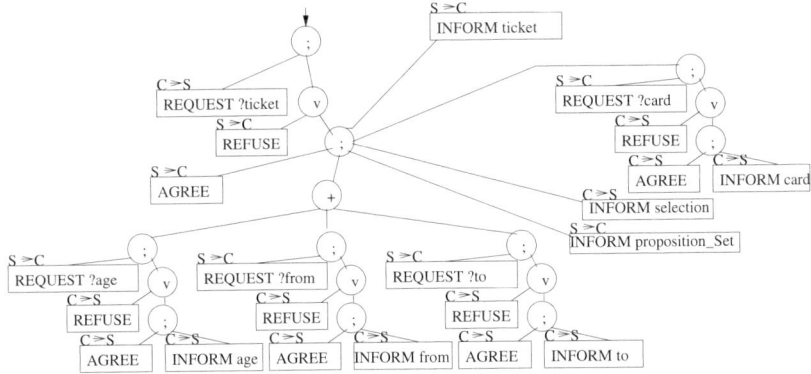

Fig. 2. Ticket buying interaction protocol between the user agent (C) and the service (S)

delivers some propositions to the user agent. When the user agent selects an offer, he informs the service agent about it. After the payment procedure, the offer is acknowledged and the dialogue is closed. The interaction protocol is specified from the viewpoint of an external observer. Having this specification a MIC*-deployment environment is generated. This process is discussed more on details in [GOU03]. All autonomous agents are executed as concurrent processes that interact with the deployment environment with explicit interaction objects sent via low-level TCP/IP sockets.

5.1   Results

By following the interaction protocol, the client interacts with the service agent and succeeds in buying a ticket. In contrast, attempts of the bogus agent have never reached the perceptions of the service agent. In fact, the deployment environment checks the validity of the dialogue among agents by recognising valid sequences of messages [GOU03]. When a sequence of messages is invalid according to the specification of the interaction protocol, the interaction attempt is simply ignored. Formally, this is expressed within MIC* as an evolution with the identity application. So, the deployment environment is not modified. Consequently, the service agent is protected from erroneous interactions. This can be considered as raising the autonomy of the agent. In fact, handling erroneous interactions is an agent's behaviour that may affect other application-level behaviours by consuming agent's resources. Furthermore, when error cases are not treated correctly the whole agent activity may be altered. On the other side, without any assumption on the bogus agent's software structure, the deployment environment was able to establish its responsibility for being not compliant with the interaction norms. Finally, since the internal software structure of agents was not accessible, their behaviours and decisional process have been conserved, which was considered in this paper as a sine qua none condition in order to achieve agents' autonomy.

6 Conclusion

This paper has addressed the problem of implementing agent-based software systems with respect to agent framework fundamental concepts such as autonomy and social abilities. This works have been conducted without specifying any particular internal agent architecture. Guaranteeing the internal integrity of the agent software structure has been found as a necessary condition to correctly implement autonomous agents. Hence, autonomous agents are completely closed boxes that perceive and react to interaction objects coming from their external world. Consequently, to conduct interaction among these autonomous entities, a non-agent entity is necessary. This entity was defined as the deployment environment and had been used in order to encode some agent society laws and norms. When encoding these norms, the deployment environment establishes a referential determining the responsibility of each autonomous agent in an open context. Finally, an example of deployment environment fulfilling the requirements on agent autonomy and responsibility has been presented. Finally, a simple application has shown, how a agent-based system has been implemented to simulate economical agents. The deployment environment has encoded the interaction norms and guaranteed that illegal interactions do not disturb other agents. The autonomy and safety of agents has been conserved.

The next step of our works is to link social models describing an agent-based system to the MIC* model. Hence, starting from a semi-formal description of the multi-agent system, using AUML or another formalism, a specific deployment environment can be automatically generated and defined independently from the autonomous agents that will populate it.

References

[Cas95]   C. Castelfranchi. Guarantees for autonomy in cognitive agent architecture. *Intelligent Agents: Theories, Architectures, and Languages*, 890:56–70, 1995.

[Dem95]   Yves Demazeau. From interaction to collective behaviour in agent-based systems. In *1st European Conference on Cogntive Science*, St Malot, France, 1995.

[dL96]   M. d'Inverno and M. Luck. A formal view of social dependence networks. In Zhang and Lukose, editors, *Distributed Artificial Intelligence Architecture and Modelling: Proceedings of the First Australian Workshop on Distributed Artificial Intelligence*, pages 115–129. Springer-Verlag: Heidelberg, Germany, 1996.

[Fer95]   Jaques Ferber. *Les Systemes Multi-Agents*. InterEditions, 1995.

[FIP96]   FIPA. Foundation of intelligent and physical agents, 1996. http://www.fipa.org.

[GB99]   Zahia Guessoum and Jean-Pierre Briot. From active objects to autonomous agents. *IEEE Concurrency*, 7(3):68–76, 1999.

[GGM03]   Abdelkader Gouaich, Yves Guiraud, and Fabien Michel. Mic▪: An agent formal environment. To appear in the 7th World Multiconference on Systemics, Cybernetics and Informatics (SCI 2003), 7 2003. Orlando, USA.

[GK94]   M.R. Genesereth and S.P. Ketchpel. Software agents. *Communications of the ACM*, 37(7):48–53, 1994.

[GOU03]   Abdelkader Gouaich. Implementing interaction norms is open, distributed and disconnected multi-agent software systems. Submitted to AOSE 2003. Url: http://www.lirmm.fr/gouaich/research.html, 2003.

[JW97]   Nicholas R. Jennings and Michael Wooldridge. Agent-based software engineering. In *Software Engineering*, volume 144, pages 26–37. IEEE, 1997.

[Ld95]   Michael Luck and Mark d'Inverno. A formal framework for agency and autonomy. In Victor Lesser and Les Gasser, editors, *Proceedings of the First International Conference on Multi-Agent Systems (ICMAS-95)*, pages 254–260, San Francisco, CA, USA, 1995. AAAI Press.

[LD01]   M. Luck and M. D'Inverno. Autonomy: A nice idea in theory. *Intelligent Agents VII: Proceedings of the Seventh International Workshop on Agent Theories, Architectures and Languages*, 2001.

[OPFB02]   James J. Odell, H. Van Dyke Parunak, Mitch Fleischer, and Sven Brueckner. Modeling agents and their environment. In *AOSE 2002, AAMAS 2002*, Bologna, 2002.

[RN95]   Stuart Russell and Peter Norvig. *Artificial Intelligence: A Modern Approach*. Prentice Hall, 1995.

[SCCD94]   Jaime Simão Sichman, Rosaria Conte, Cristiano Castelfranchi, and Yves Demazeau. A social reasoning mechanism based on dependence networks. In A. G. Cohn, editor, *Proceedings of the Eleventh European Conference on Artificial Intelligence*, pages 188–192, Chichester, 8–12 1994. John Wiley & Sons.

[Spi87]   J.M. Spivey. *The Z notation A reference manual*. Prentice Hall, 1987.

[Ste95]   Luc Steels. When are robots intelligent autonomous agents? *Robotics and Autonomous Systems*, 15:3–9, 1995.

[WJ95]   Michael Wooldridge and Nicholas R. Jennings. Intelligent agents: theory and practice. *The Knowledge Engineering Review*, 10(2):115–152, 1995.

Agent Design from the Autonomy Perspective

Massimo Cossentino[1] and Franco Zambonelli[2]

[1] Istituto di Calcolo e Reti ad Alte Prestazioni
Italian National Research Council
Viale delle Scienze Ed. 11
90128 Palermo, Italy
cossentino@pa.icar.cnr.it

[2] Dipartimento di Scienze e Metodi dell'Ingegneria
Universita di Modena e Reggio Emilia,
42100 Reggio Emilia, Italy
franco.zambonelli@unimore.it

Abstract. The design and development of multiagent systems can take advantage of a *'multi-perspectives'* approach to system design, separately focusing and the design and evaluation of one (or of a few) specific features of the system-to-be. In this paper, we introduce the basic concepts underlying the multi-perspectives approach. Then, we take a specific look at agent autonomy and try to sketch a new specific perspective to deal with it.

1 Introduction

Despite of the today already demonstrated advantages of multiagent systems (MASs) and of an agent-oriented approach to software development [18,15,10,12], software developers must be aware that the design of a MAS tends to be more articulated than the design of a traditional object-oriented system. In fact, in the design of a MAS, one should take into account novel issues such as [3]: the proactive and reactive behavior of autonomous agents; dealing with social and ontological aspects of inter-agent communications; controlling dynamic interaction with a physical or computational environment; understanding the trust and security problems connected to a potentially opened system.

During the design of a MAS, the designer is forced to pass through several different levels of abstraction looking at the problem from many different points of view. The result is a series of models that reflect this multi-faced structure. As a natural consequence, the process used in affording the different models should not be linear but multi-dimensional. In other words, the most natural way of tackling the complexity of developing and representing MASs, and taking an effective advantage of agent-based paradigm, consists in a *'multi-perspectives'* approach to the system design [14,5].

Such an approach implies representing the system-to-be according to several different *'perspectives'*; each one of them promoting an abstract representation of the system, and enabling the evaluation of one or a few features of the system,

M. Nickles, M. Rovatsos, and G. Weiss (Eds.): AUTONOMY 2003, LNAI 2969, pp. 140–150, 2004.

thus highlighting some features and aspects of current interest, while hiding some others that are not interesting from that specific point of view.

In a previous paper [5] we already presented some perspectives that could effectively support and reduce the complexity of a MAS design process. In this paper, after having better introduced the concepts underlying the '*multi-perspectives*' approach to the system design, we take a specific look at agent autonomy. In particular, we will try to outline the characteristics of a new specific perspective explicitly conceived to deal with agent autonomy and related issues.

2 Different Perspectives for MAS Design

Other works already exist where agents and MASs have been looked at from a multi-level or multi-perspective point of view.

In [5], we discussed five different perspectives in analyzing the system design. Two of them (knowledge and computer) come from the Newell's classification [14], later expanded by Jennings [11] with the inclusion of the social level. Two further perspectives (architectural and resource) come from classical software engineering concepts. However, in our opinion, looking at a MAS from different perspectives should not simply result in a series of different not-related subsystems, or in a partial descriptions of the whole system. Rather, the real outcome should be a more detailed description of the system in terms of a well-defined aspect.

When the designer looks at the system in order to study some specific problem, (s)he thinks about it as 'something' of specific. He can be concerned about the distribution of the software in the available hardware platforms in order to optimize the performances, or can be interested in defining the rules of interaction of the agent society. For this reason we intentionally call these different points of view 'perspectives' and not 'levels' or 'abstractions' because we want to stress the concept that the perspective is the representation of the system when the spectator is interested only in a specific conceptual area of the multi-faced agent system.

In order to better express our concepts we will now briefly present the key issues of some perspectives, followed by a discussion on the 'structure' that is behind a perspective. This will serve as introduction to the section where the autonomy perspective is presented and analyzed.

2.1 Review of Some Perspectives

To clarify our concepts, let us briefly sketch a few known perspectives that can be of use in MAS design.

The *architectural perspective* looks at the software as a set of functionalities to be implemented in a classical, software engineering approach [1]. This is clearly quite an abstract perspective: its elements exist in the mind of the designer, since they are abstractions of the system representing the functionalities and their logical implementation.

The *social perspective* is characteristics – although not exclusive – of MASs. Many authors adopt a social or an organizational metaphor to describe MASs, and accordingly exploit some kind of social perspective in MAS design [13,20]. Such perspective focuses on agents as individuals of a society (or members of an organization) that interact with each other to pursue social or selfish goals, as imposed by the designer.

The *knowledge perspective* is a highly detailed point of view. In such a perspective, the single agent and its functional and behavioral details (that induce some specific implementation), are conceived as entities that are able to manipulate some sort of external knowledge, with the goal of procuding/spreading new knowledge [14,9].

The *resource perspective* is oriented toward the reuse of existing design or implementation resources (e.g. architectural patterns in software architectures or detailed design patterns for components and their tasks [8]). It represents agents/components/architectures in terms of some kind of mental bottom-up process, and deals with the process of recycling an existing agent (or parts of it) and eventually adapting it to match the necessity of a new problem.

The *computer perspective* is the more physical, touchable point of view. It relates to the spreading of files that constitute the software in the available hardware platforms and to the computational and storage load imposed by the processes in charge of executing the software system. This perspective therefore considers the deployment issues arising from the interplay between the hardware and the software system [14].

Depending on the specific application problem and on the specific characteristics of the MAS to be designed, only a portion of the above perspectives are likely to be of use. As it is common in software engineering problems, the designer should find the correct trade-off between the number of different descriptions of the system, the need to ensure their coherence and the consequent increasing number of concepts that (s)he has to manage. While introducing a new perspective could allow the identification and tracking of a potential risk in the system development, the uncontrolled use of this technique could produce the undesirable relapse of inducing the designer to consider so much variables that he could loose the perception of the system as a whole.

2.2   General Outline of a Perspective

In our work we refer to the concept of perspective instead of level also because we want to emphasize the unity of the system thought as a representation of the problem-solution couple that evolves from the early stages of the requirements elicitation to the final coding and deployment activities. The system can be represented in the sequence of models and phases of a design methodology with their resulting artifacts. Looking at this unit with different scopes we obtain a perspective of it that shows some elements (under one of their possible facades) hiding what is out of the particular focus. In very general terms (and abstracting from the presence of agents and MASs) one can characterize a perspective as made of design *elements* that are composed abiding to some *constraints* in order to build a system conceived to operate in a specific *context*. The designer assembles these elements according to a (design) *rationale* that establishes the composition strategy by processing the *inputs* required by the perspective. Inputs, elements, context, rationale, and constraints are defined as follows:

1. *Inputs.* They defines the information that will be evaluated by the designer at design-time according to the prescribed design rationale. This will also be likely processed by the system (at run-time) in order to achieve its design objectives. Typically, these are static elements of the design either introduced and engineered by the designer or pre-existing in the environment. These inputs belong to two different categories: goals/requirements/features of the system, and input data available for the system. Depending on the adopted perspective, the first type of inputs can be design goals, architectural concerns, cooperation/collaboration paradigms and so on. Input data (the second category) could be files, records, computational resources, or any type of abstract knowledge.

2. *Elements.* These are the elementary computational components of the perspective (e.g., depending on the perspective, these could be functions, agents, behaviors, software components, etc.). These elements are defined/refined by the designer acting according to the design rationale. Using them like bricks the designer will composes new pieces of the system completing the definition of its appearance from the specific perspective point of view.

3. *Context.* Each element of the perspective is thought to be applied in some operating scenarios (and bringing elements outside their natural operational context could cause a system failure). Environmental considerations affecting the system design can be enumerated among context concerns; for example we should consider specific characteristics or constraints of the environment that could influence the computational capabilities of the systems component.

4. *Rationale.* What the overall system will be obviously depends on the motivation underlying how each of the system element is composed with the others. Some design choices depend on specific strategies (for example respecting the holonic architecture) that can easily be formalized, others come from the designer skills and experience (it is a matter of fact that a system designed by a student is commonly less effective than the solution provided by a senior designer). In this work we will refer to the design rationale trying to include this untouchable contributions coming from experience and skill in form of guidelines. As a result, we think about the rationale always as a set of (someway) formalized rules, algorithms, conventions, best practices and guidelines that will guide the designer work. Of course, identifying and representing such a set of motivations strictly depends on the adopted perspective, on its inputs and elements.

5. *Constraints.* These defines the rules according to which the various elements of the system can be assembled for instance to compose a complex service or reach a global application goal. These rules are of course particularly important in all contexts where a complex service or a global application

goal derives from the composition/interplay of the activities of the various components of the system, and where such a global goal can be obtained only by strictly respecting some composition rules. Also in this case, the specific adopted perspective influences the way in which these constraints are identified and expressed.

3   The Autonomy Perspective

A very distinguishing characteristic of MAS is their being composed of autonomous components, capable of proactive actions and of decisional capabilities. For this reason, and because the autonomy dimension in not something that 'traditional' software engineers are used to deal with, we think that the adoption of a specific autonomy perspective in MAS design may be needed.

Here we will try to sketch what an autonomy perspective in MAS could look like, by discussing it (according to the characterization of the Subsection 2.2) in terms of inputs, elements, context, rationale and constraints.

3.1   Inputs

The presence of some initial hypothesis, '*inputs*', is a common element of all the design activities. Such inputs will help the designer to devise and represent an appropriate architecture for the system-to-be. When adopting a specific perspective to design, of course, a limited set of '*inputs*' will be of interests.

In an autonomy perspective, whose focus in on analyzing a system from the viewpoint of the autonomous capabilities of a number of proactive, task-oriented and decision-making components, the design inputs of most interest are: '*system goals*' and '*domain ontology*'. They correspond, under the autonomy point of view, to the two fundamental needs of each perspective, i.e., the functionalities affecting the achievement of the design objectives in the specific perspective and the data to be processed by the system, respectively. In fact, the analysis of system goals guides the actual design activity in identifying the basic elements (i.e., autonomous goal-oriented agents) of the system-to-be. The analysis of the domain ontology helps identifying what knowledge will be available to system elements for them to achieve their functionalities (i.e., their goals).

While a variety of other '*inputs*' can be available to the designer of a MAS (e.g., specific non-functional requirements or specific models of knowledge acquisition) this will not play a role in an autonomy perspective, and have to be taken into account in other specific perspective (e.g., a computer perspective or a knowledge perspective).

3.2   Elements

Clearly, the elements of interests in an autonomy perspective for MAS design are autonomous agents, intended as proactive decision-making components combining specific proactive abilities (behaviors/roles) with available knowledge in order to reach goals inspired by their vocation.

The factors that mostly affect the autonomous behaviours of an agents and that should be taken into when modeling agents from the autonomy perspective are:

- Agent's vocation
- Agent's knowledge
- Agent's (behavioral) abilities
- Available resources

The agent's vocation is characterized by two interesting aspects (external and internal). The first one descends from the agent creator (the designer) point of view and it addresses the reason for which the agent has been created. This has a direct influence on the other aspect (the internal one): the will and consciousness that is put in the agent itself; in a BDI agent this could correspond to agent's desires, while in the PASSI approach [6] this is the requirement(s) that has to be fulfilled by the agent.

Agent's knowledge (at least some specific portion of it) is one of the elements of autonomy perspective since it contributes to agent autonomy by building up the consciousness the agent has of its operational scenario, and in most cases the strategy it will initially adopt is an a-priori one that is updated when new information about the situation will be available.

The expected result of the autonomous agent design is a combination of the agent's abilities to achieve some goal. Usually this is obtained with the correct coordination of some agent's behaviors and their specific duties (in terms of knowledge to be processed, options to be selected and so on). The analysis of the different agents (behavioral) capabilities is therefore one of the desiderata for the computational autonomy perspective.

The availability of specific type or resources (whether computational or physical) is another factor that influences the autonomous action of an agent. An agent could decide to adopt an alternative plan according to the possibility of using some kind of resource. Not all the resources provided in the environment are interesting for this perspective. Only their subset that has a direct influence on the agent autonomy should be included and the others should be hidden in order to limit the representation complexity.

It can be useful to remark that while the listed elements play a relevant role in characterizing the agent autonomy some others could be neglected; this is the case of agents communications (with the underlying transport mechanism, content language and interaction protocols), mechanism of knowledge updating, implementation architectures, and so on.

In order to clarify our concepts, let us abstract the execution of an autonomous agent as a movement in some abstract 'actions space' (Fig. 1). The mission of the agent (i.e., its vocation), for which it exploits all necessary knowledge, ability, and resources, is to move step by step in the action space (i.e., via a sequence of autonomous actions) until it reaches the goals positioned in this space. Clearly, depending on the actual decision-making of the agents, the goals can be reached by following various paths (i.e., via different combinations of its possible actions).

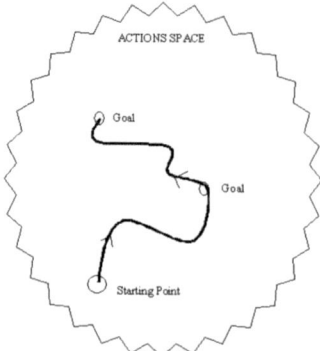

Fig. 1. An autonomous agent in an abstract trajectory towards its goal

A concrete example of this generic agent and its actions space could be a robotic agent that is devoted to the exploration of some environment in order to collect information about its topology (walls, doors) and the position of furniture elements in it. It could move around following different paths and, using its sensors (laser, infra-red, sonar or even vision), it can discover the presence of different objects that it will classify according to its (a-priori) knowledge. In this simple example, the very goal of the agent it to fully explore the building, and autonomy of the agent is mainly used to let it decide in which way to explore the building, room after room. In other words, in this example, the abstract trajectory the agent has to follow to reach its goal (as from Fig. 1) equates to a physical trajectory in space.

Thus, from the autonomy perspective, what is of interest is not the activity of collecting data and knowledge about objects in the environment, but mostly the activity of moving in such environment.

3.3   Context

Each system is designed to solve one or more problems and this situates it in the context (usually referred as the problem domain) where those problems take place. Such a characterization particularly applies to agent, which have the peculiar characteristics of being entities situated in an environment, that is, of having an explicit representation of the context and of acting in it.

A common expedient used by designers to represent the context and the system interaction is the description of some operating scenarios. This could bring an enormous number of elements to the attention but only the part of them really affecting the specific perspective should be considered and the remaining other should be neglected.

The context in which agents of a MAS situates (whether a virtual computational environment like an e-commerce marketplace or a physical one like a building to be explored) introduces in the system design some constraints. These could be rules of the environment itself (e.g., an agent should pay for the good he won in an auction) or possible environment configurations that could effect the agent activities (e.g., fog could limit vision of a robotic agent). Moreover, the data an agent can acquire from the environment can be of some relevance too.

Since autonomous agents could be not deterministic, and since the agent decisional process is something that could not always be easily deducted from a black-box external observation of the agent behavior, a perspective on resources centered around the autonomy concept should take into account this aspect. In particular, when the focus is on the autonomous actions of the agents, the characteristics of the environment (and of the data that can be found there) of interest are those that can somewhat influences the agents decisional process. In other words, by taking into account the fact that the agent executes in an environment, that may influence it and may be influenced by it, an autonomy perspective would prescribe to identify what in the environment and its data could comes to interplay with the dimension of autonomy of the agent.

Going back to our explorer robotic agent, the topological and physical characteristics of the building) to be explored could affect the activities of the agent. A very large open site may enable an agent to explore in detail all possible objects in it. A site with objects in not accessible position will prevent him to do his work in a complete satisfactory way.

However, from the autonomy perspective, the data that are really of interest are the topological information about the environment, because these will influence the way the robot will find its way through the building. For instance, the presence of a ground slope that the agent cannot safely walk through may require him to take specific exploration choices or, which it the same, to modify the physical (and abstract) trajectory of its autonomous decisions. Other information like the color of walls or the style of furniture are not generally relevant for this perspective. Sometimes, it may be the case that the system requirements calls for a *curious* robot, capable of deciding to explore some specific objects more in detail using all of its sensing capabilities and possibly requiring it to step back to analyze relations with already analyzed objects. In such a case, further characteristics other than the topology of the environment may come into play influencing the agent autonomous actions.

3.4   Rationale

We already discussed that, during his activity, the designer aims at reaching some goals for the system; all of his choices will be guided by a precise strategy (one of the many possible ones) that he considers the best solution to the problem. In the context of an architectural perspective (looking at the best architectural solution) of a system devoted to provide the control of some active network routers this mean using many small well specialized agents that will not overload the network traffic rather than multiple instances of the same big all-purpose agent. In the case of an autonomy perspective, the rationale that is behind all the design activity is the decisional process used by agents (and imposed to them by the designer) to reach their goals. In the context of cooperative agents like the ones used in Adelfe [2], this means looking for a cooperative solutions while in other approaches and agent could prefer to face the problems by itself. At the end the decisional process will decide in which way the perspective elements (tasks, roles, ...) will be composed to satisfy the agent's vocation (another element of this perspective).

Considering the robot example, the decisional process is first of all character-ized by the chosen cognitive architecture that will select the plan (for example a subsumption machine) and the strategy imposed to it by the designer. Let us suppose that the robot is not exploring the environment in order to collect new data about its topology but it is looking for bags forgotten by public in an air-port. The mission is almost the same (finding new objects in the environment) but the decisional process could be different. For example, a new line of chairs is not considered an interesting element while a bag left alone in a crowded place could be a potential danger and therefore it requires an immediate attention by the robot that could even warn security personnel of the discovery.

In this case, the rationale determines how the path to the goal (as described in Fig. 1) is to be formed, and how it can be influenced by external factors.

In other words, it determines the way an agent finds a path toward its goal, and may also determine the way an agent may not be allowed to find a path, because this clashes with some requirements (giving attention to a not dangerous new line of chairs slows the surveillance of the assigned area) or because this is not made possible by the structure of the available resources.

3.5   Constraints

In a MAS, several agents execute in the same environment, towards the achieve-ment of individual goals that may either contribute to a global application goal or that may be selfish goals. Whatever the case, the autonomy of agents does not imply that agents can do whatever they want independently of the actions of other agents. Rather, since agents live in the same universe, the 'trajectories' they follow should be disciplinated and not 'clash' with each other. In other words, in a MAS, the autonomy of agents should be somewhat reduced or 'adjusted' in order for the whole system to proceed correctly, by disciplining the abstract trajectories that each agent would follow toward the achievement of the task.

A typical example of this is in the concept of 'social laws' introduced by Moshe and Tennenholtz [16,13], which perfectly suit our example of mobile robots ex-ploring an environment. There, each robot in a group of mobile robot – each having the selfish goal of exploring an environment – is disciplined in its move-ments (that is, in its autonomy) via the superimposition of social laws (traffic laws in the specific example) that prevent it for planning motion actions that would somewhat disturb the movements of other robots.

Another example is the concept of 'organizational rules' introduced in the latest version of the Gaia methodology [19]. There, in the analysis phase of a MAS, the modeling of the internal activity of each agent in a MAS (including the goals and tasks of each agent) has to be coupled with an explicitly modeling of the external rules that the system as a whole has to ensure. Clearly, such 'organizational rules' have to be somewhat enacted in the subsequent agent design by limiting the autonomy of those agent that would otherwise be at risk of breaking the organizational rules.

4   Conclusions and Future Works

The number of different issues that a designer is forced to face in the development of a MAS may require adopting a multi-perspective approach to system design. In particular, among a number of perspectives that can be conceived, a specific perspective focusing the issue of agent autonomy may be required to effectively tackle the peculiar characteristics of agents and of their being autonomous entities interacting in a complex world.

Having sketched the key characteristics of an autonomy perspective for MAS system design, as we have done in this paper, is only a first step. Further work will be required to make such a perspective applicable to a variety of current agent-oriented methodologies, such as PASSI [6], GAIA [17,19], TROPOS [4]or MASE [7]. In addition, it will be important to verify on real-world applications the extent of applicability of the autonomy perspective and its possible limitations.

References

1. L. Bass, P. Clements, and R. Kazman. *Software Architectures in Practice (2nd Edition)*. Addison Wesley, Reading (MA), 2003.
2. C. Bernon, M.P. Gleizes, S. Peyruqueou, and G. Picard. Adelfe, a methodology for adaptive multi-agent systems engineering. In *Proceedings of the Third International Workshop Engineering Societies in the Agents World (ESAW-2002)*, Madrid, Spain, September 2002.
3. G. Booch. *Object-oriented Analysis and Design (second edition)*. Addison Wesley, Reading (MA), 1994.
4. J. Castro, M. Kolp, and J. Mylopoulos. Towards requirements-driven information systems engineering: The tropos project. In *To appear in Information Systems*, Elsevier, Amsterdam, The Netherlands, 2002.
5. M. Cossentino. Different perspectives in designing multi-agent systems. In LNCS, editor, *Proceedings of the AGES '02 workshop at NODe02*, Erfurt, Germany, October 2002.
6. M. Cossentino and C. Potts. A case tool supported methodology for the design of multi-agent systems. Las Vegas (NV), USA, June 24–27 2002. The 2002 International Conference on Software Engineering Research and Practice, SERP'02.
7. S. A. DeLoach, M. F. Wood, and C. H. Sparkman. Multiagent systems engineering. *International Journal on Software Engineering and Knowledge Engineering*, 11(3):231–258.
8. E. Gamma, R. Helm, R. Johnson, and J. Vlissides. *Design Patterns*. Addison Wesley, 1995.
9. C. Iglesias, M. Garijo, J. C. Gonzales, and J.R. Velasco. Analysis and design of multi-agent systems using mas-commonkads. In *Intelligent Agents IV: Agent Theories, Architectures, and Languages*, volume 1365, pages 313–326. Springer Verlag, 1998.

10. N. R. Jennings. An agent-based approach for building complex software systems. *Communications of the ACM*, 44(4):35–41, April 2001.

11. N.R. Jennings. On agent-based software engineering. *Artificial Intelligence*, 117, 2000.

12. J. Kephart. Software agents and the route to the information economy. *Proceedings of the National Academy of Science*, 99(3):7207–7213, May 2002.

13. Y. Moses and M. Tennenholtz. Artificial social systems. *Computers and Artificial Intelligence*, 14(3):533–562, 1995.

14. A. Newell. The knowledge level. *Artificial Intelligence*, 18, 1982.

15. H.V.D. Parunak. Go to the ant: Engineering principles from natural agent systems. *Annals of Operations Research*, 75:69–101, 1997.

16. Y. Shoham and M. Tennenholtz. Social laws for artificial agent societies: Off-line design. *Artificial Intelligence*, 73, 1995.

17. M. Wooldridge, N. R. Jennings, and D. Kinny. The gaia methodology for agent-oriented analysis and design. *Journal of Autonomous Agents and Multi-Agent Systems*, 3(3):285–315, 2000.

18. M.J. Wooldridge and N. R. Jennings. Intelligent agents: Theory and practice. *The Knowledge Engineering Review*, 10(2):115–152, 1995.

19. F. Zambonelli, N. R. Jennings, and M.J. Wooldridge. Developing multiagent systems: The Gaia methodology. *ACM Transactions on Software Engineering and Methodology*, 12(3):417–470, July 2003.

20. F. Zambonelli, N.R. Jennings, and M.J. Wooldridge. Organizational abstractions for the analysis and design of multi-agent systems. In *Proceedings of the 1st International Workshop on Agent-Oriented Software Engineering*, volume 1957 of *LNCS*, pages 253–252. Springer Verlag, 2001.

From Individual Based Modeling
to Autonomy Oriented Computation

Xiaolong Jin and Jiming Liu

Department of Computer Sicence
Hong Kong Baptist University
Kowloon Tong, Hong Kong
{jxl, jiming}@comp.hkbu.edu.hk

Abstract. In this chapter, we generalize autonomy oriented computation. We characterize the application-independent features of two main components of AOC systems, namely, entities and environment. We identify the self-organizating process of entities as the core of AOC systems. We also address the autonomy deployed in AOC systems and compare them with those in other work. Moreover, we use three typical examples, a natural system, a man-made system, and a computational system, to show how AOC is embodied and how AOC works in different cases.

1 Introduction

Individual social animals have very limited capabilities. However, a colony of such animals can exhibit complex and interesting collective behaviors. As we know, in an ant colony, individual ants only have limited memory and simple goal-directed behaviors. They have no global knowledge about the task the colony is performing. They locally and autonomously make decisions on choosing routes to travel without control of others. However, through an indirect communication media, pheromone trails, ants in the colony can self-organize themselves and show complex collective behaviors, such as cooperatively conveying large items and finding the shortest route between their nest and a food source. These complex behaviors are also called *emergent behaviors* or *emergent intelligence.*

In the ant colony system, the autonomies of individual ants play a crucial role. Specifically, we can observe the following characteristics:

- **Autonomous:** the systems' elements are rational individuals, which act independently without control from others or a 'master' from outside. In other words, no central controller for directing and coordinating individuals exists;
- **Emergent:** Although the individual elements have only simple behaviors, the systems can exhibit unpredictable complex behaviors;
- **Adaptive:** The elements can modify their behaviors in response to changes in the environment where they are situated; and
- **Self-organized:** Through communications, often not direct, the elements are able to organize themselves to achieve the above complex behaviors.

M. Nickles, M. Rovatsos, and G. Weiss (Eds.): AUTONOMY 2003, LNAI 2969, pp. 151–169, 2004.

As inspired by the above system and its characteristics, researchers have developed artificial systems to imitate social animals' collective behaviors: Reynolds developed boids to simulate birds' flocking, navigating, and gathering [18,19]; Tu *et al.* provided an artificial fish model to emulate the schooling behavior of fishes [18,20]. Due to the success in simulating complex collective behaviors, researchers have tried to use such kind of artificial systems to solve computational problems. Dorigo *et al.* used artificial ant colony systems to solve some optimization problems [7,8] (e.g., traveling salesman problems (TSPs)); Liu *et al.* [11,12,14,15] presented autonomy oriented approaches, ERA (i.e., Environment, Reactive rules, and Agents) and MASSAT (i.e., a multi-agent-based satisfiability problem solver), to solving constraint satisfaction problems (CSPs) and satisfiability problems (SATs), respectively. By generalizing the above work, Liu in [13] has proposed *autonomy oriented computation* (AOC) as a novel paradigm for characterizing the behaviors of a complex system and for solving hard computational problems. Later, in [16,17] Liu *et al.* identified two main goals of AOC:

1. To understand the underlying mechanism of a real-world complex system through hypothesizing and repeated experimentation. The end product of these simulations is a better understanding of or explanations to the real working mechanism of the modeled system.
2. To reproduce life-like behaviors in computation. With detailed knowledge of the underlying mechanism, simplified life-like behaviors can be used as models for problem solving techniques. Note that here replication of behavior is not the end, but rather the means, of these computational algorithms.

To continue the work in [13,16,17], this chapter will address the following two issues:

1. How to generalize autonomy oriented computation?
2. How to identify and characterize the problem-independent features of an AOC system, especially its two main components, entities and environment?

The remainder of the chapter is organized as follows. Section 2 surveys the related work on autonomy. Section 3 presents our generalization of autonomy oriented computation. We emphasize the autonomy in AOC systems. Sections 4, 5, and 6 provide examples to detailedly show how autonomy oriented computation works in different cases. Section 7 concludes the chapter.

2   Related Work on Autonomy

Originally, autonomy was of interest in sociology [9] and biology [21]. Since the proposal of *multi-agent systems* (MAS), autonomy has quickly becomes its central issue. In MAS, autonomy aims at studying how autonomous agents make decisions and how they operate on the decisions [2]. Autonomy has been widely studied in various contexts, and defined in different senses. In [1], Barber and Martin defined that "autonomy is an agent's active use of its capabilities to pursue some goal, without intervention by any other agent in the decision-making process used to determine how that goal should be pursued." Castelfranchi [3] defined autonomy as the degree to which an agent depend its decisions on others and external sources.

Besides definitions, researchers have also done some work on representations, classifications, and measurements of autonomy. In [22], Weiss *et al.* proposed a schema, called RNS (*Roles*, *Norms*, and *Sanctions*), to formally and precisely specify agents' autonomies. Specifically, in their schema in order to achieve goals, agents will act as owners of certain roles, for which there are some norms, such as permissions, obligations, and interdictions. If agents behave while obeying the norms, they will get certain positive sanctions, i.e., *rewards*. Otherwise, they will receive negative sanctions, i.e., *punishments*. They further developed a tool, XRNS, to facilitate developers to generate RNS-based autonomy specifications in XML format.

Hexmoor studied agent autonomy from a relativistic point of view based on a formal BDI (*Believes*, *Desires*, and *Intentions*) model [10]. He defined that if an agent is free of other agents' influences of control and power, it is autonomous with respect to them. To determine the autonomy of an agent, it should be considered that to what extent the agent prefers to cooperate with other agents and how the others contribute to its freedom to make decision and perform. Moreover, Hexmoor suggested that given a goal, there are three stages to determine the autonomy of an agent, namely *potential determination* stage, *utility analysis* stage, and *enactment* stage. In the first stage, agents determine potential autonomies. In the second stage, agents weight potential autonomies and make decisions among alternatives based on utilities. Lastly, in the third stage, agents perform action selection according to selected autonomies.

In [2], Brainov and Hexmoor argued the autonomies of agents should contain two aspects, *action autonomy* and *decision autonomy*. Action autonomy concerns the way agents behave in their environment and interact with each other. Decision autonomy focuses on the capabilities of agents to make consistent decisions. Depending on the influencer, they categorized autonomy into three types, i.e., autonomy with respect to an agent's user, autonomy with respect to a physical environment, and autonomy with respect to other agents. They further proposed methods to quantitatively measure the degree of autonomy in the contexts of agent-user and agent-environment interactions. Here, the degree of autonomy means to what extent the influencer affects the object (e.g., action, goal, task) of autonomy. Moreover, they introduced a measurement for group autonomy.

There is some other work on autonomy. Davis and Lewis proposed computational models of emotion for underpinning computational autonomy [5,6]. Costa and Dimuro defined internal states, in which agents should try to permanently be as *needs*. They regarded agent needs as one of the foundations of agent autonomy. Based on this notion, they further proposed a formal definition of agent autonomy [4].

2.1   Autonomy Oriented Computation

In [13], Liu first proposed the notion of *autonomy oriented computation* (AOC). Later, Liu *et al.* in [16,17] gave their definitions of autonomies in AOC. They identified and defined *synthetic autonomy, emergent autonomy*, and autonomy in a computational system in AOC systems. They defined autonomy oriented computation as the computational approaches, which take autonomy as the core model of complex system behaviors. They further proposed three different autonomy oriented computation approaches, namely, *AOC-by-fabrication, AOC-by-prototyping*, and *AOC-by-self-discovery*. For more details, readers are referred to [16,17]. In this chapter, we will not focusing on the definitions of autonomies, but on how to generalize AOC and how AOC employs autonomy as a core model to work.

3   Generalization of Autonomy Oriented Computation

In an AOC system, there are mainly two components, namely, an environment and a group of computational entities. In the following, we will discuss these two components in detail as well as the process of self-organization of entities.

3.1   Environment

As one of the main components of an AOC system, environment E usually play two roles. Firstly, it is the place where computational entities reside and behave. This role is absolutely necessary for all applications of AOC. Secondly, in some applications (e.g., ant colony systems), environment E acts as a media for indirect communications among computational entities. To implement this role, E will:

1. record its own state information (e.g., the intensity of pheromone); and/or
2. record the information of its entities (e.g., the positions of its entities).

In order to explain the working mechanism of the indirect communication media, let's consider a simple system containing two entities, A and B, and an environment E. While A is performing its behavior, it will change the state of environment E; Or, after A has performed its behavior, its own state will accordingly be changed. Environment E will record these changes. Later, B will consider these changes in its decision making. Consequently, A implements the indirect communication with B. This process can easily be extended to a parallel version and a multi-agent version.

In the sense of the second role, we can define environment E as follows:

Definition 1. *Environment E is a tuple $\langle ES, \{S\} \rangle$. $ES = \{es_1, \cdots, es_n\}$ is an application-dependent data structure to record the state information of E. Each es_i characterizes a property of E. D_{es_i} is its domain. $\{S\}$ is a set of state information of entities resided in E.*

Remark 1. In an application, if environment E only plays the first role. It will have no much influence on the process of self-organization of its entities. On the other hand, if it also plays the second role, the process of self-organization will depend much on it. Because, in this application, entities can interact with each other through the environment. Later, in Sections 4, 5, and 6, we will see this point.

3.2   Computational Entities

Definition 2. *In AOC, a computational entity e is a tuple $\langle G, B, S \rangle$, where G and B are two sets of goals and behaviors, respectively; S describes the state information of e.*

At any time, an entity is in a certain state. Its goals are to achieve certain targets with respect to its state. To do so, the entity will select and perform different behaviors at different time steps. While doing so, it need to interact with other entities or its environment. In the following, let's address states, goals, behaviors, and interactions of an entity in detail.

States

Definition 3. *State S is an application-dependent data structure to describe the dynamic and static properties of entity e, i.e., $S = \{s_1, \cdots, s_n\}$ $(n \geq 0)$, where each s_i describes a dynamic or a static property of e and has domain D_{s_i}.*

Remark 2. S records the values of properties which are used to characterize an entity. The goals of an entity are to reach certain special states. The goal of an AOC system is to make all entities to organize themselves and reach certain states.

Goals

Definition 4. *Goal set $G = \{g_1, \cdots, g_n\}$ where $n \geq 0$. Each g_i $(0 \leq i \leq n)$ is an individual goal of entity e. g_i is usually an application-dependent constraint to be satisfied or an application-dependent function to be optimized (i.e., maximized or minimized), with respect to state S of entity e.*

Goals G of entity e describes its local targets. In an AOC system, each entity e is a simple individual. Usually, each entity has only one goal and all entities have the same goal.

Remark 3. Each entity has its own goals. And, its simple behaviors are goal-directed. However, an entity does not know the global goal of their colony and the task their colony is performing.

Behaviors

Definition 5. *Behavior set $B = \{b_1, \cdots, b_n\}$ where $n > 0$. Behavior b_i is usually a mapping: $b_i : D_S \times D_{ES} \rightarrow D_S \times D_{ES}$, where $D_S = D_{s_1} \times \cdots \times D_{s_n}$ and*

$D_{ES} = D_{es_1} \times \cdots \times D_{es_n}$ are Cartesian products of domains of s_i and es_i, respectively.

It should be pointed out that in some applications, behaviors consists of actions with finer granularity. In other words, there is a set of actions $\{a_1, \cdots, a_m\}$. Each behavior b_i corresponds to a unique sequence of actions $\{a_1^i, \cdots, a_j^i\}$. Performing an action may not cause the state change of entity e. Only after performing a sequence of actions, which corresponds to a behavior, the state of entity e may be changed.

In the application, where the goal of an entity is to optimize a function with respect to its state, an entity can usually have three kind of behaviors, *greediest behavior*, *greedier behavior*, and *random behavior*.

Greedist behavior The greediest behavior b_{gst} is a behavior that will cause the maximum change in the state of the entity. Namely, $b_{gst} \doteq b$ *iff* b can maximize $(S(t+1) - S(t))$.

Greedier behavior a greedier behavior b_{ger} is a behavior that will cause the positive change in the state of the entity, i.e., $b_{ger} \doteq b$ if we have $S(t+1) - S(t) > 0$ after performing b.

Random behavior a random behavior b_{ran} is a behavior that will cause any change (positive or negative) in the state of the entity. Namely, $b_{ran} \doteq b$ if we have $S(t+1) - S(t) \neq 0$ after performing b.

Remark 4. At each step, by performing the greediest behavior an entity can get the currently best result; by performing a greedier behavior, it can reach a state better than its current one; by performing a random behavior, an entity may reach a state worse than its current one. Here, the question is since performing the greediest behavior an entity can reach the currently best state, why it is necessary to have greedier behaviors and random behaviors. This is because of the following two reasons:

- To perform the greediest behavior, an entity has to make decisions on lots of possible states and determine which one will maximize $(S(t+1) - S(t))$. While to perform a greedier behavior, an entity only makes decisions on a few possible states. That means the complexity of the greediest behavior is much greater than that of a greedier behavior. In this sense, performing greedier behaviors can reduce the cost.
- To performing a random behavior, an entity only needs to randomly select state. Therefore, its complexity is less than the greediest behavior and an greedier behavior. The more important reason is that only performing the greediest behavior AOC systems can easily get stuck in local optima, where all individual entities are at their currently best states and cannot transit to other states while AOC systems have not reached their goals. Through performing random behaviors, AOC systems can escape from local optimal states. Model details can be found in [11,12,15].

3.3   Interactions

There are two kinds of interactions in AOC systems, namely, *interactions between entities and their environment* and *interactions among entities*. Generally speaking, in an application, only one kind of interactions is employed.

Interactions Between Entities and Their Environment. In an AOC system, the interaction between an entity and its environment is implemented through the state change caused by the entity's behaviors. Specifically, we can define it as follows:

Definition 6. *The interaction between entity e and its environment E is a group of mappings $\{I_{eE}\}$:*

$$I_{eE} : D_{ES} \rightarrow_{b_i} D_{ES}, \tag{1}$$

where '$\rightarrow_{b_i}$' denotes that this mapping is a part of mapping b_i, i.e., the jth behavior of entity e.

Interactions Among Entities. Different AOC systems may have different fashions of interactions among their entities. Those interactions can be categorized into two types, *direct interactions* and *indirect interactions*. Which type of interactions will be used in an AOC system is dependent on the specific applications.

Direct interactions are implemented through direct information exchanges among entities. Here, the information refers to the state information of entities. In AOC systems, each entity can usually interact with some, not all, other entities. Those entities are called its *neighbors*.

Indirect interactions are implemented through the communication media role of an environment. Specifically, it can be separated into two stages: (1) through the interactions between an entity and its environment, it will 'inform' its information to the environment; (2) later, while other entities behave, if necessary, they will consider the information of that entity stored in their environment. That entity can obtain the information of others via the same way.

3.4   The Process of Self-organization

Why AOC systems can successfully emulate social animals' collective behaviors (e.g., the flocking behavior of birds and the schooling behavior of fishes) and solve some computational problems (e.g., constraint satisfaction problems and satisfiability problems)? The key point is the self-organization of entities in AOC systems. All complex collective behaviors in AOC systems come from the self-organization of their entities.

In order to achieve their respective goals, computational entities autonomously make decisions on selecting and performing their simple behaviors. While selecting or performing, they will not only consider their own state information, but also that of some others. To do so, they will either directly interact with each other, or indirectly interact via the communication media – their environment, to exchange their information. By performing behaviors, entities will change their states towards their respective goals. Because computational entities take into account of others while behaving, from a global view entities are aggregated together to achieve the global goals of AOC systems. In other words, the final collective behaviors are aggregation results of individual entities' behaviors.

3.5   Autonomy in AOC

Autonomies in AOC systems are embodied in three aspects. First, computational entities in an AOC system autonomously make decision on how to select behaviors to perform at the next steps, according to the information of their own states and the sensed environment. Second, while performing selected behaviors, entities are free of intervention or control from others. Although there are direct or indirect interactions among entities, those interactions are usually information exchanges. No 'command' is sent from one entity to another. Third, the whole AOC systems are also autonomous. It is not controlled by a 'commander' from outside. In other words, it is an open but 'command-repellent' system.

Remark 5. In Section 2, we have surveyed related work on autonomy. As compared with those work, autonomy in AOC has the following differences:

1. In other work, the subjects of autonomy are agents. While in AOC, the subjects are computational entities. We believe that agents are also entities, but they have more considerations on mental states, namely, believes, desires, and intentions. In our point of view, autonomy can also be embodied in 'lower level' entities.
2. In other work, the subjects of autonomy are only agents. But in AOC, we have addressed that autonomy can also be embodied on the whole system. The whole AOC system behaves autonomously without interventions from outside.

4   Autonomy in a Natural System: An Ant Colony

In this section, let us take a look at how AOC operates in an ant colony. To do so, we also use a common scenario in ant colony research, where ants search for food; if they really find a food resource, they will convey food to their nest through the shortest route between their nest and the food resource.

4.1   Scenario

At the beginning, ants have no knowledge about where food is. Therefore, they will randomly search the space they can reach. While foraging, they will lay down pheromone over their routes. If one of them finds a food source, it will return to its nest based on its own limited memory, at the same time it will lay pheromone on its route again therefore reinforcing its trail. After that, when other ants encounter this trail, they will have greater probability to autonomously follow this trail. Consequently, more and more pheromone will be laid on this trail. The more ants that travels through this trail, the higher the intensity of the pheromone over it. On the other hand, the higher the intensity of the pheromone over this trail, the more ants that will decide to follow it. In other words, this process will form a positive feedback. In order to show how this process leads the ant colony to find the shortest route, let's see Fig. 1, where the ant colony finds two routes ACB and ADB. Because ADB is shorter than ACB, those ants on route ADB will complete their travel more times than those on route ACB and therefore lay more pheromone on ADB. Due to this reason, those ants previously traveling through ACB will re-select ADB to travel. Fewer and fewer ants will travel on ACB. Therefore, less and less pheromone will be laid over ACB. At the same time, pheromone previously laid over ACB will evaporate gradually. Finally, no ants will select ACB and a shorter route emerges.

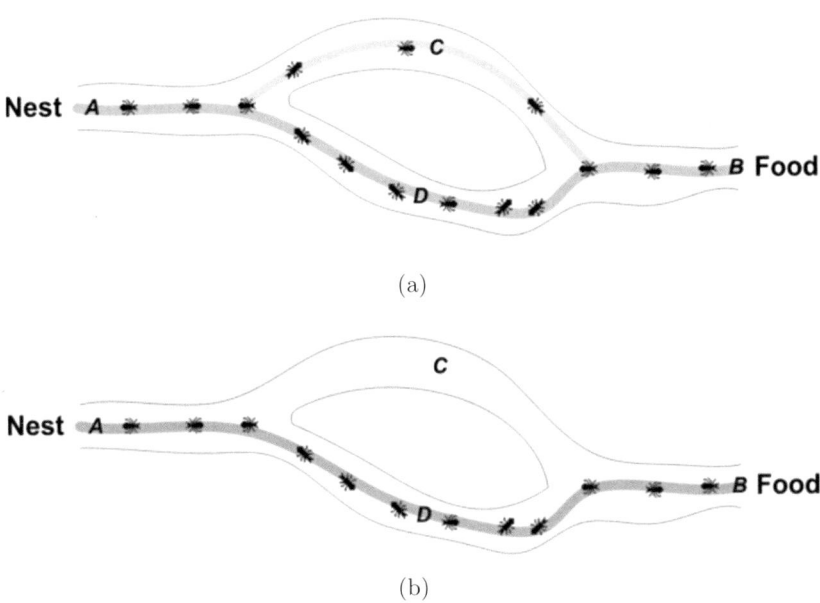

(a)

(b)

Fig. 1. An ant colony system where ants are carrying food from a food source to their nest. (a) At the beginning, the colony finds two routes from the nest to the food source, ACB and ADB. ACB is longer than ADB. (b) Finally, the colony selects ADB as a optimal route to convey food

4.2   Environment

In an ant colony, the environment E plays two roles: (1) a place where ants reside; (2) an indirect communication media among ants. To implement the second role, environment E is embodied through three matrices, d_{ij}, τ_{ij}, and p_{ij}, which respectively denote the distance between adjacent nodes i and j, the intensity of pheromone on route from i and j, and the transition probability with which an ant select to travel through path between i and j. d_{ij} is constant, while τ_{ij} and p_{ij} vary over time.

4.3   Computational Entities: Ants

In an ant colony, the autonomous entities are ants.

Ants' Goals. The goal set of ants is $G = \{foraging\}$. Specifically, ants forage for food. If finding a food source, ants will convey the food to their nest.

Ants' Behaviors. The behavior set of ants is $B = \{pickfood, walk, laypheromone, chooseroute, leavefood\}$:

- *pickfood:* At the food source, ants pick up a part of food.
- *walk:* Ants walk on the route to their nest or the food source.
- *laypheromone:* While walking, ants lay pheromone over their routes. Say, at time interval $[t, t + 1]$, ant k travels from node i to j, then it lays a mount of $\Delta\tau_{ij}^k(t, t+1)$ pheromone on the path between i and j.
- *chooseroute:* At a node, if there are multiple routes, ants autonomously make decisions according to the environment information. Say, an ant reaches node i at time t. It has a set of nodes to choose, $\{j, \cdots, k\}$. In this case, it will autonomously select one node to travel based on probabilities $\{p_{ij}, \cdots, p_{ik}\}$.
- *leavefood:* In their nest, ants put down food they are carrying.

After ants perform their *laypheromone* behaviors, τ_{ij} and p_{ij} of their environment will be updated as follows [7,8]:

$$\tau_{ij}(t + 1) = \rho\tau_{ij}(t) + \Delta\tau_{ij}(t, t + 1), \qquad (2)$$

where ρ is an constant to denote the evaporation rate of pheromone; $\Delta\tau_{ij}(t, t+1)$ denotes the amount of pheromone laid by ants that travel between cities i and j at time interval $[t, t + 1]$,

$$\Delta\tau_{ij}(t, t + 1) = \sum_{k=1}^{m} \Delta\tau_{ij}^k(t, t + 1). \qquad (3)$$

where m is the total number of ants that travel between cities i and j at time interval $[t, t+1]$; $\Delta\tau_{ij}^k(t, t+1)$ denotes the amount of pheromone laid by the kth ants.

$$p_{ij}(t) = \frac{[\tau_{ij}(t)]^\alpha[\eta_{ij}]^\beta}{\sum_{j=1}^{n}[\tau_{ij}(t)]^\alpha[\eta_{ij}]^\beta}, \qquad (4)$$

where $\eta_{ij} = 1/d_{ij}$, called *visibility* between cities i and j; α and β are two constants to control the relative importance of pheromone intensity and visibility.

4.4   Interactions

In an ant colony, ants interact with each other through their environment. Specifically, while walking ants will lay down pheromone over their routes. When other ants encounter these routes, they will have higher probabilities to follow these routes. In this sense, we can argue that in an ant colony the followers interact with their forthgoers through the pheromone trails laid by their forthgoers.

4.5   Summary

This section showed how AOC works in a natural system. In an ant colony, individual ants performs their simple behaviors – foraging for food. In doing so, they autonomously make decisions on choosing routes to travel. While walking, they will lay down pheromone on their routes. Consequently, those routes will have higher probabilities to be choosen by other ants. Since more ants choose to travel through these routes, these routes will be laid more and more pheromone. The more pheromone, the more ants will travel through these routes. The more ants traveling through certain routes, the more pheromone will be laid on them. This will form a positive feedback, through which the ant colony can gradually find the shortest route between two places (e.g., their nest and a food source).

5   Autonomy in a Man-Made System: Boids

5.1   Scenario

Flocks of birds do not have a leader to decide where to go. But they yet are able to fly as a 'mass' as if they have planned their route already. Such behaviors clearly cannot be precisely modeled by a top-down approach, e.g., differential equations. Instead, Reynolds [19] has simulated perfectly these behavior in the well-known boid model. Fig. 2 presents a snapshot of the boids system. Each individual in the model, called *boid*, is characterized by three simple steering behaviors that govern its movements. These steering behaviors are based on the positions and velocities of a boid's neighboring flockmates [18,19]:

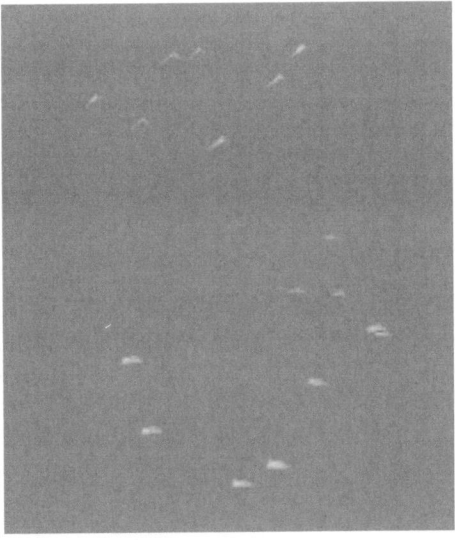

Fig. 2. A snapshot of the boids system

- **Separation**: steer to avoid crowding local flockmates;
- **Alignment**: steer towards the average heading of local flockmates;
- **Cohesion**: steer to move toward the average position of local flockmates.

In the boids system, neighborhood is defined by distance and the angle from the direction of flight. It limits the 'visibility' of boids in a way similar to the murky waters obscure the vision of fishes beyond a certain distance. Advanced features such as predictive obstacle avoidance and goal seeking are added to simulate some behaviors. The boid model has been used to produce animation in films. Similar idea has been used to visualize information cluster [18].

5.2   Environment

In the boids system, environment E plays only the first role, namely, the place where boids reside.

5.3   Computational Entities: Boids

Boids' Goal. The boids system is only designed to emulate the flocking, navigating, gathering etc. behaviors of birds. The goal of each individual boid is $G = \{$keeping in the flock while flying$\}$.

Boids' States. The state of boid b is characterized by five parameters, $S_b = \{p_b, v_b, d_b, r_b, a_b\}$. Specifically, p_b, v_b, d_b, r_b, and a_b are the position, velocity, direction, eyeshot radius, and eyeshot angle of boid b, respectively. Here, r_b and a_b are two constants used to define the eyeshot of boid b. b will ignore the boids beyond its eyeshot.

Boids' Behaviors. In the boids system, boid b performs its three steering behaviors simultaneously at each step. While doing so, its state information about position p_b and velocity v_b is changed as follows:

Velocity v_b :

$$v_b(t+1) = v_b(t) + \Delta v_S(t) + \Delta v_A(t) + \Delta v_C(t) \qquad (5)$$

where $\Delta v_S(t)$, $\Delta v_A(t)$, and $\Delta v_C(t)$ are three changes in b's velocity caused by its three behaviors, respectively.

$$\Delta v_S(t) = \alpha \cdot \sum_{i=1}^{n} p_{ib}(t), \qquad (6)$$

This change is caused by the *separation* behavior of boid b. Here, α is a coefficient; n is the total number of other boids in b's eyeshot; and

$$p_{ib}(t) = \begin{cases} p_i(t) - p_b(t) & \text{if } |p_i(t) - p_b(t)| < \delta \\ 0 & \text{otherwise} \end{cases} \qquad (7)$$

Note that δ is a threshold to judge if or not a boid is excessively close to boid b.

$$\Delta v_A(t) = \beta \cdot \left(\frac{\sum_{i=1}^{n} v_i(t)}{n} - v_b(t) \right). \qquad (8)$$

It is caused by the *alignment* behavior of boid b. Here, β is a coefficient.

$$\Delta v_C(t) = \gamma \cdot (c_b(t) - p_b(t)) \tag{9}$$

is caused by the *cohesion* behavior of boid b. Here, γ is a coefficient; $c_b(t)$ is the center of boids in b's eyeshot,

$$c_b(t) = \frac{\sum_{i=1}^n p_i(t)}{n}. \tag{10}$$

Position p_b :

$$p_b(t+1) = p_b(t) + v_b(t) \cdot \Delta(t) \tag{11}$$

5.4   Interactions

In the boids system, boids can interact with others in their neighboring regions: They can directly get the state information of their neighbors.

5.5   Summary

In this section, we showed how AOC works in the boids system, to emulate the flocking behavior of birds. In the system, boids use three simple behaviors to keep in the flock while flying. They can directly interact with others in their neighboring regions to obtain their information. Because of the interactions, boids can self-organize themselves to exhibit some complex collective behaviors, such as gathering and homing.

6   Autonomy in a Computational System: MASSAT

The previous two sections provided two perfect examples, which employ different kinds of interactions among entities, to show that AOC can be used to simulate complex collective behaviors in social animal colonies. As inspired by these, one can ask whether or not AOC can be used to solve some computational problems, such as constraint satisfaction problems and satisfaction problems. In fact, this question has been answered by Liu *et al.* in [11,12,14,15]. In this section, we will use an example, i.e., MASSAT to show how AOC works in solving computational problems.

As inspired by the ERA approach [14,15], Jin and Liu [11,12] have presented an autonomy oriented approach, called MASSAT, to solving satisfiability problems (SATs). This approach is intended to provide a multi-agent-based SATs solver. The following is a brief, but formal, introduction to SATs.

Definition 7. A *Satisfiability Problem* (SAT), P, consists of:

1. A finite set of propositional variables, $\mathbf{X} = \{X_1, X_2,..., X_n\}$;
2. A domain set, $\mathbf{D} = \{D_1, D_2,..., D_n\}$, for all $i \in [1, n]$, $X_i \in D_i$ and D_i $=\{True, False\}$;
3. A clause set, $\mathbf{CL} = \{Cl(R_1), Cl(R_2), ..., Cl(R_m)\}$, where R_i is subset of X, and clause $Cl(R_i)$ is a conjunction or disjunction of literals corresponding to the variables in R_i.

Definition 8. The solution, S, to a SAT is an assignment to all variables such that, under this assignment, the truth values of all given clauses are true, i.e.,

1. S is an ordered set, $S = \langle v_1, v_2, ..., v_n \rangle$, for all $i \in [1, n]$, v_i is equal to $True$ or $False$, $S \in D_1 \times D_2 \times ... \times D_n$;
2. $\forall j \in [1, m], T(Cl(R_j)) = True$, where $T(\cdot)$ is a function that returns the truth value of a clause.

6.1   Scenario

MASSAT is essentially to create a multiagent system for solving SATs. In this system, each agent represents a group of variables. These variables in turn construct a local space where this agent resides. Each position in the local space specifies values for these variables. An agent freely moves within its own local space. To solve a given SAT problem, all agents need some kind of coordination among each other. In the MASSAT formulation, we use a coordination mechanism, called *coordination without communication*, for agents to achieve their coordination. To do so, in our approach, we employ an information sharing mechanism, i.e., a shared blackboard. Each agent can write its position information onto the blackboard and read that of other agents from it. At each step, an agent will first select a certain behavior. If the selected behavior needs the agent to evaluate its position(s), the agent will read the position information of other agents from the blackboard. Next, it will evaluate its current position. If it is not a 'good' position, it will select another one in its local space according to the selected behavior, and then move to it. All agents will proceed the above process until they all find a 'good' position, i.e., the system reaches a solution state. In this situation, combining the positions of all agents, we can obtain a solution to the given SAT problem. Accordingly, agents stop moving.

In the following descriptions, we assume that we divide n variables into u groups. $G_i = \{X_{i1}, X_{i2}, ..., X_{ik}\}$ is the ith variable group. The domain set corresponding to G_i is denoted by $\{D_{i1}, D_{i2}, ...D_{ik}\}$.

6.2   Environment

In MASSAT, all agents are homogeneous. Each agent represents a group of variables, say, agent a_i represents variable group G_i. Therefore, the number of agents is equal to the number of variable groups, u. Agent a_i inhabits in the corresponding *local space* s_i.

Definition 9. The *local space* s_i of agent a_i is a row vector $\langle p_1, p_2, ..., p_{|D_{i1} \times D_{i2} \times ... \times D_{ik}|} \rangle$ which is a permutation of all elements in the Cartesian product of G_i. $\forall j \in [1, |D_{i1} \times D_{i2} \times ... \times D_{ik}|]$, p_j is a cell, which is called a *position*.

Local space s_i only belongs to agent a_i. It is the environment for agent a_i to inhabit. By combining the local spaces of all agents together, we get the global environment E of the whole system.

Definition 10. An *environment*, E, is a column vector composed by the local spaces of all agents, i.e.,

$$E = \begin{pmatrix} s_1 \\ s_2 \\ ... \\ s_u \end{pmatrix}$$

In E, we use (j, i) to index the jth position in local space s_i. Positions in E are not only used for an agent to stay, but also used to record *domain values* and *evaluation values*.

Definition 11. The *domain value* recorded in position (j, i), denoted with $(j, i).value$, is the jth element in the Cartesian product of $D_{i1} \times D_{i2} \times ... \times D_{ik}$.

If $(j, i).value = \langle x_{i1}, x_{i2}, ..., x_{ik} \rangle$ (where $\forall x_{iq}, q \in [1, k], x_{iq} \in D_{iq}$), and agent a_i stays at (j, i), it means: $X_{i1} = x_{i1}$, $X_{i2} = x_{i2}$,..., and $X_{ik} = x_{ik}$.

Definition 12. The *evaluation value* related to a position (j, i), denoted by $(j, i).evaluation$, is a number that reflects the desirability for agent a_i to stay at position (j, i).

Between two evaluation values, there exists a relationship, $\succeq$. If $x \succeq y$, it means evaluation value x is better than or equal to evaluation value y. Here, 'better than' is a notion depending on the specific meaning of the evaluation value.

To evaluate a position, an agent computes an evaluation function $\Phi(\mathbf{X})$. In local search, one of the most important issues is how to select the next variable to flip based on an evaluation function. To do so, an appropriate function is necessary to evaluate the current state. Like the local search, in the MASSAT approach, how to select the next position for an agent to move to is also important. A good evaluation function $\Phi(\mathbf{X})$ is crucial, because it can make the solution space search process more efficient.

A commonly used evaluation function is the number of satisfied clauses. In this case, $(x, i).evaluation \succeq (y, i).evaluation$ means that if agent a_i stays at position (x, i), more clauses will be satisfied than at position (y, i), or at least the numbers of satisfied clauses will be equal.

6.3   Computational Entities: Agents

Agents' Actions. After all agents are put into their own local spaces, in order to find a solution for a given SAT problem, all agents will be activated. Basically, we can reduce the behaviors of an agent to the following basic actions:

1. **Select** a behavior;
2. If the selected behavior needs the position information of other agents, **Read** the position information from a blackboard;
3. **Evaluate** all (in the case of best-move) or some (in the case of better-move) positions in its local space, or skip this step if it selects a random-move;
4. **Select** a best-position (in the case of best-move) or a better-position (in the case of better-move) or a random position (in the case of random-move);
5. **Move** to the new position;
6. **Write** its own position information onto the blackboard.

All agents will repeat the above actions, until the system finds that all clauses are satisfied, or a certain predefined termination condition is reached.

Agents' Behaviors. Now, let us consider the autonomous behaviors of agents in the MASSAT approach. First, we give the definitions of *best-position* and *better-position*.

Definition 13. Position (j, i) in local space s_i is called:

- *best-position*: if $\forall x \in [1, |D_{i1} \times D_{i2} \times ... \times D_{ik}|], (j, i).evaluation \succeq (x, i).$ *evaluation*;
- *better-position*: if (x, i) is agent a_i's current position, and $(j, i).evaluation \succeq (x, i).evaluation$.

In MASSAT, we have designed three autonomous behaviors: *best-move*, *better-move*, and *random-move*. If necessary, these three basic behaviors can be combined into more complicated behaviors [15]. At each step, an agent can probabilitically choose one to perform.

A best-move means that an agent moves to a *best-position* with a probability *best-p*. To perform a best-move, an agent needs to evaluate all positions in its local space to find the best one. If there exists more than one best-position, the agent will randomly choose one. Generally speaking, this strategy is instinctive to all agents, because, in the views of individual agents, this strategy can more efficiently lead them to a solution.

If an agent selects a better-move, which means it will move to a *better-position* with a probability *better-p*, this agent will first randomly select a position and then compare its evaluation value with that of its current position to decide whether or not to move to this new position. Although a better-move may not be the best choice for an agent, the computational cost required for this strategy is much less than that of a best-move, because only two operations are involved in deciding this movement, i.e., randomly selecting a position and performing a comparison.

With the autonomous behaviors of best-move and better-move alone, the system may readily get stuck in local optima and cannot escape from them. In the state of local optima, all agents are at best-positions, but it is not a solution state. And in this case, no agent will move to a new position. The agents will lose their chance for finding a solution. To solve this problem, we introduce random-move to our system. A random-move means that an agent randomly selects a position with a probability *random-p* and moves to it. It is somewhat like a random-walk in local search. Obviously, the probability of random-move, i.e., *random-p*, should be relatively smaller than the probabilities of selecting best-move and better-move behaviors, i.e., *best-p* and *better-p*.

6.4   Interactions

In MASSAT, agents interact with each other via their environment, in particular, the blackboard. The movement of each agent will change evaluation values of the positions in its local space, and also change the values of the variables it represents. All these changes will be considered by some other agents at their next steps. In this way, an agents transit its 'information' to others and consequently influences their movements at the next steps.

6.5   Summary

This section provided MASSAT as an example to show how AOC solves satisfiability problems. In fact, AOC has been successfully employed to solve other computational problems, such as traveling salesman problems and constraint satisfaction problems. These experimentally proves that besides emulating complex collective behaviors of social animals, AOC can also be used to solve computational problems.

7   Conclusion

Liu in [13] proposed autonomy oriented computation as a novel paradigm for characterizing collective behaviors of a complex system and solving some hard computational problems. In this chapter, we have tried to generalize autonomy oriented computation. Specifically, we identified and characterized the application-independent features of the main components of AOC systems, i.e., their entities and environments. We argued that the environment of an AOC system plays two roles: a place where entities resides and a communication media among entities. We defined an entity in an AOC system as an tuple of state S, goal set G, and behavior set B. Further, we identified the self-organizing process of entities in AOC systems as the core of the systems. All collective behaviors of a complex system come from the self-organization of entities. We also addressed the autonomy embodied in AOC systems and compared them with those in other work. We argued that in AOC systems, computational entities autonomously make decisions and behave in their environments. And, the whole AOC systems are also autonomous. After that, we using three examples, a natural system (i.e., ant colonies), a man-made system (i.e., boids), and a computational system (i.e., MASSAT) showed (1) how AOC is embodied in specific applications? and (2) how AOC emulates collective behaviors of social animals and solves hard computational problems.

References

1. K. S. Barber and C. E. Martin. Agent autonomy: Specification, measurement, and dynamic adjustment. In *Proceedings of the Autonomy Control Software Workshop at Autonomous Agents 1999 (Agents99)*, pages 8–15, Seattle, USA, May 1999.
2. S. Brainov and H. Hexmoor. Quantifying relative autonomy in multiagent interaction. In *The IJCAI-01 Workshop on Autonomy, Delegation, and Control: Interacting with Autonomous Agents*, pages 27–35, Seatlle, USA, 2001.
3. C. Castelfranchi. Guaranties for autonomy in cognitive agent architecture. In N. Jennings and M. Wooldridge, editors, *Agent Theories, Architectures, and Languages*, pages 56–70. Spinger-Verlag, 1995.
4. A. C. da Rocha Costa and G. P. Dimuro. Needs and functional foundation of agent autonomy.
5. D. N. Davis. Cellular automata, computational autonomy and emotion. In *Proceedings of the International Conference on Computational Intelligence for Modeling, Control, and Automata (CIMCA 2001)*, Las Vegas, USA, July 2001.
6. D. N. Davis and S. C. Lewis. Computational models of emotion for autonomy and reasoning. *Informatica*, 27(2):159–165, 2003.
7. M. Dorigo, G. D. Caro, and L. M. Gambardella. Ant algorithms for discrete optimization. *Artificial Life*, 5(2):137–172, 1999.
8. M. Dorigo, V. Maniezzo, and A. Colorni. The ant system: Optimization by a colony of cooperating agents. *IEEE Transactions on Systems, Man, and Cybernetics*, 26(1):1–13, 1996.
9. G. Dworkin. *The Theory and Practice of Autonomy*. Cambridge University Press, 1988.
10. H. Hexmoor. Stages of autonomy determination. *IEEE Transactions on Systems, Man, and Cybernetics*, 31(4):509–517, November 2001.
11. X. Jin and J. Liu. Multiagent SAT (MASSAT): Autonomous pattern search in constrained domains. In *Proceedings of the Third International Conference on Intelligent Data Engineering and Automated Learning (IDEAL'02)*, pages 318–328, 2002.
12. X. Jin and J. Liu. An autonomy-oriented, distributed approach to satisfiability problems. To be submitted, 2003.
13. J. Liu. *Autonomous Agents and Multi-Agent Systems: Explorations in Learning, Self-Organization and Adaptive Computation*. World Scientific, 2001.
14. J. Liu and J. Han. Alife: A multi-agent computing paradigm for constraint satisfaction problems. *International Journal of Pattern Recognition and Artificial Intelligence*, 15(3):475–491, 2001.
15. J. Liu, H. Jing, and Y.-Y. Tang. Multi-agent oriented constraint satisfaction. *Artificial Intelligence*, 136(1):101–144, 2002.
16. J. Liu and K. C. Tsui. Autonomy oriented computation. Technical Report COMP-03-003, Department of Computer Science, Hong Kong Baptist University, 2003.
17. J. Liu, K. C. Tsui, and J. Wu. Introducing autonomy oriented computation. In *Proceedings of the First International Workshop on Autonomy Oriented Computation*, pages 1–11, 2001.
18. G. Proctor and C. Winter. Information flocking: Data visualisation in virtual worlds using emergent behaviours. In *Proceedings of First International Conference on Virtual Worlds*, pages 168–176, Berlin, Germany, 1998. Springer-Verlag.
19. C. W. Reynolds. Flocks, herds, and schools: A distributed behavioral model. *Computer Graphics*, 21(4):25–34, 1987.

20. X. Tu and D. Terzopoulos. Artificial fishes: physics, locomotion, perception, behavior. In *Proceedings of the 21st annual conference on computer graphics and interactive techniques*, pages 43–50, New York, USA, 1994.
21. F. J. Varela. *Principles of Biological Autonomy*. The Netherlands: North-Holland, 1979.
22. G. Weiss, M. Rovatsos, and M. Nickles. Capturing agent autonomy in roles and XML. In *Proceedings of the Second International Joint Conference on Autonomous Agents and Multi-Agent Systems (AAMAS'03)*, pages 105–112, Melbourne, Australia, 2003. ACM Press.

Toward Quantum Computational Agents

Matthias Klusch

German Research Center for Artifical Intelligence, Deduction and
Multiagent Systems, Stuhlsatzenhausweg 3, D-66123 Saarbrücken, Germany
klusch@dfki.de

Abstract. In this chapter, we provide some first thoughts on, and preliminary answers to the question how intelligent software agents could
take most advantage of the potential of quantum computation and communication, once practical quantum computers become available in foreseeable future. In particular, we discuss the question whether the adoption of quantum computational and communication means will affect the
autonomy of individual and systems of agents. We show that the ability
of quantum computing agents to perform certain computational tasks
more efficient than classically computing agents is at the cost of limited
self-autonomy, due to non-local effects of quantum entanglement.

1   Introduction

Quantum computing technology based on quantum physics promises to eliminate
some of the problems associated with the rapidly approaching ultimate limits
to classical computers imposed by the fundamental law of thermodynamics. According to Gordon Moore's first law on the growth rate of classical computing
power, and the current advances in silicon technology, it is commonly expected
that these limits will be reached around 2020. By then, the size of microchip components will be on the scale of molecules and atoms such that quantum physical
effects will dominate, hence irrevocably require effective means of quantum computation.

Quantum physics has been developed in the early 1920's by physicists and
Nobel laureates such as Max Planck, Niels Bohr, Richard Feynman, Albert Einstein, Werner Heisenberg, and Erwin Schrödinger. It uses quantum mechanics
as a mathematical language to explain nature at the atomic scale. In quantum
mechanics, quantum objects including neutrons, protons, quarks, and light particles such as photons can display both wave-like and particle-like properties that
are considered as complementary. In contrast to macroscopic objects of classical
physics, any quantum object can be in a superposition of many different states
at the same time that enables for quantum parallelism. In particular, it can exhibit interference effects during the course of its unitary evolution, and can be
entangled with other spatially separated quantum objects such that operations
on one of them may cause non-local effects that are impossible to realize by
means of classical physics.

It has been proven that quantum computing can simulate classical computing.
However, the fundamental raison d'être of quantum computation is the fact

M. Nickles, M. Rovatsos, and G. Weiss (Eds.): AUTONOMY 2003, LNAI 2969, pp. 170–186, 2004.
© Springer-Verlag Berlin Heidelberg 2004

that quantum physics appears to allow one to transgress the classical boundary between polynomial and exponential computations [25]. Though there is some evidence for that proposition, only very few practical applications of quantum computing and communication have been proposed so far including quantum cryptography [5].

Quantum computing devices have been physically implemented since the late 1990's by use of, for example, nuclear magnetic resonance [43], and solid state technologies such as that of neighbouring quantum dots implanted in regions of silicon based semiconductor on the nanometer scale [27]. As things are now, they work for up to several tens of qubits. Whether large-scale fault-tolerant and networked quantum computers with millions of qubits will ever be built remains purely speculative at this point. Though, rapid progress and current trends in nanoscale molecular engineering, as well as quantum computing research carried out at research labs across the globe could make it happen to let us see increasingly sophisticated quantum computing devices in the era 2020 to 2050. This leads, in particular, to the question how intelligent software agents [46,45] could take most advantage of the potential of quantum computation and communication, once practical quantum computers are available. Will quantum computational agents be able to outperform their counterparts on classical von-Neumann computers? What kinds of architectures and progamming languages are required to implement them? Does the adoption of quantum computational and communication means affect the autonomy of individual and systems of agents? This chapter provides some first thoughts on, and preliminary answers to these questions based on known fundamental and recent results of research in quantum computing and communication. It is intended to help bridging the gap between the agent and quantum research community for interdisciplinary research on quantum computational intelligent agents.

In sections 2 and 3, we briefly introduce the reader to the basics of quantum information, computation and communication in terms of quantum mechanics. For more comprehensive and in-depth introductions to quantum physics, and quantum computation we refer the interested reader to, for example, [12], respectively, [33,23,42,1]. [17] provides a well-readable discussion of alternative interpretations of quantum mechanics. Readers who are familiar with the subjects can skip these sections. In section 4, we outline an architecture for a hybrid quantum computer, and propose a conceptual architecture and examples of quantum computational agents for such computers in section 5. Issues of quantum computational agent autonomy are discussed in section 6.

2   Quantum Information

Quantum computation is the extension of classical computation to the processing of quantum information based on physical two-state quantum systems such as photons, electrons, atoms, or molecules. The unit of quantum information is the quantum bit, the analogous concept of the bit in classical computation.

2.1   Quantum Bit

Any physical two-state quantum system such as a polarized photon can be used to realize a single *quantum bit* (qubit). According to the postulates of quantum mechanics, the state space of a qubit ψ is the 2-dimensional complex Hilbert space $H_2 = \mathbb{C}^2$ with given orthonormal computational basis in which the state $|\psi >$ is observed or measured[1]. The standard basis of qubit state spaces is $\{|0 >, |1 >\}$ with coordinate representation $|0 >= (1,0)^t$, and $|1 >= (0,1)^t$. Any *quantum state* $|\psi >$ of a qubit ψ is a coherent *superposition* of its basis states

$$|\psi >= \alpha_0|0 > +\alpha_1|1 > \tag{1}$$

where the probability *amplitudes* $\alpha_1, \alpha_2 \in \mathbb{C}$ satisfy the normalization requirement $|\alpha_0|^2 + |\alpha_1|^2 = 1$ for classical probabilities $p(|\psi >= |0 >) \equiv p(0) = |\alpha_1|^2$, respectively, $p(|\psi >= |1 >) \equiv p(1) = |\alpha_2|^2$ of the occurrence of alternative basis states[2]. The decision of the physical quantum system realizing the qubit on one of the alternatives is made non-deterministically upon irreversible measurement in the standard basis. It reduces the superposed qubit state to the bit states '0' and '1' in classical computing. This transition from the quantum to the observable macroscopic world is called *quantum decoherence*.

2.2   Quantum Bit Register

A *n-qubit register* $\psi = \psi_1...\psi_n$ of n qubits $\psi_i, i \in \{1, ..n\}$ is an ordered, composite n-quantum system. According to quantum mechanics, its state space is the n-folded *tensor (Kronecker) product* $H_2^{\otimes n} = \overbrace{H_2 \otimes ... \otimes H_2}^{n}$ of the (inner product) state spaces H_2 of its n component qubits. Each of the 2^n n-qubit basis states $|x_i >, x_i \in \{0,1\}^n$ of the register can be viewed as the binary representation of a number k between 0 and $2^n - 1$. Any *composite state* of a n-qubit register is in a superposition of its basis states

$$|\psi >= |\psi_1\psi_2...\psi_n >= \sum_{k=0}^{2^n-1} \alpha_k|k >, \quad \sum_{k=0}^{2^n-1} |\alpha_k|^2 = 1 \tag{2}$$

As the state of any n- and m-qubit register can be described by 2^n, respectively, 2^m amplitudes, any distribution on the joint state space of the $n + m$-qubit register takes 2^{n+m} amplitudes. Hence, in contrast to classical memory, quantum

---

[1]   Paul Dirac's bra-ket notation $< \psi| = (\alpha_1, ..., \alpha_k)^T$ (bra) and $|\psi >= (\alpha_1^\square,, ..., \alpha_k^\square)$ (ket) with complex conjugates $\alpha_i^\square, i \in \{1, .., k\}$ is the standard notation for system states in quantum mechanics. The inner product of quantum state vectors in H_k is defined as $< \psi_1|\psi_2 >= (\alpha_i^\square)_{i\square\ \{1,..,k\}} \otimes (\beta_i)_{i\square\ \{1,...,k\}} = \sum_{i=1}^{k} \alpha_i^\square\beta_i$. The orthonormal basis of H_k can be chosen freely, but if fixed refers to one physical observable of the quantum system ψ such as position, momentum, velocity, or spin orientation of a polarized photon, that can take k values.

[2]   In contrast to physical probabilistic systems, a quantum system can destructively interfere with itself which can be described by negative amplitude values.

memory increases exponentially in the size of the number of qubits stored in a quantum register. It can be doubled by adding just one qubit.

2.3   Measurement of Qubits

Measurement of a n-qubit register ψ in the standard basis yields a n-bit post-measurement quantum state $|\psi_k>$ with probability $|\alpha_k|^2$. Measurement of the first $z < n$ qubits corresponds to the orthogonal measurement with 2^z projectors $M_i = |i><i| \otimes I_{2^{n-z}}, i \in \{0,1\}^z$ which collapses it into a probabilistic classical bit vector, yielding a single state randomly selected from the exponential set of possible states[3]. Measurement of the individual qubit ψ_m of a n-qubit register $\psi = \psi_1...\psi_m...\psi_n, n \geq m$ in compound state $|\psi> = \sum_{i=0}^{2^n-1} c_i |i_1..i_n>$ with measurement operator M_m will give the classical outcome $x_m \in \{0,1\}$ with probability $p(x_m) = \sum_{i_1..i_n} |c_{i_1..i_{m-1}xi_{m+1}..i_n}|^2 = <\psi|M_m^* M_m|\psi>$, and post-measurement state is

$$|\psi>' = \frac{1}{\sqrt{p(x_m)}} \sum_{i_1..i_{m-1}i_{m+1}..i_n} c_{i_1..i_{m-1}xi_{m+1}..i_n} |i_1..i_{m-1}xi_{m+1}..i_n>$$

where $c_{i_1..i_{m-1}xi_{m+1}..i_n}$ denote the amplitudes of those 2^n alternatives for which x could be observed as state value of the m-th qubit of ψ upon measurement. In general, the post-measurement quantum state $|\psi_k>'$ of $|\psi_k>$ is $\frac{M_m|\psi_k>}{\sqrt{<\psi|M_m^* M_m|\psi>}}$.

2.4   Unitary Evolution of Quantum States

According to the postulates of quantum mechanics, the time evolution of any n-qubit register, $n \geq 1$, is determined by any linear, unitary[4] operator U in the 2^n-dimensional Hilbert space $H_2^{\otimes n}$. The size of the unitary matrix of a n-qubit operator is $2^n \times 2^n$, hence exponential in the physical size of the system. Since any unitary transformation U has an inverse $U^{-1} = U^*$, any non-measuring quantum operation is reversible, its action can always be undone. Measurement of a qubit ψ is an irreversible operation since we cannot reconstruct its state $|\psi>$ from the observed classical state after measurement.

2.5   Entangled Qubits

Entangled n-qubit register states cannot be described as a tensor product of its component qubit states. Central to entanglement is the fact that measuring one of the entangled qubits can affect the probability amplitudes of the other

---

[3]   According to the *standard interpretation of quantum mechanics* it is meaningfully to attribute a definite state to a qubit only *after* a precisely defined measurement has been made. Due to Heisenberg's *uncertainty principle* complementary observables such as position and momentum cannot be exactly determined at the same time.

[4]   *Unitarity* preserves the inner product ($<\phi|U^\bullet U|\psi>$), similar to a rotation of the Hilbert space that preserves angles between state vectors during computation.

entangled qubits no matter how far they are spatially separated. Such kind of non-local or holistic correlations between qubits captures the essence of the *non-locality principle of quantum mechanics* which has been experimentally verified by John Bell in 1964 [3] but is impossible to realize in classical physics.

Example 2.1: *Entangled qubits*
Prominent examples of entangled 2-qubit are the *Bell states*

$$|\psi^+> = \frac{1}{\sqrt{2}}((|01> +|10 >), |\phi^+> = \frac{1}{\sqrt{2}}((|00 > +|11 >),$$

$$|\psi^\square> = \frac{1}{\sqrt{2}}((|01 > -|10 >), |\phi^\square> = \frac{1}{\sqrt{2}}((|00 > -|11 >)$$

The Bell state $|\phi^+> = (\square\frac{1}{2}, 0, 0, \square\frac{1}{2})$ is not decomposable. Otherwise we could find amplitudes of a 2-qubit product state $(\alpha_{11}|0 > +\alpha_{12}|1 >)(\alpha_{21}|0 > +\alpha_{22}|1 >)$ $= \alpha_{11}\alpha_{21}|00 > +\alpha_{11}\alpha_{22}|01 > +\alpha_{12}\alpha_{21}|10 > +\alpha_{12}\alpha_{22}|11 >$ such that $\alpha_{11}\alpha_{21} = \square\frac{1}{2}$, $\alpha_{11}\alpha_{22} = 0$, $\alpha_{12}\alpha_{21} = 0$ and $\alpha_{12}\alpha_{22} = \square\frac{1}{2}$ which is impossible. We cannot reconstruct the total state of the register from the measurement outcomes of its component qubits. $|\phi^+ >$ can be produced by applying the conditioned-not 2-qubit operator $M_{cnot} = ((1,0,0,0), (0,1,0,0), (0,0,0,1), (0,0,1,0))$ to the separable register state $|\psi_1\psi_2 >= \square\frac{1}{2}(|00 > +|10 >)$.

Suppose we have measured 0 as definite state value of the second qubit in state $|\phi^+ >\equiv a|00 > +b|01 > +c|10 > +d|11 >\equiv (a, b, c, d)$ with amplitudes normalized to 1. The corresponding measurement operator is the self-adjoint, non-unitary projector $M_{2:0} = ((1,0,0,0), (0,0,0,0), (0,1,0,0), (0,0,0,0))$, which yields the outcome 0 or 1 with equal probability, for example, $p(0) = < \phi^+|M_{2:0}^\square M_{2:0}|\phi^+ >= (\square\frac{1}{2}, 0, 0, \square\frac{1}{2})(\square\frac{1}{2}, 0, 0, 0)^t = \frac{1}{2}$, and the post-measurement state $|\phi^+ >\square= \frac{M_{2:0}|\psi>}{\sqrt{<\psi|M_{2:0}|\psi>}}$

$$= \frac{(\frac{1}{\sqrt{2}}, 0, 0, 0)}{\sqrt{1/2}} = \sqrt{2}(\square\frac{1}{2}, 0, 0, 0) = (1, 0, 0, 0) = |00 >\neq a|00 > +c|10 >.$$

That means, measurement of the second qubit caused also the entangled first qubit to instantaneously assume a classical state without having operated on it.   o

Pairs of entangled qubits are called *EPR pairs* with reference to the associated Einstein-Podolsky-Rosen (EPR) thought experiment [20]. The *non-local effect of instantaneous state changes* between spatially separated but entangled quantum states upon measurement belongs to the most controversial issue and debated phenomenon of quantum physics, and caused interesting attempts of developing a quantum theory of the humand mind and brain [39,38]. Entanglement links information across qubits, but does not create more of it [22], nor does it allow to communicate any classical information faster than light.

Entangled qubits can be physically created either by having an EPR pair of entangled particles emerge from a common source, or by allowing direct interaction between the particles, or by projecting the state of two particles each from different EPR pairs onto an entangled state without any interaction between them (*entanglement swapping*) [12]. Entanglement of qubits is considered as one essential feature of, and resource for quantum computation and quantum communication [25,11].

3   Quantum Computation and Communication

The quantum Turing machine model [37], and the quantum circuit model [18] are equivalent models of quantum computation. In this paper, we adopt the latter model.

3.1   Quantum Logic Gates and Circuits

A *n-qubit gate* is a unitary mapping in $H_2^{\otimes n}$ which operates on a fixed number of qubits (independent of n) given n input qubits. Most quantum algorithms to date are described through a *quantum circuit* that is represented as a finite sequence of concatenated quantum gates. Basic quantum gates are the 1-qubit *Hadamard* (H) and *Pauli* (X, Y, Z) gates, and the 2-qubit XOR, called *conditioned not* (CNOT), gate. These operators are defined by unitary matrices as follows.

$$M_H = \frac{1}{\sqrt{2}}((1,1),(1,-1))$$
$$M_{CNOT} = ((1,0,0,0),(0,1,0,0),(0,0,0,1),(0,0,1,0))$$
$$M_X = ((0,1),(1,0)), M_Z = ((1,0),(0,-1)), M_Y = ((0,-i),(i,0))$$

The Hadamard gate creates a superposed qubit state for standard basis states, demonstrates *destructive quantum interference* if applied to superposed quantum states $(M_H(\frac{1}{\sqrt{2}}(|0>+|1>)) = |0>)$, and can be physically realized, for example, by a 50/50-beamsplitter in a Mach-Zehnder interferometer [12]. The CNOT gate flips the second (target) qubit if and only if the first (control) qubit is in state $|1>$. The quantum circuit consisting of a Hadamard gate followed by a CNOT gate creates an entangled *Bell state* for each computational basis state. The X gate is analogous to the classical bit-flip NOT gate, and the Z gate flips the phase (amplitude sign) of the basis state $|1>$ in superposition. Other common basic qubit gates include the NOP, S, and T gates for quantum operations of identity, phase rotation by $\pi/4$, respectively $\pi/8$. The set {H, X, Z, CNOT, T} is universal [33].

3.2   Quantum vs. Classical Computation

The constraint of unitary evolution of qubit states yields a generalization of the restriction of classical (Turing machine or logic circuit based) models of computation to unitary, hence reversible computation [4]. It has been shown that each classical algorithm computing a function f can be converted into an equivalent quantum operator U_f with the same order of efficiency [49,1], which means that quantum systems can imitate all classical computations. However, the fundamental raison d'être of quantum computation is the expectation that quantum physics allows one to do even better than that.

The linearity of quantum mechanics gives rise to *quantum parallelism* that allows a quantum computer to simultaneously evaluate a given function $f(x)$

for all inputs x by applying its unitary transformation $U_f : |x > |0 >\mapsto |x > |0 \oplus f(x) >= |x > |f(x) >$ to a suitable superposition of these inputs such that

$$U_f \left(\frac{1}{\sqrt{2^n}} \sum_{x \in \{0,1\}^n} |x > |0 > \right) = \frac{1}{\sqrt{2^n}} \sum_{x \in \{0,1\}^n} |x > |f(x) > \qquad (3)$$

Though this provides, in essence, not more than classical randomization, if combined with the effects of *quantum interference* such as in the Deutsch-Josza algorithm ([33], p.36) and/or quantum entanglement [11] it becomes a fundamental feature of many quantum algorithms for speeding-up computations. The basic idea is to compute some global property of f by just one evaluation based on a combination of interfered alternative values of f, whereas classical probabilistic computers only can evaluate different but forever mutually excluding alternative values of f with equal probability.

In general, *quantum algorithms* appear to be best at problems that rely on promises or *oracle* settings, hence use some hidden structure in a problem to find an answer that can be easily verified through, for example, means of amplitude amplification. Prominent examples include the quantum search developed by Grover (1996) for searching sets of n unordered data items [21], and the quantum prime factorization of n-bit integers developed by Shor (1994) [41] with complexity of $O(\sqrt{n})$, respectively, $O(n^3)$ time, which is a quadratic and exponential speed-up compared to the corresponding classical case. It is not known to date whether quantum computers are in general more powerful than their classical counterparts[5]. However, it is widely believed that the existence of an efficient solution of the NP-hard problem of integer prime factoring using the quantum computation model [41], as well as the quadratically speed up of classical solutions of some NP-complete problems such as the Hamiltonian cycle problem by quantum search ([33], p.264), provides evidence in favor of this proposition.

3.3 Quantum Communication Models

In this paper, we consider the following models of quantum based communication between two quantum computational agents A and B.

1. **QCOMM-1**. Agents A and B share entangled qubits and use a classical channel to communicate.
2. **QCOMM-2**. Agents A and B share entangled qubits and use a quantum channel to communicate.
3. **QCOMM-3**. Agents A and B share no entangled qubits and use a quantum channel to communicate.

---

[5] In terms of the computational complexity classes P, BPP, NP, and $PSPACE$ with $P \subseteq NP \subseteq PSPACE$, it is known that $P \subseteq QP$, $BPP \subseteq BQP$, and $BQP \subseteq PSPACE$ [10]. QP and BQP denote the class of computational problems that can be solved efficiently in polynomial time with success probability of 1 (exact), or at least 2/3 (bounded probability of error), respectively, on uniformly polynomial quantum circuits.

QCOMM-1: Quantum teleportation of n qubits with 2n bits. The standard process of teleporting a qubit ϕ from agent A to agent B based on a shared EPR pair $\psi_1\psi_2$ and classical channel works as follows [7]. Suppose agent A (B) keeps qubit ψ_1 (ψ_2). A entangles ϕ with ψ_1 by applying the CNOT, and the Hadamard gate to the 2-qubit register $[\phi\psi_1]$ into one of four Bell states $|\phi\psi_1>$. It then sends the measurement outcome (00, 10, 01, or 11) to agent B through a classical communication channel at the cost of two classical bits. Only upon receipt of A's 2-bit notification message, agent B is able to create $|\phi>$ by applying the identity or Pauli operator gates to its qubit ψ_2 depending on the content of the message (00: I, 01: X; 10: Z; 11: XZ)[6].

QCOMM-2: Quantum dense coding of n-bit strings in n/2 qubits. Agent A dense codes each of consecutive pairs of bits $b_1 b_2$ at the cost of one qubit as follows [8]. Suppose agent A (B) keeps qubit ψ_1 (ψ_2) of shared EPR pair in entangled Bell state $|\psi> = |\psi_1\psi_2> = \frac{1}{\sqrt{2}}(|00> + |11>)$. According to prior coding agreement with B, agent A applies the identity or Pauli operators to its qubit depending on the 2-bit message to be communicated (for example, 00: $I \otimes I|\psi>$, 01: $X \otimes I|\psi>$, 11: $Z \otimes I|\psi>$, 10: $(XZ)^t \otimes I|\psi>$) which results in one of four Bell states $|\psi>'$ and physically transmits the qubit ψ_1 to B. Upon receipt of ψ_1, agent B performs $M_{CNOT}|\psi>'$ yielding separable state $|\gamma_0\gamma_1>$, applies the Hadamard operation to the first qubit $M_H|\gamma_0> = |\delta_0>$ and decodes the classical 2-bit message depending on measured states of $\delta_0\gamma_1$ (e.g., $\delta_0\gamma_1 = 00$: 00, 01: 01, 11: 10, 10: 11).

A fundamental result in quantum information theory by Holevo (1973) [24] implies that by sending n qubits one cannot convey more than n classical bits of information. However, for every classical (probabilistic) communication problem [48] where agents exchange classical bits according to their individual inputs and then decide on an answer which must be correct (with some probability), quantum protocols where agents exchange qubits of communication are at least as powerful [31].

4   Quantum Computers

All known quantum algorithms require the determinism and reliability of classical control for the execution of suitable quantum circuits consisting of a finite sequence of quantum gates and measurement operations. Figure 1 shows a master-slave architecture of a *hybrid quantum computer* based on proposals in [35] and [9] in which classical signals and processing of a classical machine (CM) are used to control the timing and sequence of quantum operations carried out in a *quantum machine* (QM).

---

[6] Due to (Bell state) measurement of $|\phi\psi_1>$ agent A lost the original state $|\phi>$ to be communicated. However, since ψ_1 and ψ_2 were entangled, this measurement instantaneously affected the state of B's qubit ψ_2 (cf. Ex. 2.1) such that B can recover $|\phi>$ from $|\psi_2>$.

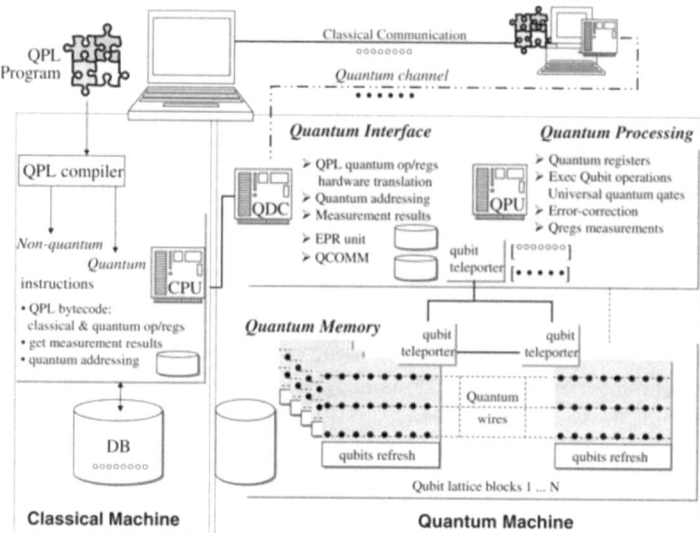

Fig. 1. Master-slave architecture of a hybrid quantum computer

The QM consists of *quantum memory*, *quantum processing unit* (QPU) with error correction, *quantum bus*, and *quantum device controller* (QDC) with interface to the classical machine (CM). The classical machine consists of a CPU for high-level dynamic control and scheduling of the QM components, and memory that can be addressed by both classical and quantum addressing schemes (e.g., [9] p.20, [33] p.268). Quantum memory can be implemented as a lattice of static physical qubits, which state is factorized in tensor states over its nodes[7]. Qubit states can be transported within the QM along point-to-point quantum wires either via teleportation (cf. section 3.3), or chained quantum swapping and repeaters [36][8].

A few *quantum programming languages* (QPL) for hybrid quantum computers exist, such as the procedural QCL [34], and QL [9], and the functional qpl [40]. A QPL program contains high-level primitives for logical quantum operations, interleaved with classical work-flow statements. The QPL primitives are compiled by the CPU into low-level instructions for qubit operators that are passed to and then translated by the QDC to physical qubit (register) operations which are executed by the QPU. The QPU performs scheduled sequences of measurement and basic qubit operations from a universal set of 1- and 2-qubit quantum

---

[7] According to the *no-cloning theorem* of quantum computing [47], a qubit state cannot be perfectly copied unless it is known upon measurement. Thus, no backup copies of quantum data can be created in due course of quantum computation.

[8] In short *quantum wires* a qubit state can be progressively swapped between pairs of qubits in a line, where each qubit is represented, for example, by the nuclear spin of a phosporus atom implanted in silicon (quantum dot). Each swap operation along this line of atoms is realized by three back-to-back CNOT gates.

gates (cf. section 3.1) with error correction[9] to minimize quantum decoherence caused by imperfect control over qubit operations, measurement errors, number of entangled qubits, and the pysical limits of the quantum systems such as nuclear spins used to realize qubits [19]. The QM returns only the results of quantum measurements to the CM.

5   Agents on Quantum Computers

5.1   QC Agents

A *quantum computational agent* (QCA) extends an intelligent software agent by its ability to perform both classical, and quantum computing and communication on a quantum computer to accomplish its goals individually, or in joint interaction with other agents. QC agents on hybrid quantum computers are coded in an appropriate QPL. The deliberative component of a QC agent uses sensed input, beliefs, actions, and plans that are classically or quantum coded depending on the kind of respective QPL data types and statements. The QPL agent program is executed on both the classical and the quantum machine in an interleaved master-slave fashion using the QPL interface of the quantum machine (cf. section 4).

QC agents are supposed to exploit the power of quantum computing to reduce the computational complexity of certain problems, where appropriate. For example, a *quantum computational information agent* (QCIA) is a special kind of QC agent which extends an intelligent information agent on a classical computer [29] by its ability to perform quantum computation and communication for information search and management tasks. How can a QCIA exploit oracle-based quantum search algorithms for searching local data or knowledge bases?

Local quantum based search. Suppose a QCIA has to search its local unstructured classical database LDB with $N = 2^n$ l-bit data entries d_x each of which is indexed by value $x = 0...N - 1$ for given l-bit input s and search oracle O with $1 \leq M \leq N$ solutions. The oracle is implemented by an appropriate quantum circuit U_f that checks whether the input is a solution to the search problem ($f(x) = 1$ if $d_x = s$, else $f(x) = 0$). No further structure to the problem is given. Any classical search would take an average of $O(N/M)$ oracle calls to find a solution. Using Grover's quantum search algorithm [21] the QCIA can do the same in $O(\sqrt{N/M})$ time. The basic idea is that (a) the search is performed on a $logN$-qubit index register $|x>$ which state is in superposition of all $N = 2^n$ index values x[10],and (b) the oracle O marks the M solutions ($|x> \rightarrow (-1)^{f(x)}|x>$ with $f(x) = 1$ if $d_x = s$, 0 else) which are amplified to increase the probability that

---

[9] According to the *threshold theorem* of quantum computing [28,2], scalable quantum computers with faulty components can be built by using quantum error correction codes as long as the probability of error of each quantum operation is less than 10^{-4}.

[10] The initial superposed index state $|x>$ is created by n-folded Hadamard operation $(H^{\otimes n}|0> = \frac{1}{2^n} \sum_{i \in \{0,1\}^n} |i>)$.

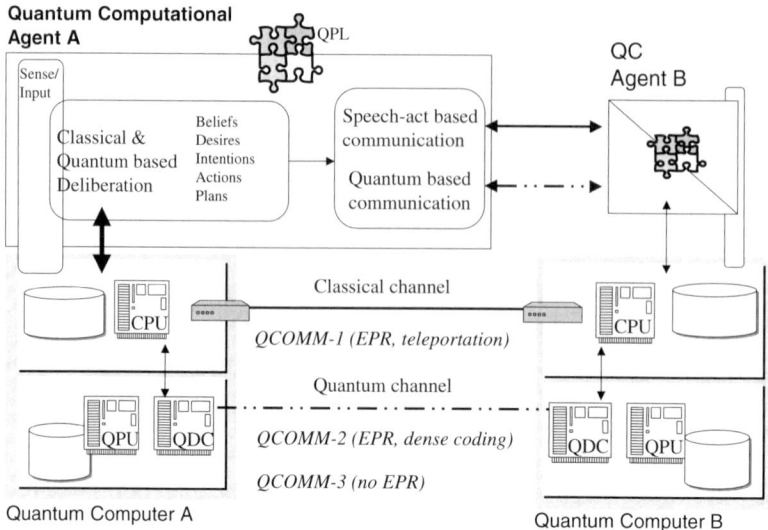

Fig. 2. Conceptual scheme of a quantum computational agent

they will be found upon measurement of the index register after $O(\sqrt{N/M})$ iterations. Type and cost of each oracle call (matching operation U_f) depends on the application. Implementation of the search uses n-qubit index, l-qubit data and input, and 1-qubit oracle register of the QPU. Like in the classical search, we need a quantum addressing scheme ([33] p.268) with $O(logN)$ per operation to access, load, and restore indexed data d_x to the data register, and recreate respectively measured index states $|x >$ for further processing.

Local quantum based matchmaking is a special case of local quantum search for binary coded service descriptions. The type of service matching depends on the implemented search oracle. We assume that both service requests and service ads are encoded in the same way to allow for meaningful comparison by a *quantum computational matchmaker agent*. Componentwise quantum search with bounded rather than exact success probability for partial matching could be used for syntactic but not semantic service matching such as in LARKS [44].

5.2   QC Multi-Agent Systems

A *quantum computational multi-agent system* (QCMAS) is a multi-agent system that consists of both classical and quantum computing agents which can interact to jointly accomplish their goals. A *pure QCMAS* consists of QC agents only. QCMAS which members cannot interact with each other using a quantum communication model (cf. section 3.3) are called *type-I QCMAS*, and *type-II*

QCMAS otherwise. QC agents of type-I or type-II QCMAS are called *type-I QC agents*, respectively, *type-II QC agents*.

Inter-agent communication between type-I QC agents bases on the use of classical channels without sharing any EPR pairs. None of the quantum communication models is applicable. As a consequence, quantum computation is performed locally at each individual agent. In addition, type-II QC agents can use quantum communication models which cannot be simulated in any type-I QCMAS. It is assumed that type-II QC agents share a sufficient number of EPR pairs, and have prior knowledge on used quantum coding operations for this purpose. Any QC agent communicates appropriate speech-act based messages via classical channels to synchronize its actions, if required. Messages related to quantum communication between type-II QC agents concern, for example, the notification in quantum teleportation (QCOMM-1), the prior agreement on the order of operations in quantum dense coding (QCOMM-2), and the semantics of qubits (QCOMM-3).

What are the main benefits of QCMAS? In certain cases, QCMAS can be computationally more powerful than any MAS by means of properly designed and integrated QC agents. The main challenge is the development of application-specific quantum algorithms that can do better than any classical algorithm.

Quantum-based communication between type-II QC agents is inherently secure. Standard quantum teleportation (QCOMM-1) ensures data integrity, since it is impossible to deduce the original qubit state from eavesdropped 2-bit notification messages of the sender without possessing the respective entangled qubit of the receiver. Quantum dense coding (QCOMM-2) is secure, since any quantum operation on the physically transmitted qubit in any of the four Bell states takes the same value. The physical transmission of qubits via a quantum channel (QCOMM-3) is secure due to the no-cloning theorem of quantum mechanics, a fact that is also used in quantum key distribution [5]. Any attempt of eavesdropping will reckognizably interfere with the physical quantum states of transmitted qubits.

Finally, certain communication problems [48], that is the joint computation of some boolean function f minimizing the number of qubits to communicate for this purpose, can be solved more efficiently by type-II QC agents. In general, it has been proven in [15] that the gap between bounded-error (zero-error) classical and (exact) quantum communication complexity is near quadratic (exponential), and that each quantum communication model is at least as powerful than a classical one for every communication problem on n-bit inputs [31]. More interesting, they can do even better for certain communication problems such as the computation of inner product, equality, and disjointness of boolean functions $f(x), g(x)$ according to individual n-bit inputs $x \in \{0, 1\}^n$. Quantum based solutions to latter problems can be applied to quantum based collaborative search, and matchmaking [30], with respective quadratic or exponential reduction of communication complexity.

5.3   Examples of Type-I and Type-II QCMAS

Quantum based collaborative search in type-I QCMAS. Upon receipt of a multi-casted l-bit request s from QCIA A_1, each agent $A_j, j = 2..n$ locally computes $M_j \geq 1$ solutions to the given search problem $LQS(s, O, LDB_j)$ in $O(\sqrt{N_j/M_j})$ time, instead of $O(N)$ in the classical case, and returns the found data items to A_1. Due to non-quantum based interaction, both requests and replies have to be binary coded for transmission via classical channel, and binary requests are directly1 quantum coded prior to quantum search (cf. section 5.1).

Quantum based collaborative search in type-II QCMAS. Suppose two QCIAs A_1, A_2 want to figure out whether a n-bit request s matches with data item $s' \in LDB_2$ ($N = 1$) with the promise that their Hamming distance is $h(s, s') = 0$ else $n/2$. In this case, it suffices to solve the corresponding equality problem with $O(logn)$ qubits of communication, instead of $O(n)$ in the classical case [15]. Basic idea is that A_1 prepares its n-bit s in a superposition of $logn+1$ qubits such that A_2 can test, upon receipt of s, whether $s_i \oplus s'_i = 0, i = 1..n$ by applying the known oracle-based Deutsch-Josza quantum algorithm ([33], p.34) to $|s> |o \oplus s \oplus s'>$, followed by Hadamard operations ($H^{\otimes(logn+1)}$), and measurement of the final state yields the desired result.

QC matchmaking in type-I and type-II QCMAS. As in the classical case, quantum based service matchmaking can be directly performed by pairs of QC agents in both types of QCMAS. In fact, it is a special case of the collaborative search scenario where two QC agents can both advertise and request a set of N (N') l−bit services from each other. For example, QC service agents A of a type-II QCMAS can physically send a set of (QCOMM-2: dense coded) n-bit service request each of size $n/2$ qubits to a QC matchmaker A^* via a quantum channel (QCOMM-3). In cases where only classical channels are available (QCOMM-1), A can teleport the qubit request to A^* at the cost of $2n$ bits. In any case, upon receipt of the request, A^* quantum searches its classical database of N service ads, and returns those that matches it according to the given search ("matching") oracle. Using quantum search, the disjointness of sets of quantum coded service descriptions interpreted as ads and/or requests can be decided with just $logN + 1$ qubits of communication [15], instead of at least N bits in the classical randomized setting [26].

6   Autonomy of QC Agents

Following the classification of different types of agent autonomy in [16], we define a QC agent A autonomous from QC agent B for given autonomy object o in the context c, if, in c, its behaviour regarding o is not imposed by B. The ability of an individual QC agent in type-I QCMAS to exhibit autonomous behaviour is not affected by its local quantum computation, since non-local effects are restricted to local quantum machine components. Hence, the self-autonomy of individual type-I QC agents in terms of the ability to autonomously reason about sets of goals, plans, and motivations for decision-making remains intact.

That is independent from the fact that the computational complexity of deliberative actions could possibly be reduced by, for example, quantum searching of complex plan libraries. Regarding user autonomy, any external physical interaction with the quantum machine by the user will cause massive quantum decoherence which puts the success of any quantum computational process and associated accomplishment of tasks and goals of individual type-I QC agents at risk.

A type-II QC agent shall be able to adjust its behaviour to the current quantum computing context of the overall task or goal to accomplish. It can freely decide on whether and with which agents to share a sufficient number of EPR pairs, or to make prior coding agreements according to the used quantum communication model. However, both its adjustable interaction and computational autonomy, turn out to be limited to the extent of entanglement based joint computation and communication with other type-II QC agents. Any type-II QC agent can change the state of non-local qubits that are entangled with its own qubits by local Bell state measurements. This way, if malevolent, it can misuse its holistic correlations with other type-II QC agents to corrupt their computations by manipulating their respective entangled quantum data. Even worse, there is no way for these agents to avoid such kind of influence.

For example, suppose that agents A and B share EPR pairs to interact using quantum teleportation (QCOMM-1). Since the change of B's entangled qubits caused by A's local Bell state measurements is instantaneous, B cannot avoid it at all. B does not even know that such changes occurred until it receives A's 2-bit notification messages (cf. section 3.3). B is not able to clone its entangled qubits, and measuring their state prior to A's notification would let communication fail completely. The same situation occurs when entanglement swapping is used to teleport qubit states along a path of correlated QC agents in a type-II QCMAS; in fact, it holds for any kind of entanglement based computation in general.

To summarize, the use of entanglement as a resource for computation and communication requires type-II QC agents to strictly trust each other. The ability of individual type-II QC agents to influence other type-II QC agents is inherently coupled with the risk of being influenced in turn by exactly the same agents in the same way. Though, for an individual agent the degree of its influence can be quantified based on the number, and the frequency of respective usage of its entangled quantum data.

7   Conclusions

In essence, quantum computational agents and multi-agent systems are feasible to implement on hybrid quantum computers, and can be used to solve certain problems in practical applications such as information search and service matchmaking more efficiently than with classically computing agents. Type-II QC agents can take most computational advantages of quantum computing and communication, but at the cost of limited self-autonomy, due to non-local effects of quantum entanglement. Quantum-based communication between type-II agents is inherently secure.

Ongoing and future research on QC agents and multi-agent systems focuses on appropriate integration architectures for QCMAS of both types, type-II QC information and matchmaker agents, as well as potential new applications such as secure quantum based distributed constraint satisfaction, and qualitative measures and patterns of quantum computational autonomy in type-II QCMAS.

References

1. D. Aharonov. Quantum Computation. LANL Archive quant-ph/981203, 1998.
2. D. Aharonov, M. Ben-Or. Polynomial Simulations of Decohered Quantum Computers. Proc. 37th Ann. Symp. Foundations of Computer Science (FOCS), 1996.
3. J.S. Bell. Speakable and Unspeakable in Quantum Mechanics. Cambridge University Press, 1987.
4. C.H. Bennett. Time/Space Trade-offs for Reversible Computation. SIAM Journal of Computing, 18(4):766–776, 1989.
5. C.H. Bennett, G. Brassard. Quantum Cryptography: Public Key Distribution and Coin Tossing. Proc. IEEE Intl. Conference on Computers, Systems, and Signal Processing, pp 175–179, 1984.
6. C.H. Bennett, D.P. DiVincenzo. Quantum Information and Computation. Nature, 404(6775):247–254, 2000.
7. C.H. Bennett, G. Brassard, C. Crepau, R. Josza, A. Peres, W.K. Wootters. Phys. Reviews Letters, 70, 1895, 1993.
8. C.H. Bennett,, S.J. Wiesner. Communication via one- and two-particle operators on EPR states. Phys. Review Letters, 69(20), 1992.
9. S. Betteli, T. Calarco, L. Serafini. Toward an Architecture for Quantum Programming. LANL Archive cs.PL/0103009, March 2003.
10. E. Bernstein, U. Vazirani. Quantum complexity theory. SIAM Journal of Computing, 26(5):1411–1473, 1997.
11. E. Biham, G. Brassard, D. Kenigsberg, T. Mor. Quantum Computing Without Entanglement. LANL Archive quant-ph/0306182, 2003.
12. D. Bouwmeester, A. Ekert, A. Zeilinger. The Physics of Quantum Information. Springer, Heidelberg, 2000.
13. G. Brassard, I. Chuang, S. Lloyd, and C. Monroe. Quantum Computing. Proc. National Academy of Sciences, USA, vol. 95, p. 11032–11033, 1998.
14. G.K. Brennen, D. Song, C.J. Williams. A Quantum Computer Architecture using Nonlocal Interactions. LANL Archive quant-ph/0301012, 2003.
15. H. Buhrman, R. Cleve, A. Wigderson. Quantum vs. Classical Communication and Computation. Proc. 30th Ann. ACM Symp. Theory of Computing (STOC 98), 1998.
16. C. Carabelea, O. Boissier, A. Florea. Autonomy in Multi-agent Systems: A Classification Attempt. Proc. Intl. Autonomous Agents and Multiagent Systems Conference Workshop on Computational Autonomy, M Rovatso, M Nickles (eds.), Melbourne, Australia, 2003.
17. P.C.W. Davies, J.R. Brown (Eds.). The Ghost in the Atom. Cambridge University Press, Canto edition reprint, 2000.
18. D. Deutsch. Quantum Theory, the Church-Turing Principle, and the Universal Quantum Computer. Proc. Royal Society London A, 400:97, 1985

19. D.P. DiVincenzo. The Pysical Implementation of Quantum Computation. LANL Archive quant-ph/0002077, 2000.
20. A. Einstein, B. Podolsky, N. Rosen. Can quantum mechanical description of physics be considered complete?. Phys. Review, 47:777–780, 1935.
21. L. Grover. A Fast Quantum Mechanical Algorithm for Database Search. Proc. 28th Annual ACM Symposium on Theory of Computation, ACM Press, NY USA, pp 212–219, 1996.
22. Gruska, Imai. Power, Puzzles and Properties of Entanglement. M. Margenster, Y. Rogozhin (eds.), Lecture Notes in Computer Science LNCS, 2055, Springer, 2001.
23. M. Hirvensalo. Quantum Computing. Natural Computing Series, Springer, 2001.
24. A.S. Holevo. Some estimates of the information transmitted by quantum communication channels. Problems of Information Transmission, 9:177–183, 1973.
25. R. Josza. Entanglement and Quantum Computation. Geometric Issues in the Foundations of Science, S. Huggett et al. (eds.), Oxford University Press, 1997.
26. B. Kalyanasundra, G. Schnitger. The probabilistic communication complexity of set intersection. SIAM Journal on Discrete Mathematics, 5(4), 1992.
27. B. Kane. A Silicon-Based Nuclear Spin Quantum Computer. Nature, 393, 1998
28. E. Knill, R. Laflamme, W.H. Zurek. Resilient Quantum Computation. Science, 279:342–345, 1998
29. M. Klusch. Information Agent Technology for the Internet: A Survey. Data and Knowledge Engineering, 36(3), Elsevier Science, 2001.
30. M. Klusch, K. Sycara. Brokering and Matchmaking for Coordination of Agent Societies: A Survey. Coordination of Internet Agents, A. Omicini et al. (eds), Springer, 2001.
31. I. Kremer. Quantum Communication. Master Thesis, Hebrew University, Jerusalem, Israel, 1995.
32. Los Alamos National Lab Archive, USA: http://xxx.lanl.gov/archive/
33. M.A. Nielsen, I.L. Chuang. Quantum Computation and Quantum Information. Cambridge University Press, Cambrige, UK, 2000.
34. B. Oemer. Quantum Programming in QCL. Master Thesis, Technical University of Vienna, Computer Science Department, Vienna, Austria, 2000.
35. M. Oskin, F.T. Chong, I.L. Chuang. A Practical Architecture for Reliable Quantum Computers. IEEE Computer, 35:79–87, January 2002.
36. M. Oskin, F.T. Chong, I.L. Chuang, J. Kubiatowicz. Building Quantum Wires: The Long and the Short of it. Proc. 30th Intnl Symposium on Computer Architecture (ISCA), 2003.
37. M. Ozawa. Quantum Turing Machines: Local Transitions, Preparation, Measurement, and Halting Problem. LANL Archive quant-ph/9809038, 1998.
38. R. Penrose. The Large, the Small, and the Human Mind. Cambridge University Press, 1997.
39. A. Pereira. The Quantum Mind/Classical Brain Problem. NeuroQuantology, 1:94–118, ISSN 1303–5150, Neuroscience & Quantum Physics, 2003.
40. P. Selinger. Towards a Quantum Programming Language. Mathematical Structures in Computer Science, 2003.
41. P. Shor. Algorithms for Quantum Computation: Discrete Logarithms and Factoring. Proc. 35th Annual Symposium on Foundations of Computer Science, Los Alamitos, USA, 1994.
42. A. Steane. Quantum computing. LANL Archive quant-ph/9708022, 1997.
43. M. Steffen, L.M.K. Vandersypen, I.L. Chuang. Toward Quantum Computation: A Five-Qubit Quantum Processor. IEEE Micro, March/April, 2001.

44. K. Sycara, S. Widoff, M. Klusch, J. Lu. LARKS: Dynamic Matchmaking Among Heterogeneous Software Agents in Cyberspace. Autonomous Agents and Multi-Agent Systems, 5(2), 2002.
45. G. Weiss. Introduction to Multiagent Systems. MIT Press, 1999.
46. M. Wooldridge. An Introduction to Multiagent Systems. John Wiley & Sons, Chichester, UK, 2002.
47. W.K. Wootters, W.H. Zurek. A Single Quantum Cannot be Cloned. Nature, 299:802–803, 1982.
48. A.C-C. Yao. Quantum Circuit Complexity. Proc. 33rd Annual Symposium on Foundations of Computer Science (FOCS), pp. 352–361, 1993.
49. P. Zuliani. Logical Reversibility. IBM Journal of Res. & Devel., 45, 2001.

Adjustable Autonomy Challenges in Personal Assistant Agents: A Position Paper

Rajiv T. Maheswaran[1], Milind Tambe[1], Pradeep Varakantham[1],
and Karen Myers[2]

[1] University of Southern California
Salvatori Computer Science Center, Los Angeles, CA 90089
[2] SRI International, 333 Ravenswood Avenue, Menlo Park, CA 94025

Abstract. The successful integration and acceptance of many multi-agent systems into daily lives crucially depends on the ability to develop effective policies for adjustable autonomy. Adjustable autonomy encompasses the strategies by which an agent selects the appropriate entity (itself, a human user, or another agent) to make a decision at key moments when an action is required. We present two formulations that address this issue: user-based and agent-based autonomy. Furthermore, we discuss the current and future implications on systems composed of personal assistant agents, where autonomy issues are of vital interest.

1 Introduction

Increasingly, researchers have focused on deploying multi-agent systems to support humans in critical activities, revolutionizing the way a variety of cooperative tasks are carried out, at home, at the office, in a large organization, or in the field. For example, multi-agent teams could support rapid response to natural (or man-made) disasters, such as earthquakes or assist coordination of autonomous spacecraft. One promising application of interest to many organizations is the development of personal assistant agents [2,3,10]. These entities would be capable of performing a variety of tasks such as scheduling meetings, gathering information or communicating on behalf of its user. Unburdened by routine or mundane tasks, humans would free up time for more productive pursuits.

To be able to perform their roles as assistants, agents will need to be endowed with a measure of autonomy. This raises many issues as agents may not have the information or ability to carry out a required action or the human user may not want agents to make certain decisions. These concerns have led to an emergence of the study of *adjustable autonomy*, i.e. agents dynamically varying their own autonomy, transferring decision making control to other entities (typically human users) in key situations [9,12]. One personal assistant agent project where adjustable autonomy issues are being addressed is CALO [2]. The goal is to create an agent that will observe and learn while interacting with users, other agents and the environment enabling it to handle a vast set of interrelated decision-making tasks that remain unaddressed by today's technology. This next-generation personal assistant agent and many similar multi-agent

M. Nickles, M. Rovatsos, and G. Weiss (Eds.): AUTONOMY 2003, LNAI 2969, pp. 187–194, 2004.
© Springer-Verlag Berlin Heidelberg 2004

systems will critically depend on the ability of today's research community to answer the challenges of adjustable autonomy.

Arguably, the fundamental research question is determining whether and when transfers of control should occur from the agent to other entities. Answering this question is very important for the development of personal assistant agents (along with other multi-agent systems with similar autonomy issues). An effective autonomous agent must be able to obtain the trust of a user through it's action policies. These policies must balance the ability to make competent and timely decisions without treading into domains where the user wants sole control.

In this position paper, we present two approaches in dealing with issues of transfer of control in adjustable autonomy: user-based and agent-based adjustable autonomy. We also discuss future directions of research and how they will be applied to a next-generation personal assistant agent.

2   User-Based Autonomy

User-based adjustable autonomy is driven by the need to support user control over the activities of an agent at runtime. In situations where activities are routine and decisions are straightforward, a human may be content to delegate all problem-solving responsibility to an agent. However, in situations where missteps could have undesirable consequences, the human needs the ability to control the problem-solving process to ensure that appropriate actions are taken. The ideal in such situations is not to devolve all responsibility to the human. Rather, the human should be able to guide the agent in much the same way that a human manager would direct a subordinate on tasks that were beyond his or her capabilities. Thus, what is required is a form of supervised autonomy for the agent [1].

The requirement for user-driven adjustable autonomy stems from two main issues: capability and personalization.

- *Capability:* For many applications, it will be difficult to develop agents whose problem-solving capabilities match or even approach those of humans. Agents can still play a valuable role in such applications by performing routine tasks that do not require user intervention and by working under user supervision on more complex tasks. For example, an office assistant agent may be capable of scheduling meetings that do not involve changes to the users current commitments, but require human guidance if commitments need to be revised.
- *Personalization:* Many decisions do not have a clear "best" answer, as they depend on multiple factors that can interact in complex ways. An individual user may have his own preference as to how to respond in different situations, making it impossible to precode appropriate responses in an off-the-shelf agent. Additionally, those preferences may change over time. While automated learning techniques present one possible mechanism for addressing customization, they are a long way from providing the sophisticated

adaptability needed for agent customization. Furthermore, they are inherently limited to projecting preferences based on past decisions.

The central issue for user-based adjustable autonomy agent is the design of mechanisms by which a user can dynamically modify the scope of autonomy for an agent. Such mechanisms should be natural, easy to use, and sufficiently expressive to enable fine-grained specifications of autonomy levels.

One approach to user-based adjustable autonomy is oriented around the notion of policies [9]. A policy can be considered a declarative statement that explicitly bounds the activities that an agent is allowed to perform without user involvement. Policies can be asserted and retracted throughout the scope of an agents operation, thus providing the means to tailor autonomy dynamically in accord with a users changing perspective on what he or she feels comfortable delegating to the agent. This comfort level may change as a result of the user acquiring more confidence in the agents ability to perform certain tasks, or the need for the user to focus his problem-solving skills on more important matters.

Policies could be defined from the perspective of either explicitly enabling or restricting agent activities. We are interested in domains where agents will generally need to operate with high degrees of autonomy. For this reason, we assume a permissive environment: unless stated otherwise, agents are allowed to operate independently of human interaction. In this context, policies are used to limit the scope of activities that can be performed autonomously. Activities outside that set may still be performable, but require some form of interaction with the user.

Our policies assume a Belief-Desire-Intention (BDI) model of agency [11] in which an agent has a predefined library of parameterized plans that can be applied to achieve assigned tasks or respond to detected events. Within this framework, we consider two classes of policies: permission requirements for action execution and consultation requirements for decision making.

- *Permission Requirements*: Permission requirements declare conditions under which an agent must obtain authorization from the human supervisor before performing activities. For example, consider the directive *Obtain permission before scheduling any meetings after 5:00.*
- *Consultation Requirements*: Consultation requirements designate a class of decisions that should be deferred to the human supervisor. These decisions can relate to either the selection of a value for parameter instantiation (e.g., *Let me choose airline flights*) or the selection of a plan for a goal (e.g., *Consult me when deciding how to respond to requests to cancel staff meetings*).

These adjustable autonomy policies are part of a more general framework for agent guidance that enables high-level user management of agent activities [9]. An alternate class of policies, called strategy preference guidance, supports the specification of recommendations on how an agent should accomplish tasks. These preferences could indicate specific plans to employ restrictions on plans that should not be employed, as well as constraints on how plan parameters can be instantiated. For example, the directive *Use Expedia to find options for flying to Washington next week* expresses a preference over approaches to finding flight information for planning a particular trip. On the other hand, the directive *Dont schedule project-related meetings for Monday mornings* restricts the choice for instantiating parameters that denote project meeting times.

Given that strategy preference guidance can be used to restrict the set of problem-solving activities of an agent, it can be viewed as a mechanism for limiting agent autonomy. However, this form of guidance does not support the explicit transfer of decision-making control to the user, as is the case with most work in the area of adjustable autonomy. Rather, the users influence on the decision-making process is implicit in the guidance itself.

These policies are formulated in terms of high-level characterizations of classes of activities, goals, and decisions over which an agent has autonomy. The language used to encode policies makes use of a logical framework that builds on both the underlying agent domain theory, and a domain metatheory. The domain metatheory is an abstract characterization of the agent domain theory that specifies important semantic attributes of plans, planning parameters, and instances. For instance, it could be used to identify parameters that play certain roles within agent plans (e.g., parameters for project meeting times), or to distinguish alternative procedures according to properties such as cost or speed. This abstraction allows a user to express policies in compact, semantically meaningful terms without having to acquire detailed knowledge of the agents internal workings or constituent knowledge.

One nice feature of this policy-based approach to adjustable autonomy is that it has minimal set-up costs, requiring only the formulation of the domain metatheory for defining policies. As argued in [8], the metatheory should be a natural by-product of a principled approach to domain modeling. Enforcement of the policies for adjustable autonomy can be done via a simple filtering mechanism that overlays standard BDI executor models. This filtering mechanism adds little to the computation cost of agent execution, as it requires simple matching of policies to plan and task properties.

3   Agent-Based Autonomy

In settings with agent-based adjustable autonomy, domains of authority are not prescribed by the user. Rather, an agent explicitly reasons about whether and when to transfer decision-making control to another entity. If control is to be transferred, an agent must choose the appropriate entity to which to yield responsibility. A policy to transfer control for a decision or action needs to balance the likely benefits of giving control to a particular agent or human with the potential costs of doing so, thus the key challenge is to balance two potentially conflicting goals. On one hand, to ensure that the highest-quality decisions are made, an agent can transfer control to a human user (or another agent) whenever that entity has superior decision-making expertise. On the other hand, interrupting a user has high costs and the user may be unable to make and communicate a decision, thus such transfers-of-control should be minimized.

The work so far in agent-based AA has examined several different techniques that attempt to balance these two conflicting goals and thus address the transfer-of-control problem. For example, one technique suggests that decision-making control should be transferred if the expected utility of doing so is higher than the expected utility of making an autonomous decision [7]. A second technique uses uncertainty as the sole rationale for deciding who should have control, forcing the agent to relinquish control to the user whenever uncertainty is high [6]. Yet other techniques transfer control to a user if an erroneous autonomous decision could cause significant harm [4] or if the agent lacks the capability to make the decision [5].

Previous work has investigated various approaches to addressing this problem but has often focused on individual agent-human interactions, in service of single tasks. Unfortunately, domains requiring collaboration between teams of agents and humans reveals at least two key shortcomings of these previous approaches. First, these approaches use rigid one-shot transfers of control that can result in unacceptable coordination failures in multi-agent settings. Second, they ignore costs (e.g., in terms of time delays or effects on actions) to an agent's team due to such transfers of control.

To remedy these problems, we base a novel approach to adjustable autonomy on the notion of *transfer-of-control strategy*. A transfer-of-control strategy consists of a conditional sequence of two types of actions: (i) actions to transfer decision-making control (e.g., from the agent to the user or vice versa) and (ii) actions to change an agent's pre-specified coordination constraints with team members, aimed at minimizing miscoordination costs. The goal is for high quality individual decisions to be made with minimal disruption to the coordination of the team. Strategies are operationalized using Markov decision processes (MDPs) which select the optimal strategy given an uncertain environment and costs to individuals and teams. A general reward function and state representation for an MDP have been developed, to help enable implementation such strategies [12].

An agent strategy can be an ordering, composed of authority owners and constraint changes, which outlines a particular sequence of responsibility states interconnected by temporal limits. For instance, a strategy denoted ADH implies that an agent (A) initially attempts to make an autonomous decision. If the agent makes the decision autonomously the strategy execution ends there. However, there is a chance that it is unable to make the decision in a timely manner, perhaps because its computational resources are busy with higher priority tasks or a high quality decision cannot be made due to lack of information. To avoid miscoordination, the agent executes a D action that changes the coordination constraints on the activity. For example, a D action could be to inform other agents that the coordinated action will be delayed, thus incurring a cost of inconvenience to others but buying more time to make the decision. If the agent still cannot make the decision, it will eventually transfer decision-making control to the human (H) and wait for a response.

Transfer-of-control strategies provide a flexible approach to adjustable autonomy in complex systems with many actors. By enabling multiple transfers-of-control between two (or more) entities, rather than rigidly committing to one entity (i. e., A or H), a strategy attempts to provide the highest quality decision, while avoiding coordination failures. Thus, a key AA problem is to select the right strategy, i.e., one that provides the benefit of high quality decisions without risking significant costs such as interrupting the user or miscoordination with the team. Furthermore, an agent must select the right strategy despite significant uncertainty. Markov decision processes (MDPs) (Puterman, 1994) are a natural choice for implementing such reasoning because they explicitly represent costs, benefits and uncertainty as well as doing look ahead to examine the potential consequences of sequences of actions.

This agent-based autonomy approach has been tested in the context of a real-world multi-agent system, called Electric Elves (E-Elves) [3, 10] that was used for over six months at the University of Southern California and the Information Sciences Institute. Individual user proxy agents called Friday (from Robinson Crusoe's servant Friday) act in a team to assist with rescheduling meetings, ordering meals, finding presenters and other day-to-day activities. Figure 1 describes the interactions and architecture of the E-Elves project.

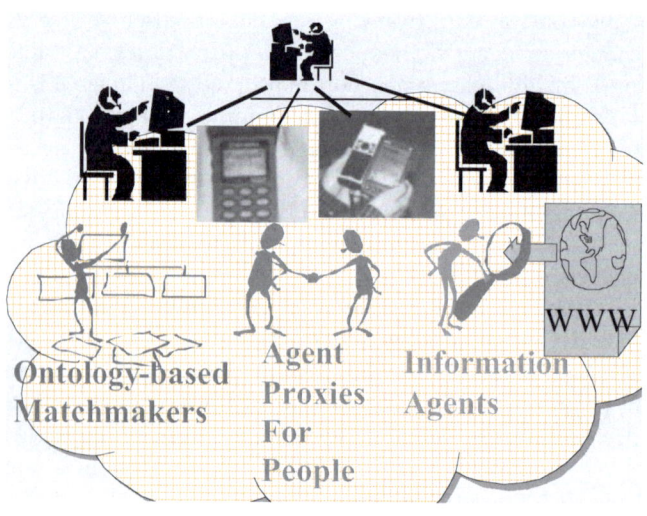

Fig. 1. Model for Interaction and Architecture of Electric Elves Project

MDP-based AA reasoning was used around the clock in the E-Elves, making many thousands of autonomy decisions. Despite the unpredictability of the user's behavior and the agent's limited sensing abilities, the MDP consistently made sensible AA decisions. Moreover, the agent often performed several transfers-of-control to cope with contingencies such as a user not responding.

4   Future of Adjustable Autonomy

The greatest challenges in adjustable autonomy involve bridging the gaps in the user-based and agent-based methodologies to create agents that can leverage the benefits of both systems. An ideal system would be able to include the situational sensitivity and personalization capabilities of the user-based autonomy while incorporating the autonomous adaptability, modeling of uncertainty in decision making and constraint modification aspects of the agent-based model. The result should be an agent customizable with low set-up cost, yet be able to factor in multi-agent issues when considering autonomy issues. A key to developing such a system will be resolving a fundamental formulation in the BDI and MDP modeling. This can be approached by applying hybrid schemes that weave both structures into their policy space. Another method may be to discover mapping from BDI plans to decision theoretic structures that yield congruent plans.

One lesson learned when actually deploying the system was that sometimes users wished to influence the AA reasoning, e.g., to ensure that control was transferred to them in particular circumstances. To enable users to influence the AA reasoning, safety constraints were introduced that allowed users to prevent agents from taking particular actions or ensuring that they do take particular actions. These safety constraints provide guarantees on the behavior of the AA reasoning, making the basic approach more generally applicable and, in particular, making it more applicable to domains where mistakes have serious consequences. Creating structures and models for safety constraints can be considered a step towards more effective adjustable autonomy systems that infuse personalization into agents.

Another area for exploration are situations where the underlying state is not known explicitly. One can envision scenarios where the current information available to a personal assistant agent does not allow it to deduce a single state from which to apply a policy. One way to approach this problem might be to assign a most probable state given every possible information set *a priori*, and apply a plan for the resulting probable state. This would become extremely cumbersome for many problem settings where the underlying state space or the information sets are large. Another approach is to apply Partially Observable Markov Decision Processes (POMDPs), into the models for autonomy decisions, allowing for uncertainty in both observation and action.

Acknowledgements

This work was sponsored by a subcontract from SRI International based on the DARPA CALO project.

References

[1] K.S. Barber and C.E. Martin. Dynamic adaptive autonomy in multi-agent systems: Representation and justification. *International Journal of Pattern Recognition and Artificial Intelligence*, 15(3), 2001.

[2] CALO, 2003. http://www.ai.sri.com/project/CALO.

[3] HH. Chaulpsky, Y. Gil, C. Knoblock, J. Oh, K. Lerman Dnd D. Pynadath, T. Russ, and M. Tambe. Electric elves: Applying agent technology to support human organizations. In *International Conference on Innovative Applications of AI*, pages 51–58, 2001.

[4] G. Dorais, R. Kortenkamp, B. Pell, and D. Schreckenghost. Adjustable autonomy for human-centered autonomous systems on mars. In *proceedings of the First International Conference of the Mars Society*, pages 397–420, 1998.

[5] G. Ferguson, J. Allen, and B. Miller. Towards a mixed-initiative planning assistant. In *proceedings of the Third conference on Artificial Intelligence Planning Systems*, pages 70–77, 1996.

[6] J. Gunderson and W. Martin. Effects of uncertainty on variable autonomy in maintenance robots. In *Workshop on Autonomy Control Software*, pages 26–34, 1999.

[7] E. Horvitz, A. Jacobs, and D. Hovel. Attention-sensitive alerting. In *proceedings of Conference on Uncertainty and Artificial Intelligence*, pages 305–313, Stockholm, Sweden, 1999.

[8] K.L. Myers. Domain metatheories: Enabling user-centric planning. In *proceedings of AAAI Workshop on Representational Issues for Real-World Planning Systems*, 2000.

[9] K.L. Myers and D.N. Morley. *Agent Autonomy*, chapter The TRAC Framework for Agent Directability. Kluwer Academic Publishers, 2002.

[10] D. Pynadath, M. Tambe, Y. Arens, H. Chalupsky, Y. Gil, C. Knoblock, H. Lee, J. Oh, K. Lerman, S. Kamachandran, P. Rosenbloom, and T. Russ. Electric elves: Immersing an agent organization in a human organization. In *Proceedings of the AAAI Fall symposium on socially intelligent agents - the human in the loop*, 2000.

[11] A.S. Rao and M.P. Georgeff. Bdi agents: From theory to practice. In *proceedings of International Conference on Multi-Agent Systems*, San Francisco, CA, 1995.

[12] Paul Scerri, David V. Pynadath, and Milind Tambe. Towards adjustable autonomy for the real world. *Journal of Artificial Intelligence Research*, 17, 2003.

Autonomy in an Organizational Context

Olga Pacheco

University of Minho, Dep. of Informatics, Braga, Portugal
omp@di.uminho.pt
http://www.di.uminho.pt/~omp

Abstract. In this paper it is discussed how organizations deal with autonomy of agents that constitute them. Based on human organizations and on their legal characterization, it is proposed a normative and role-based model for organizations (human or not), that assumes autonomy of agents as a natural ingredient. It is discussed how an organization can work without regimenting agents behavior, but simply by describing their expected (ideal) behavior (through the deontic characterization of the roles agents hold) and fixing sanctions for agents that deviate from what is expected of them. Interaction between agents is ruled through contracts that agents are free to establish between each other. A formal model, supported by a deontic and action logic, is suggested. Although this model is in a preliminary stage, it might be an useful approach to incorporate autonomy as a natural property of agents in an organizational context.

1 Introduction

Organizations are entities that exist with specific purposes. They are constituted by a set of autonomous agents. The achievement of an organization objectives depends on agents' actions. But, as agents are autonomous, the organization cannot regiment their actions to assure that agents do exactly what they should do. Autonomous agents take their own decisions, according to their own objectives. Thus, a natural question is: *Why should an agent give up of their own objectives and adopt those of an organization?* or *How can an organization convince an agent to act for it?*[1].

If we look into human organizations a possible answer might be: *The agent must be convinced that it is better for its own objectives to act for the organization.* It is the agent that decides to play a role in the organization, in fist place, and after, it is up to him to act in accordance to what has been stated in the role or not.

The commitment of an agent to act for an organization is formalized through a contract. An organization, an agent itself – as discussed in [12], establishes contracts with each of the agents that constitute it, stating what they should do in the organization (what are their roles), what they receive if they do what is

---

[1] M.Gilbert in [4] and R. Tuomela in [15] discuss in detail similar questions.

M. Nickles, M. Rovatsos, and G. Weiss (Eds.): AUTONOMY 2003, LNAI 2969, pp. 195–208, 2004.
© Springer-Verlag Berlin Heidelberg 2004

expected of them and what sanctions are applied to them if they deviate from the expected behavior. But an organization can never be sure that agents do what they should. An agent may,in each moment, decide to act in accordance with what has been contracted or not, depending on his evaluation of what is better for him: to respect the contract or to suffer the sanctions of violating it. The only thing an organization can do to make autonomous agents act toward organization objectives, is to establish adequate contracts that reward adequately when agents fulfill their obligations and define sanctions that agents would not want to suffer.

The perspective we have just presented does not restrict autonomy of agents, but assumes it as a natural ingredient. This is what happens in human organizations and we think that this perspective can be transposed to any kind of organization formed by autonomous agents (humans or not)[2]. We will continue to use humans as a metaphor for autonomous agents, because they are the most complex autonomous entities we know. But we believe that the analysis we present can be useful to computational agents, where similar questions arise.

We do not assume any particular kind of agent. The only assumption we use is autonomy (ability of action choice and act under self-control) and knowledge about the relationships with other agents an its deontic characterization (e.g. obligations, permissions or prohibitions that apply to them). We do not discuss in this paper what are the motivations of an agent to accept a role in an organization or how he decides to follow a norm or deviate from it.

In the rest of the paper we first present the model we adopt for organizations (c.f. [12] for more details). Then we briefly refer to the action and deontic logic that we have proposed in [2] and [12] to capture the concepts involved in the organization model. This logic is then explored in the formal specification of organizations, societies of agents and normative relationships between agents, and in the rigorous analysis of them.

2 A Role-Based Model of Organizations

2.1 Organizations as Institutional Agents

An organization is created when a group of agents have some common interests and need to act as a whole in order to pursue their interests. This collective entity has its own identity (different from the set of agents that has created it), and for many purposes it is seen by the external environment as acting by itself and as being responsible for its acts. For instance, in the legal system, most organizations are classified as *artificial persons* (in addition to natural persons – the human beings), and have *juridical personality* and *legal qualification*, which means, respectively, that they may be the subject of obligations and rights and that they can exercise their rights and be responsible for the unfulfillment of their obligations. Thus, an organization may be seen itself as an agent. In [12]

---

[2] This work may be extended to other kind of collective entities, as far as they can be structured by roles and norms.

we called those agents *institutional agents*. Organizations interact in the society like any other agent: they can establish contracts or other normative relationship with other agents, they can hold roles, they may be the subject of obligations, permissions or other normative concepts, and may be responsible for the non-fulfilment of obligations or any other "non ideal" situation.

But organizations are abstract entities, in the sense that they cannot act directly. So, to achieve their objectives, organizations must act through concrete agents, that act on their behalf. Agents (human or not) that integrate an organization must contribute with their actions to achieve organization's objectives.

When an organization is created its statutory norms are defined, stating the purpose of the organization and how it should be attained. An organization has a stable structure formed by a set of roles and a set of norms stating what should be done in each role – the deontic characterization of each role (obligations, permissions and prohibitions, of the holders of each role)[3]. This information is described in the statutes of an organization.

This abstract structure is supported by other agents: the holders of the roles. The holders of the roles of an organization are not necessarily human agents. They may be artificial agents or even other organizations[4].

For an organization to be able to interact with other agents in a society, it must have defined the representative roles of the organization and its respective scope of representation, stating who is authorized to act on behalf of the organization and to what extent. Another aspect that needs to be defined is a mechanism of transmission of the obligations of an organization to the roles of its structure (and indirectly to the holders of those roles), stating who will be responsible for fulfilling those obligations.

2.2   Autonomous Agents in an Organization

The agents that hold a particular role inherit the deontic characterization of that role. Hence, when acting in that role, his behavior will be evaluated according to that deontic characterization[5]. The deontic characterization of roles describes the expected behavior of agents and what are the consequences of deviating from it, trying to assure that agents actions contribute to achieve the aims of the organization.

We may be tempted to say that autonomy of agents within an organization is restricted, because they are under the norms defined in the roles they hold.

---

[3] In this paper, whenever we mention norms or rules, we refer to obligations, permissions or prohibitions of agents in a role.

[4] For instance, in some companies the role of *single auditor* is held by other company. Another example is ATM that may be seen as an employee of a bank (or other companies).

[5] For example, I may have permission to drive a company car when I am on duty but I may not be allowed to drive it when I am on holiday. So, by doing the same action, and depending on the role I am playing, I may be having an ideal or non-ideal behavior.

But norms are just guidelines, and agents are always free to follow or not the norms. In that sense, agents remain autonomous.

However, if an agent decides not to follow a norm, he will be responsible for that decision. This means that when a violation occurs, after identifying the agent that caused it, and the role he was playing, sanctions may be applied to that agent (or to others that have relationships with him[6]). Sanctions may be new obligations, new prohibitions, or permissions that are withdraw, applied to the agent in the role he was playing or even to the agent itself.

Agents must be aware of the norms that apply to them, and of the consequences of deviating from them. We assume that when an agent establishes a contract with an organization, assuming a particular role, he knows its deontic characterization.

But deontic characterization of roles are not the only source of obligations, permissions or prohibitions that apply to an agent in a role, in an organizational context. When an agent establishes a contract with an organization accepting to hold a particular role, other obligations or permissions (distinct from the ones of the deontic characterization of the role), may be attributed to the agent in that role. For example, in an organization, it may be defined that administrators are permitted to use a company car when they are on duty, but they don't have permission to drive it home; but a particular administrator may negotiate with the organization to use a company car all the time. That permission must be stated in the contract he established with the organization.

Another source of obligations, permissions or prohibitions, in a role, are the effects of actions of agents, particularly, sanctions to actions that unfulfilled some other obligation. For instance, if an employee has permission to use an equipment but does not obey the instructions of use, he will loose that permission (he will be forbidden to use it). Another possible sanction for the same situation, would be the obligation to pay for the damage caused. But in this case, that obligation would not be applied to the agent in the role, but to the agent itself. Sanctions that are applied to agents because they have a non-ideal behavior, deviating from what was expected of them in some role, must also be defined in the contract, or in some global norms of the organization, known to all of its members.

In the organizational context just presented, there are several concepts that have an imprecise and often divergent meaning, like the concept of role, action in a role, obligation, permission, representation, violation, sanction, contract, among others. Formal specification is needed in order to remove ambiguity and vagueness, defining precisely what is the meaning we assume for each of them.

3   A Deontic and Action Logic

Logic can be an useful tool in a formal specification process, fixing in a rigorous way the meaning of the concepts relevant to the problem under consideration,

---

[6] For instance, to the external society, responsibility for "non-ideal" behavior of agents that hold roles in an organization is, usually, also attributed to the organization.

relating them with each other and characterizing its properties. This formal description can then be used to specify the entities involved and reason about them.

The formal characterization of the concepts presented in the previous section has been done through the definition of deontic and action modal logics, following the tradition initiated by S. Kanger, I. Pörn and L.Lindahl of combining deontic and action logics as basic building blocks to describe social interaction and complex normative concepts (see e.g. [8], [9], [13], [14], [10]). In [2] and [12] a first-order, many sorted and multi-modal logic, including a new action operator that captures the notion of *action of an agent in a role*, has been proposed.

We will not present here the logical details nor discuss the need for a new logic. We will briefly try to give some intuition for the modal operators of the logic and present some of their properties. For more details and for the technical issues see [2] and [12].

Naturally, we assume that all tautologies are axioms of our logic, and that we have the rule of Modus Ponens (in the sense that the set of theorems of our logic is closed under Modus Ponens). With respect to the first-order component, we have the usual properties of quantifiers (e.g. see [5]).

3.1 Action of an Agent in a Role

We need an action operator able to distinguish situations where an agent brings about a state of affairs in different roles. This is fundamental because the consequences of the action of an agent and its deontic classification (e.g. permitted or not) depends on the role the agent was playing when he did that action.

For these reasons, in [2] we have proposed an action operator of the form $E_{a:r}$ (for a an agent and r a role), where expressions of the form $E_{a:r}B$ are read "agent a, playing the role r (or acting in the role r) brings it about that B".

This relativized modal operator relates an agent playing a role, and the state of affairs he brings about, omitting details about the specific actions that have been performed (and setting aside temporal aspects).

Thus, these action operators are suited to support a first-level of specification of organizations, where we want to describe agent's behavior at an abstract level, without yet considering specific actions and tasks.

The formal properties of this action operator $E_{a:r}$ are described bellow:

Axioms:	
(T_E)	$E_{a:r}B \rightarrow B$
(C_E)	$E_{a:r}A \wedge E_{a:r}B \rightarrow E_{a:r}(A \wedge B)$
(Qual)	$E_{a:r}B \rightarrow qual(a:r)$
(Itself)	$(\forall_x)qual(x:itself)$
Proof rule:	
(RE_E)	If $\vdash A \leftrightarrow B$ then $\vdash E_{a:r}A \leftrightarrow E_{a:r}B$

The (T) schema captures the intuition that if agent a, when acting in a role r, brings it about that B, then B is indeed the case (is a "success" operator).

We impose the axiom (Qual), where $qual(a : r)$ is a predicate read as "agent a is qualified to act in the role r" (i.e. the agent holds the role), because for an agent to bring about something, playing some role, he must have the correspondent qualification. The axiom (Itself) is also imposed, in order to capture that any agent is qualified to play the role of *itself*.

3.2   Obligation, Permission and Prohibition of an Agent in a Role

As deontic operators we have considered $O_{a:r}$, $P_{a:r}$ and $F_{a:r}$. $O_{a:r}B$ means that *agent a is obliged to bring about B by acting in role r*; $P_{a:r}B$ means that *agent a is permitted to bring about B when acting in role r*; and $F_{a:r}B$ means that *a is forbidden to bring about B when acting in role r*.

Only $O_{a:r}$ and $P_{a:r}$ are primitives. $F_{a:r}$ will be defined as $\neg P_{a:r}$.

With respect to the formal properties of the deontic operators, and of the relationships between each other and with the action operator, we consider the following axioms and proof-rules:

Axioms:	
(C_O)	$O_{a:r}A \land O_{a:r}B \rightarrow O_{a:r}(A \land B)$
$(O \rightarrow P)$	$O_{a:r}B \rightarrow P_{a:r}B$
$(O \rightarrow \neg P\neg)$	$O_{a:r}B \rightarrow \neg P_{a:r}\neg B$
$(O \land P)$	$O_{a:r}A \land P_{a:r}B \rightarrow P_{a:r}(A \land B)$
Proof rules:	
(RE_O)	If $\vdash A \leftrightarrow B$ then $\vdash O_{a:r}A \leftrightarrow O_{a:r}B$
(RM_P)	if $\vdash A \rightarrow B$ then $\vdash P_{a:r}A \rightarrow P_{a:r}B$
(RM_{EP})	If $\vdash E_{a_1:r_1}A \rightarrow E_{a_2:r_2}B$ then $\vdash P_{a_1:r_1}A \rightarrow P_{a_2:r_2}B$

More details can be found in [2] and [12].

4   Formal Specification of Organizations

The logic referred above is now going to be used in the formal specification of organizations and of normative interactions between agents.

Lets introduce some abbreviations and extensions to the logic, in order to increase its expressiveness.

4.1   Some Abbreviations

The deontic characterization of a role in an organization is part of the identity of the organization and does not depend on the agent that hold that role in a particular moment. To capture this idea, deontic notions will be attached to roles, but they are actually interpreted as applied to the holders of such roles, when acting in such roles (deontic notions are only meaningful when applied to agents). Thus, we do not introduce new operators, but just new abbreviations:

$$O_r B \stackrel{abv}{=} (\forall_x)(qual(x:r) \to O_{x:r} B)$$
$$P_r B \stackrel{abv}{=} (\forall_x)(qual(x:r) \to P_{x:r} B)$$
$$F_r B \stackrel{abv}{=} (\forall_x)(qual(x:r) \to F_{x:r} B)$$

For example, if we want to characterize the role of *administrator of a company* x, we could write:

Role: administrator-of(x)

$O_{administrator-of(x)}$ pay the salaries
$P_{administrator-of(x)}$ sign contracts
. . .

We are also able to express the transmission of obligations of an organization (resultant from the interactions of the organization with other agents) to specific roles of its structure (and indirectly, to the holders of those roles):

$O_{x:itself} A \to O_r A$ (for r a role of the structure of organization x).

An example is $O_{i:i} A \to O_{President-of-Administration(i)} A$ meaning that whenever i is under the obligation to bring about A, that obligation is attributed to the role *President-of-Administration(i)* (so, the holder of that role will be obliged to bring about A).

4.2   Representative Roles

As we said before, for an organization to be able to interact with other agents in a society, it must be defined the representative roles of the organization and its respective scope of representation, stating who is authorized to act on behalf of the organization and to what extent. We will classify some of the roles of the structure of the organization as *representative roles*, introducing the following abbreviation:

$$r : REP(x, B) \stackrel{abv}{=} (\forall_y)(E_{y:r} B \to E_{x:itself} B)$$

$r : REP(x, B)$ means that the role r is a representative role of x, for B. Thus, any agent that holds role r and brings it about that B when acting in that role, acts on behalf of x.

When an agent acts as representative of another agent he does not act on his own behalf. So, it is natural to impose that:

$$E_{a:r} B \wedge r : REP(x, B) \to \neg E_{a:itself} B$$

In our model of organizations we will have to identify what are the *representative roles* and to specify the *scope of representation* of each of those roles.

Representative roles are not necessarily roles of the structure of an organization. They may result from contracts or other normative relations that agents are free to establish between each other.

4.3   Non-ideal Situations and Sanctions

We have referred before that agents may violate norms and that they must be responsible for that non-ideal behavior, having sanctions applied to them.

So, we have to define what we mean by violation, by sanction and how they are applied to agents[7].

Some situations that we classify as non-ideal, i.e., that correspond to violations of norms, are:

- when an agent does not fulfill an obligation:

  $O_{x:r}A \wedge E_{x:r}\neg A.$

- when an agent brings about a state of affairs, acting in a role, that he is not permitted to bing about in that role:

  $E_{x:r}A \wedge \neg P_r A$ or $E_{x:r}A \wedge \neg P_{x:r}A^8.$

There are other kinds of non-ideal situations that we don't consider here (e.g. agents playing roles that they are not qualified to play or agents pretending to be other agents).

When a violation occurs, sanctions may be applied to the agents responsible for that non-ideal situation. The identification of the agent that caused the violation and the role he was playing is important to detect the violation, in first place, and then to attribute responsibilities and to apply sanctions to the agents involved. Responsible agents might be not only the ones that directly caused the violation but also other agents that are related with them. For example, if an agent is acting as representative of another agent, the represented agent may also be responsible for the actions of the representative.

We will consider that sanctions applied to a non-ideal situation are new obligations or prohibitions applied to the agents involved[9].

Possible representation of a violation and of sanctions applied to the agents involved are:

1) $(O_{x:r}A \wedge E_{x:r}\neg A) \rightarrow O_{x:r}B$ or
2) $(O_{x:r}A \wedge E_{x:r}\neg A) \rightarrow \neg P_{x:r}B.$

In some kind of violations, sanctions may be extended to other roles (including the role *itself*):

3) $(O_{x:r}A \wedge E_{x:r}\neg A) \rightarrow O_{x:r1}B$ or
4) $(O_{x:r}A \wedge E_{x:r}\neg A) \rightarrow \neg P_{x:r2}B$ (assuming qual(x:r1) and qual(x;r2));

or even to all the roles held by the agent in the organization:

---

[7] These are complex issues, specially if we take the legal domain as a reference. We do not have any pretension of discussing them in detail in this paper. We just refer to some aspects that arise naturally in the model we adopt.

[8] A permission that has been withdraw from an agent in a role (e.g. through a sanction).

[9] It is a simplification. Representation of contrary-to-duties is a complex subject (see [1]about this issue).Moreover, in the classification of an action as being a violation, and in determining what sanctions apply to the agents involved, there are several factors that we deliberately ignore, such as: mental attitudes of agents, knowledge about the norms, possibility of avoiding violation, and many others.

5) $(O_{x:r}A \wedge E_{x:r}\neg A) \rightarrow O_x B$ or
6) $(O_{x:r}A \wedge E_{x:r}\neg A) \rightarrow \neg P_x B,$

where $O_x B$ (or $P_x B$) means that the obligation (or permission) applies to agent x when he acts in any of the roles he holds in the organization i[10], and are defined as follows:

$$O_x B \stackrel{abv}{=} (\forall_r)((qual(x:r) \wedge is-role-str(i,r) \rightarrow O_{x:r}B)$$

$$P_x B \stackrel{abv}{=} (\forall_r)((qual(x:r) \wedge is-role-str(i,r)) \rightarrow P_{x:r}B)^{11}$$

This representation of sanctions may rise some problems. Notice that the new obligations or prohibitions applied to an agent in a role might be in conflict with the deontic characterization of the role. We may have for instance $P_r A$ and $\neg P_{x:r}A$[12]. In this paper we just assume that sanctions do not contradict what has been stated in the roles.

Another consequence of representing sanctions in this way, is that we have to register the "history" of agents: the effects of actions done by the agents. Two different agents, acting in the same role, and producing the same state of affairs, but having different "histories", may have different deontic qualifications of their acts (c.f. next section).

4.4   Contracts

Autonomous agents in a society are free to establish arbitrary normative relationships between each other. A particular kind of those relationships are *contracts*. As we have already discussed, contracts play a crucial role in societies of autonomous agents.

When two agents[13] establish a contract between each other, they may attribute roles to each other and they deontically characterize those roles, that is, they define what are the obligations, permissions or prohibitions associated to each role. Some of the roles may be representative roles of one of the agents. In that case, it must be also defined in the contract, the scope of representation for that role. Frequently, contracts also include conditional obligations (or conditional permissions). In particular, in legal contracts it is usual to include conditional obligations describing the effects of the fulfillment or violation(unfulfillment) of other obligations in the contract. For instance, besides an

---

[10] The predicate *is-role-str(i,r)* means that the role r is part of the structure of organization i.

[11] The count-as operator proposed by A.Jones and M.Sergot in [7], would be more appropriate in this definition, instead of the material implication, because this information is local to the organization.

[12] If we want to consider this possibility we must define priorities among norms, stating which norm prevails.

[13] For simplicity reasons we only consider contracts between two agents.

obligation $O_{x:r_1}A$ on x a contract $C(x,y)$ may include an obligation on y on the condition that x fulfills the previous obligation

$$E_{x:r_1(y)}A \rightarrow O_{y:r_2(x)}B,$$

or another obligation on x if he does not fulfill it

$$\neg E_{x:rg_1(y)}A \rightarrow O_{x:rg_1(y)}B.$$

Using $C(x,y)$ to denote (the content of) a contract between agents x and y, we may say that a contract $C(x,y)$ can be a formula similar to the following one:

$qual(x:r1) \wedge qual(y:r2) \wedge$	*Attribution of roles to agents*
$P_{x:r_1}A_1 \wedge P_{x:r_1}A_2 \wedge \ldots$	*Deontic characterization of r_1*
$O_{x:r_1}B_1 \wedge O_{x:r_1}B_2 \ldots$	
$P_{y:r_2}C_1 \wedge P_{y:r_2}C_2 \wedge \ldots$	*Deontic characterization of r_2*
$O_{y:r_2}G_1 \wedge O_{y:r_2}G_2 \wedge \ldots$	
$r_1 : REP(y, H_1) \wedge \ldots$	*Scope of representation of r_1*
$r_2 : REP(x, I_1) \wedge \ldots$	*Scope of representation of r_2*
$E_{x:r_1}J_1 \rightarrow O_{y:r_2}K_1$	*Conditional obligation*
$E_{x:r_1}\neg L_1 \rightarrow O_{x:r_1}M_1$	*Sanction to violations of $O_{x:r_1}L_1$*
$\ldots$	

An example of a mandate contract is:

$$C(a,b) = qual(a{:}mandatory(b)) \wedge qual(b{:}manager(a)) \wedge$$
$$O_{a:mandatory(b)}A \wedge P_{a:mandatory(b)}B \wedge$$
$$E_{a:mandatory(b)}B \rightarrow O_{b:manager(a)}C \wedge mandatory(b) : REP(b, A)$$

where:
A = *buy a house*;
B = *pay more than it was previously agreed*;
C = *give the necessary amount of money*.

Another example of a titularity contract, where agent a accepts to hold role r in the organization i, is:

$$C(a,i) = qual(a{:}r) \wedge$$
$$O_{a:r}B \wedge P_{a:r}C \wedge O_{i:itself}D$$
$$E_{a:r}\neg B \rightarrow O_{a:r}F$$

In this contract, specific obligations and permissions are attributed to agent a when acting in role r. These obligation and permission will be added to the deontic characterization of role r, inherited by agent a because he will become holder of r. Sanctions for unfulfilled obligations are also stated.

Finally, an example of a contract established by organization i and an agent b:

$$
\begin{aligned}
C(i,b) &= O_{i:itself}A \;\wedge\; O_{b:itself}B \\
E_{i:itself}\neg A &\to O_{i:itself}J \\
E_{b:itself}\neg B &\to O_{b:itself}G
\end{aligned}
$$

4.5 Institutional Agents in a Society of Autonomous Agents

We are now going to propose a formal model for organizations based on the concepts we have formalized above. The specification of an organization involves a name, i, and a structure ST_i: $\quad ST_i =< R_i, DCR_i, TO_i, RER_i >$.

ST_i, the *structure* of the organization i, is formed by:

R_i : a *set of roles*, associated to the organization. It is constituted by a finite set of atomic formulas of the form *is-role-str(r, i)*, stating that the role r is a role of the structure of the organization i (is a structural role).

DCR_i : the *deontic characterization of each role* - obligations, permissions or prohibitions that are intrinsic to the role. It is a (finite) set of formulas of the form $O_r A$, $P_r A$ or $F_r A$, where r is a role of the structure of the organization i. It may also include sanctions to be applied when violations occur, in any of the forms discussed before (e.g. $(\forall x).qual(x : r) \to E_{x:r}\neg A \to O_{x:r}B$, knowing that $O_r A$ is part of the deontic characterization of r).

TO_i : *transmission of obligations* from the organization to specific roles of its structure. It is formed by a set of formulas of the form $O_{i:itself}A \to O_r A$ (for r a role of the structure of i).

RER_i : contains information about the *representative roles* of the organization and its respective scope of representation. It is constituted by a set of formulas of the form $r : REP(i, B)$.

The specification of an organization i may also include other components, not considered here.

The description of $< i, ST_i >$ contains those aspects that do not change and define the identity of the organization. We need also to include in the specification of i, information describing the agents that in the present moment hold the roles r of the structure of i. Since this component corresponds to relationships between i and other agents, we have decided to include it in component NR (normative relationships) of the specification of the society of agents (see below).

A society of agents, SA, is constituted by:

$$SA =< IA, nIA, NR, GK >$$

where:

IA : Specification of each institutionalized agent of the society. So it is formed by a set of pairs $< x, ST_x >$, as explained above.

nIA : The component nIA contains the identification of the other agents that exist in the society.

NR : Contains normative relationships that agents have established between each other, and in particular the contracts $C(a, b)$ that are actually in force. Contracts between organizations and agents, attributing to the agents titularity of roles of its structure, are also included in this component.

GK : Contains general knowledge about the society[14].

5   Normative Analysis of Autonomous Agents in an Organizational Context

It is now possible to use the logic proposed to analyze, in a rigorous way, the effects of an action of an agent (in the actions of other agents and in the attribution of new obligations or permissions to agents), or in the detection of non-ideal behavior (unfulfillment of obligations).

A specification of a society of agents SA defines a particular language (e.g. with a constant for each agent) and a set of formulas of such a language. Let us call the theory defined by SA, denoted by $\mathcal{T}(SA)$, the logic obtained by adding such formulas to the underlying logic as new axioms. Then, supposing that Δ is a sequence of action formulas, we can analyze if some formulas B can be deduced from Δ in $\mathcal{T}(SA)$, which can be represented by:

$$\Delta \vdash_{\mathcal{T}(SA)} B$$

We assume that the notions of theorem, and of deduction with hypotheses, are defined as in, for example, Chellas [3].

To illustrate the type of reasoning supported by the proposed logic, we will present two examples of inferences that can be made from a specification of a society of agents.

Lets consider the following situation:

- We have an organization i, containing a role r as part if its structure, which deontic characterization is:
  $O_r A \ \wedge \ r : REP(i, A) \ \wedge \ (\forall_x).(E_{x:r} \neg A \rightarrow (O_{x:r} G \ \wedge \ F_{x:r} H))$
- The organization i has established a contract with agent b, as specified in section 4.4.
- The organization i has established the titularity contract with agent a described in section 4.4.:

Due to space limitations, we do not present the formal proofs. We just present some inferences and informally indicate how they can be deduced.

Case 1: Agent a acting in role r brings it about: B, then $\neg A$ and finally H.

1. $E_{a:r} B$: agent a fulfills the obligation of the titularity contract he has established with i.

---

[14] For example, if violations are typified, sanctions to general types of violations might be included here.

2. $E_{a:r} \neg A$: agent a violates the obligation $O_{a:r} A$, resultant from the fact that he holds role r (and we have $O_r A$). So he is sanctioned with $O_{a:r} G \wedge F_{a:r} H$, as stated in the deontic classification of r.

3. Moreover, as r is a representative role of i for A, we have $E_{a:r} \neg A \rightarrow E_{i:itself} \neg A$, and so, $E_{i:itself} \neg A$. Thus, the organization violates the contract it has with b, and is sanctioned with $O_{i:itself} J$.

4. $E_{a:r} H$: agent a is again in a non-ideal situation, because as a result of his past actions he is under the following prohibition $F_{a:r} H$.

Case 2: Agent a acting in role r brings it about: A, then $\neg B$ and finally H.

1. $E_{a:r} A$: agent a fulfills the obligation resultant from the fact that he holds role r ($O_r A$).Moreover, as r is a representative role of i for A, we have $E_{a:r} A \rightarrow E_{i:itself} A$, and so, $E_{i:itself} A$ holds, and i fulfills his obligation resultant from the contract with b.

2. $E_{a:r} \neg B$: agent a violates the obligation $O_{a:r} B$ of the titularity contract he has established with i. He will be sanctioned with $O_{a:r} F$.

3. $E_{a:r} H$: this is a regular action of a.

Suppose that agent a has to choose if he does A or B. He will have to do an analysis similar to the one presented above, in order to decide what is better for him.

6 Conclusion

In this paper we have discussed how organizations deal with autonomy of agents that constitute them. We have been specially interested in understanding how relationships between agents (and in particular between an agent and an organization that he is part of) may restrict or not their autonomy. We have used an high-level, role-based model of organizations, that does not regiment agents behavior, but just describes their expected behavior, fixing sanctions for agents that deviate from what is expected of them. An autonomous agent can always decide to follow or not the norms that apply to them. But in that latter case, he will be responsible for violating the rules, having sanctions applied to him. As organizations can never be sure that autonomous agents behave according to what is expected of them, they have to admit that violations may occur and must be prepared to deal with them. By sanctioning deviant behavior, organizations may try to influence agents not to violate norms. Contracts are a central notion to express interaction between autonomous agents.

The proposed model is supported by a first-order,many-sorted, action and deontic modal logic, allowing a formal representation of the concepts presented and reasoning about them. Part of this work have been presented in [11], [2] and [12]. Here, we adapt and extend it to the discussion of autonomy questions. We use a simplified version of the logic proposed, we omit some of the components of an organization and extend others.The notion of contract is extended and the notions of violation and sanction are discussed in more detail.

Although the discussion and the model proposed is based on human organizations, taking human persons as a paradigm of autonomous agents, we think it can be transposed to computational autonomy.

This research is in a preliminary stage, and the model used is yet very simple. In spite of that, we believe the introduced approach might be useful to understand autonomy of agents.

References

1. J. Carmo and A. Jones: "Deontic Logic and Contrary-to-Duties", D. Gabbay and F. Guenthner (eds.) *Handbook of Philosophical Logic - Second Edition*, **4**, pp. 287–363, Kluwer, 2001.
2. J. Carmo and O. Pacheco: "Deontic and action logics for organized collective agency, modeled through institutionalized agents and roles", *Fundamenta Informaticae*, Vol. 48 (No. 2, 3), pp. 129–163, IOS Press, November, 2001.
3. B. J. Chellas: *Modal Logic - an Introduction*, Cambridge University Press, 1980.
4. M. Gilbert: *On Social Facts*, Princeton University Press, 1989.
5. A.G. Hamilton: *Logic for Mathematicians*, Cambridge University Press, 1988.
6. R. Hilpinen (ed.), *Deontic Logic: Introductory and Sistematic Readings*, Dordrecht: D. Reidel, 1971.
7. A. J. I. Jones and M. J. Sergot: "A Formal Characterization of Institutionalized Power", *Journal of the IGPL*, **4(3)**, pp. 429–445, 1996. Reprinted in E. Garzón Valdés, W. Krawietz, G. H. von Wright and R. Zimmerling (eds.), *Normative Systems in Legal and Moral Theory*, (Festschrift for Carlos E. Alchourrón and Eugenio Bulygin), Berlin: Duncker & Humblot, pp. 349–369, 1997.
8. S. Kanger: *New Foundations for Ethical Theory*, Stockholm, 1957. (Reprinted in [6].)
9. S. Kanger: "Law and Logic", *Theoria*, **38**, 1972.
10. L. Lindahl: *Position and Change - A Study in Law and Logic*, Synthese Library **112**, Dordrecht: D. Reidel, 1977.
11. O. Pacheco and J. Carmo: "Collective Agents: From Law to AI". *Proceedings of 2nd French-American Conference on Law and Artificial Intelligence*, Nice, 1998.
12. O. Pacheco and J. Carmo: *A Role Based Model for the Normative Specification of Organized Collective Agency and Agents Interaction*, Journal of Autonomous Agents and Multi-Agent Systems, Vol. 6, Issue 2, pp. 145–184, Kluwer, March 2003.
13. I. Pörn: *The Logic of Power*. Oxford : Blackwell, 1970.
14. I. Pörn: *Action Theory and Social Science: Some Formal Models*, Synthese Library, **120**, Dordrecht: D. Reidel, 1977.
15. R. Tuomela: *The Importance of Us: A Philosophical Study of Basic Social Notions*, Stanford series in Philosophy, Stanford University Press, 1995.

Dynamic Imputation of Agent Cognition

H. Van Dyke Parunak and Sven A. Brueckner

Altarum Institute, 3520 Green Ct. Suite 300, Ann Arbor, MI 48105 USA
{van.parunak, sven.brueckner}@altarum.org

Abstract. People can interact much more readily with a multi-agent system if they can understand it in cognitive terms. Modern work on "BDI agents" emphasizes explicit representation of cognitive attributes in the design and construction of a single agent, but transferring these concepts to a community is not straightforward. In addition, there are single-agent cases in which this approach cannot yield the desired perspicuity, including fine-grained agents without explicit internal representation of cognitive attributes, and agents whose inner structures are not accessible. We draw together two vintage lines of agent be imputed externally to a system irrespective of its internal structure, and the insight from situated automata that dynamical systems offer a well-defined semantics for cognition. We demonstrate this approach in both single agent and multi agent examples.

1   Introduction

Communities of agents are engineered artifacts. They are constructed by people, in order to solve problems for people, and to be effective they must be understood and controllable by people. This simple observation accounts for the popularity of cognitive metaphors in motivating and designing agents. People deal with one another in cognitive terms such as beliefs, desires, and intentions. Our intuitions and predictive instincts are tuned by generations of experience to manipulate epistemological, praxiological, and axiological categories, and we find it most natural to interact with computer systems in the same terms. This reality explains the growing popularity of so-called "BDI agents," those whose internal representations include structures that map explicitly onto cognitive characteristics (for examples, see [12]).

In some cases, such a design approach to cognition is not feasible. If a system is a collection of agents with emergent properties, there is no central location where such a representation could live. In addition, the internal structure of some agents (e.g., neural networks or reactive architectures) may not lend themselves to interpretation as cognitive primitives, or may be unknown. Still, people must interact with such systems. How can the cognitive gap be bridged in these cases?

This paper draws together two vintage lines of thought in addressing this question. The first, dating back more than thirty years, is the recognition that cognitive categories may legitimately be used for *analysis* even if they are inappropriate for *design*. This insight tells us what we want to do: impute cognition to individual agents and

M. Nickles, M. Rovatsos, and G. Weiss (Eds.): AUTONOMY 2003, LNAI 2969, pp. 209–226, 2004.

communities externally, based on their behavior, independent of their internal structure. The second line of thought is the notion that the semantics of cognition can be grounded in dynamical systems theory, a position advocated in the situated automata community. In that community, this notion is used as a design principle, but we argue that it can also give us a disciplined way to carry out the analysis agenda suggested by the first line of thought.

Section 2 reviews the need for a cognitive understanding of agents and the cases in which BDI architectures are not the answer. Section 3 reviews the two lines of thought summarized above (cognitive categories for analysis, dynamics as a semantics for design) and proposes a synthesis (dynamics as a semantics to support cognitive analysis) and supporting tools. Section 4 applies this scheme to several examples. Section 5 concludes.

2   The Problem

The growing popularity of BDI architectures attests to the intuitive need that people have to relate to their agents in cognitive terms. The motivation for this paper is the recognition that this need persists even when it is difficult or impossible to satisfy it through explicit representations of cognitive constructs in an agent or multi-agent system. We summarize three increasingly common circumstances that can frustrate the design approach to cognition, and discuss the contribution that an analytic approach to cognition can make to research in multi-agent systems.

When does the design approach fail?

The design approach to cognition fails in at least three cases: decentralized agent communities with emergent properties, non-symbolic agents, and black-box agents.

Emergent Behavior of Swarms.—A human stakeholder's interest in a multi-agent system is often at the level of the community as a whole rather than the individual agents, and cognition at this level can be difficult to represent. Consider, for example, resource allocation. A single agent seeks to maximize its own utility, but the system as a whole needs to maximize the productivity of the entire set of agents, which usually requires that some individual agents receive less than they want. A human relying on a multi-agent resource allocation system needs to be assured that the system (*qua* system) wants to maximize overall productivity, and may inquire concerning actions it intends (again, as a system) to take. Team coordination is an active area of research. Some efforts take the design approach to cognition by maintaining explicit cognitive constructs such as team plans and intentions, for example in a team infrastructure to which all participants have access [30]. In other cases, coordination may take place emergently through the shared problem domain [25] or through a non-symbolic infrastructure such as digital pheromones [4], and in these cases another approach is needed to provide human stakeholders with the cognitive interface they require.

Non-symbolic Agents.—Explicit representation of cognitive constructs is usually grounded in a logical model of these constructs (a "BDI logic," e.g., [21, 31]), which in turn guides the construction and manipulation of symbolic structures in the agent's reasoning. There is increasing interest in models of agent reasoning that are numerical rather than symbolic (for example, in the swarm intelligence community [2, 23]). Reasoning in such agents may consist of neural networks, evaluation of polynomial or transcendental functions, or even weighted stochastic choice, with no clear mapping between the internal representation and useful cognitive concepts.

Black-box Agents.—Agent internals may be invisible to the humans who need to interact with an agent or agent system, and thus not helpful in reasoning about the system's behavior. This "black box" status can occur in two ways. First, the population of an open system [14] may be constructed by many different people, using different models. As long as the sensors and effectors of individual agents are compatible with the environment shared by the agents, their internals can be inaccessible (and should be, to take full advantage of the modularity that agents can offer). Second, it is increasingly difficult to distinguish between synthetic and natural agents. A distributed military simulation, for example, may include both carbon-based and silicon-based agents, and a trainee interacting with the system through a computer terminal may not be able to tell the difference. The internal representations and processing of people (carbon-based agents) are in general not accessible to outside observers. Even introspection is notoriously unreliable.

Why should agent researchers care?

The insights we develop will be useful to researchers and developers in MAS at several levels.

Human interfaces to the three cases outlined in the previous section depend on the ability to represent these systems cognitively. Our insights will enable the development of interfaces that make non-symbolic or black-box agents, or swarms of agents, more understandable to the human user. They will also reduce the bandwidth required to communicate between operators and systems, by permitting both status reports and commands to be expressed in cognitive terms rather than by transmitting large segments of the system state. This economy is particularly critical for swarming systems, in which the information needed to characterize the system's state is typically not localized spatially.

Hybrid agent systems combining conventional BDI agents with the kinds of agents and systems described in the previous section are becoming more common, and require interoperability between BDI agents and these other agents and systems. Our methods facilitate agent-to-agent interaction in two ways: as a tool used by a BDI agent to interpret the behavior of an agent or swarm without designed cognition, or as a "wrapper" around such agents or swarms to enable them to communicate with BDI agents on their own terms.

Introspection is the ability of an agent to reflect on its own behavior and that of groups in which it is involved, and is directly supported by an analytic approach to cognition. Even BDI agents may make use of such methods, for two reasons. First, full prediction of the dynamics of interaction is difficult and sometimes formally impossible, so that agents in a MAS must resort to interpretation of the group's actual dynamics in order to understand the full effect of their actions. Second, while a BDI agent may rationally plan its interactions with the real world, the world may constrain the effects of those interactions in unexpected ways [11], so that it may sometimes be helpful for such an agent to observe the effects of its own situated actions and interpret them cognitively.

3   New Life for Old Theories

A solution to the challenge of a cognitive interface to communities or non-cognitive agents lies in synthesis of two classical concepts: the legitimacy of an analytical (as opposed to a design) approach to cognition, and the usefulness of dynamics as a semantic foundation for cognitive concepts.

Legitimacy of Cognitive Analysis

Much of the inspiration for BDI architectures comes from publications of Daniel Dennett [9, 10] and John McCarthy [22]. In view of the current emphasis on BDI as an architectural framework, it is important to recall that both Dennett and McCarthy were arguing the legitimacy of using cognitive categories to describe entities and systems whose internal structure does *not* explicitly reflect those categories.

Dennett's "Intentional Stance".—Dennett describes three increasingly abstract stances that an observer can assume in predicting the behavior of a system. The *physical stance* is the grand Laplacian strategy of understanding the universe from first physical principles, based on the physical composition of the system and its environment and the laws of physics. The *design stance* presumes that the system in question has been designed by an intelligent engineer and that it will behave in accordance with the engineer's design. Thus the observer can reason at whatever levels of abstraction the engineer has employed. The *intentional stance* treats an entity as an intelligent agent, imputes to it cognitive attributes such as beliefs, desires, and intentions, and then reasons over those constructs to predict the entity's behavior.

The difference between the intentional stance and primitive animism is that the former must be warranted in some disciplined way. BDI architectures seek to warrant the intentional stance by applying cognitive constructs at the design level. This approach does in fact justify taking the intentional stance, but at the same time renders it unnecessary. If the agent design is cognitive, one can reason cognitively about the agent from the design stance, and the problem is solved. Dennett argues extensively that the intentional stance is warranted even in the absence of a cognitive design, *if* it enables accurate predictions to be made more parsimoniously than reasoning from the physical or design stance.

McCarthy's Formalisms.—McCarthy addresses the question of how to formalize the predictive warrant that Dennett requires. In line with his other contributions to the neat tradition of AI, the two constructions he suggests are both based on formal logic.

The first approach draws on the notion that one logical theory T' may be an approximation to another T, in the sense that when the world is in a certain state (according to T), a correspondence may be drawn between some states and functions of T and those of T'. T' may have some additional functions of its state variables that cannot usefully be mapped onto functions of T, but that are useful for constructing theorems within T'. Such functions might include cognitive concepts that do not have natural correlates in the base theory T. In Dennett's terms, the base theory is the physical or design reality of the system, while the approximating theory is the intentional stance, and the usefulness of the cognitive functions corresponds to predictability in Dennett's framework.

McCarthy's second approach uses second-order structural descriptions. "Structural" in this case is contrasted with "behavioral," as dealing with the system's state in contrast with its actual or potential behavior. Consider the specific cognitive function of belief. Let s be a state of a machine, p a sentence, and B a belief predicate, such that $B(s,p)$ means that the machine believes p whenever it is in state s. The applicability of such a predicate clearly depends on the individual machine M and the world W in which it is situated, so the full specification of such a first-order structural predicate would have to include these arguments: $B(W, M, s, p)$. McCarthy deems construction of such a predicate to be impossible. Instead, he suggests explicating belief by means of a second-order predicate (W, M, B), which "asserts that the first order predicate B is a 'good' notion of belief for machine M in world W," where "'good' means that the beliefs that B ascribes to M agree with our ideas of what beliefs M would have." Clearly, explanatory and predictive power will be central in such a common-sense evaluation of the appropriateness of a given ascription, again aligning with Dennett's basic position.

Jonker and her colleagues have revived the notion of cognition as something that is useful for analyzing agents (in contrast with designing them). Continuing in the logical tradition of McCarthy, they build on the BDI logics developed in mainline agent research [15]. They observe that these logics are temporally defective: "within a BDI-logic, for a given world state all beliefs, desires and intentions are derived at once, without internal dynamics." Without internal dynamics, it is difficult to specify the dynamics of the interaction between the agent's cognitive state and its environment. Their solution is a formal trace language that reasons over temporal sequences of the agent's input and output states and can express formal criteria for relating interaction histories to a specific intentional notion. They demonstrate the power of these tools by using them to attribute beliefs, desires, and intentions to the bacterium *Escherichia coli*, then use these cognitive properties to derive predictions that are judged nontrivial and interesting by cell biologists, in a way that is much more efficient than reasoning from biochemical first principles (the "physical stance").

Dennett and McCarthy argue that cognition can legitimately be imputed in *analyzing* agents even if it is not present in the *design*. The approach of Jonker *et al.* emphasizes the need for a dynamical (i.e., time-based) formalism to support such imputation, but they graft it onto the logical tradition of McCarthy, which in turn reflects the adherence of mainstream AI to Simon's "physical symbol system" hypothesis.

Kaminka, Veloso, and colleagues [3, 16] detect primitive behaviors on the basis of precompiled rules, and then analyze statistical dependencies in time series of these primitive behaviors to learn coordination and detect coordination failure. Their approach is closer in spirit to ours than is Jonker's, but grounds the semantics of the system in its primitive behaviors, not in dynamics per se.

Dynamic Semantics for Cognitive Design

In spite of the many accomplishments of the Physical Symbol System approach, some aspects of natural cognition seem better understood as the result of non-symbolic dynamical processes [29], and issues of computational limitations and the need for real-time response have prompted the exploration of a non-symbolic approach to the design and implementation of artificial agents [20, 33]. In particular, research in situated automata has laid the foundation for using dynamical systems theory to ground cognitive concepts.

Situated Automata Meets Cognition.—The situated automata community [28] has advocated an approach to agent reasoning that stands in contrast to the symbolic paradigm. Driven by the real-time constraints of robotics, this community seeks to reduce or eliminate symbolic deliberation in an agent, and construct agents as finite-state machines driven by sensors immersed in the environment and issuing directly in effectors that attempt to change the environment.

A landmark in this community's engagement with the cognitive paradigm was Rosenschein's 1985 proposal [32] that the semantics of the cognitive concept of "knowing" or "believing" be described nonsymbolically, as a correlation between states of the environment and states of the agent. He showed that such a definition satisfies the axioms of epistemic logic (the S5 system), and thus qualified as a well-founded model of knowledge. He and his colleagues used this insight as the basis for a robotic design methodology, including the languages Rex and Gapps, that specify action mappings and compile them into runtime circuitry [33].

Kiss [18] extended the use of nonsymbolic dynamics as a semantics for cognition beyond the epistemic notions considered by Rosenschein, to include praxiologic (action-related) concepts such as intentions, commitment, actions, and plans, and axiologic (value-related) concepts such as desires, goals, and values[1]. Like Rosenschein, his focus was on design of new agents, not analysis of existing ones, but his insights can be extended and applied in analysis as well.

Understanding this approach requires a brief summary of dynamical[2] systems theory and its cognitive interpretation.

Dynamical Systems Theory.—Dynamical systems theory [1, 38] is a body of theory that emphasizes the evolution of a system over time more than its end-point. It focuses on the process more than the product, the journey more than the destination. Thus its concepts are useful tools for describing ongoing agent processes. Some of these concepts are system state space, trajectory, phase change, attractor, transient, repeller, and basin of attraction.

The *state space* of a dynamical system is a collection of time-dependent variables that capture the behavior of the system over time. A system's state variables may include both independent variables (those imposed exogenously, for example by the

---

[1] We independently noted a number of these same correspondences in a more applied context in [24].

[2] The use of "dynamical" rather than the simple adjective "dynamic" has become conventional in this community.

experimenter), and dependent ones (those that vary as a result of the system's opera-tion). Independent variables are sometimes called "parameters."

Classical AI is familiar with the notion of a state space as a problem-based struc-ture within which a reasoner searches for a solution. The dynamical systems use of the concept differs in two subtle but important ways. First, we are interested, not in the state space of something external to the reasoner (the "problem" being solved), but in the state space of the reasoning system itself. Second, our focus is not on the end point of movement in the state space, but on trajectory of the movement itself.

In the situated automata agenda, the notion of state space (including both overall system state and local agent state) is the foundation for epistemic notions of **knowl-edge** and **belief**. An agent is said to believe a proposition p out its environment just when there is a correlation between its state and the environmental state predicated in p. The treatment by Rosenschein, refined by Kiss, focuses on the knowledge of a sin-gle agent. As a later example (Section 4) will make clear, these concepts need refine-ment when one moves to a multi-agent system. In their original form, they warrant ascribing knowledge to the system if any agent in the system has the knowledge, but do not distinguish between different degrees of access to this knowledge across the population. One can quantify the correlation among agents in terms of their joint in-formation [26], and this concept permits the formal distinction of concepts such as common knowledge within the dynamical framework.

The state variables vary as the system evolves over time. Their successive values as the system evolves define the system's *trajectory* through state space. For an agent-based system, this trajectory represents a trace of the system's behavior through time, and the trajectory of the agent through its state space represents a trace of the agent's history.

Not all regions of state space are equally accessible to the system, and the remain-ing concepts characterize different aspects of this uneven accessibility.

A *phase change* occurs when a trajectory that changes continuously in some vari-ables experiences a discontinuous change in others. Most commonly, the continuous change is imposed exogenously, but sometimes the system's endogenous evolution leads to a phase change [5]. We distinguish "phase change" (an experimentally ob-served discontinuity) from a "phase transition" (a point of non-analyticity, predicated on an analytic model of the dynamics). Phase changes are often symptomatic of phase transitions, but one is an experimental, the other a theoretical concept.

In many dynamical systems, the trajectory eventually enters an *attractor*, a sub-space of restricted dimensionality from which (under deterministic evolution) it never emerges. The portion of the trajectory that precedes the attractor is the system's *tran-sient*. A system may have several disjoint attractors, each accessible from different initial conditions. This set of attractors offers a natural semantics for the system's **de-sires** [19]. It is entirely natural, and predictively fruitful, to say that the system "wants" to enter one or another of its attractors. There can also be regions that repel the system trajectory, naturally termed *repellors*, and these offer a natural interpreta-tion of **aversions** or dislikes.

The set of states from which trajectories enter a given attractor is termed that attractor's *basin of attraction*. The basin of attraction is a superset of the attractor, because it includes regions traversed by transients on their way to the attractor. In cognitive terms, a system in a basin of attraction has chosen the associated attractor as its **goal**. If it persists in this choice, we say that it **intends** to pursue that goal[3].

Indecision can arise from the inevitable noise associated with an agent's sensors and effectors. As a result of this noise, an agent generally knows its location in state space and the consequences of its actions only to within a region of state space. If the volume of this region is large compared with the local dimensions of the agent's current basin of attraction, noise may carry the agent out of that basin into another one, leading to a change of goal and perhaps eventually to a change in intention. We will see an example of this in Section 4.

Noise in an agent's sensors or actuators, or the discontinuities inherent in a phase change, can sometimes carry an agent out of an attractor or a basin of attraction. In this case the agent has **changed its mind**.

Synthesis: Dynamical Systems for Cognitive Analysis

The dynamical semantics that prove so useful in situated automata for agent *design* can be also used to *analyze* agents and systems of agents. This insight extends the agenda of Dennett, McCarthy, Jonker et al., and Kaminka et al. from its original foundation in formal logic to domains where nonsymbolic methods may be more appropriate. It also extends the situated automata program, by allowing us to talk meaningfully about collections of agents as having group cognition even in the absence of explicit representations.

Implementing this agenda requires some familiarity with the basic tools used for analyzing nonlinear dynamical systems. These include reduction of dimensionality, simulation, and reconstruction from time series.

Reduction of Dimensionality.—The step from problem state space to system state space can be daunting. One important characteristic of a test problem for a problem-solving mechanism is that its state space be constrained for tractability. The state variables needed to describe most AI mechanisms are orders of magnitude larger, and it can be intimidating to contemplate representing and manipulating such a state space. Fortunately, a greatly reduced subspace is often sufficient to capture and analyze behaviors of interest in a dynamical system. For example, the state space of the human heart in terms of fundamental physical variables is very large, but important characterizations can be developed in a subspace of only two dimensions [17]. Analysis of the example in Section 4 will illustrate this technique.

Simulation.—Simulation studies are a key technique for exploring the dynamics of non-symbolic agents and of agent communities. They permit us to test the sensitivity of the dynamics on various dimensions of state space, enabling us to identify those variables that are most critical. Since many of these systems have stochastic components, simulations are essential for determining the range of variation of behavior un-

---

[3] Our choice of terms follows the definition of intention as choice with commitment proposed in [7], but the use of dynamics to provide a semantics for cognition is equally applicable to the simpler view under which intention exists once an agent has chosen a particular desire, whatever its level of commitment.

der noise. Simulations can also produce visualizations of regions of state space, from which the characteristics described in the last section can be determined by inspection. Simulation underlies the analysis of both examples discussed in Section 4.

Exhaustive simulation of a high-dimensional state space can be prohibitively expensive, particularly when multiple runs at each location are needed to determine the aggregate effect of agents with stochastic behaviors. APSE (Adaptive Parameter Search Environment) [6] enables us to allocate simulation processes to different regions of state space dynamically on the basis of runs already completed, significantly reducing the computational effort needed to explore complex systems.

Reconstruction from Time Series.—Visualization from simulations is adequate to analyze the examples we will consider in this paper, but in more complex cases the agent community will want to take advantage of a wide range of analysis techniques that have been developed for time series from nonlinear systems (e.g., [17]). Many of these methods are based on a theorem by Takens [34, 39] that permits reconstruction of the system's trajectory from any time-varying observable that meets minimal requirements. Let $x(t)$ be the system's state vector as a function of time, and $s(x)$ an accessible observable. Observations are taken at regular time intervals, $s(x(t_0))$, $s(x(t_0+\Delta t))$, $s(x(t_0+2\Delta t))$, ... One can construct a synthetic n-dimensional state vector $\xi(i)$ whose elements are successive n-tuples from this time series,

$$\xi(i) = <s(x(t_0+i\Delta t)), s(x(t_0+(i+1)\Delta t)),...s(x(t_0+(i+n-1)\Delta t))> \tag{1}$$

Informally, Takens's Theorem asserts that the topology of the system's trajectory in its native state space, projected onto an n-dimensional subspace, is completely captured by these synthetic state vectors reconstructed by time delays from an accessible observable. The conditions on the theorem are surprisingly lenient, and widely satisfied. The observational mapping $s(x)$ must be continuous through the second derivative, and either the intrinsic dynamics of the system or the observational mapping must couple all of the intrinsic state variables, thus ensuring that the single observable carries information from all dimensions of the underlying state space. In some cases the requirement for evenly spaced observations can be relaxed, permitting the use of event-based data [13, 36].

It is sufficient and often surprisingly effective to plot the derived multi-dimensional vectors and inspect the resulting plots for structure. Mathematical techniques for detecting such structure include the S-measure [35] and epsilon machines [37].

4 Examples

We illustrate how one can impute cognition externally to agents using dynamics with two examples. The first is a non-symbolic single agent, while the second is a multi-agent system with emergent system-level behavior.

Newt

Consider a simple agent, "Newt," that lives in the complex plane and executes Newton's method for finding roots of equations. This method iteratively approximates the root of a function $f(x)$, beginning with an initial estimate x_0, using the update rule

$$x_{n+1} = x_n - f(x_n)/f'(x_n) \qquad (2)$$

Our particular instance of Newt seeks the roots of $x^4 = 1$. At each time step, Newt measures its current position using its sensors, computes the next position to which it should move, and moves there using its effectors. What can we say about Newt's cognition?

Beliefs.—Newt's *beliefs* consist of its current location in state space. Ideally, Newt knows its location exactly, but in practice, due to errors in its sensors and effectors, its beliefs are approximate. However, there will still be a correlation between its actual position and the values of the variables with which it computes, warranting the imputation of beliefs.

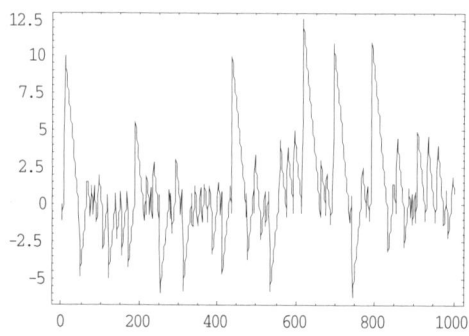

Fig. 1. Log plot of successive iterates starting at (1+i)

Desires.—Dynamically, Newt has four stable attractors: (1+0i), (−1+0i), (0+i), (0 − i). If ever it reaches any one of these points, it will stay there. It also has two unstable attractors, consisting of the diagonals where the real and imaginary parts of x are equal. Exact iteration of (2) from a point such as (1+i) will keep the real and imaginary parts of x identical, but both will oscillate erratically (Fig. 1). However, any small noise will knock the process off the diagonal, and lead it eventually to one of the four stable attractors. In cognitive terms, Newt *desires* to reach one of the four stable points, but can be *distracted* if it starts on one of the unstable ones.

Intentions.—In spite of its simplicity, Newt can have a hard time translating its beliefs about its current location and its global desires into a specific intention. Fig. 2 shows a portion of the complex plane around (0+0i). Each attractor is surrounded by a large triangular basin of attraction, and if Newt enters one of these basins, it will be drawn to the corresponding attractor. Cognitively, we say that it has adopted an *intention* to pursue the desire represented by the attractor.

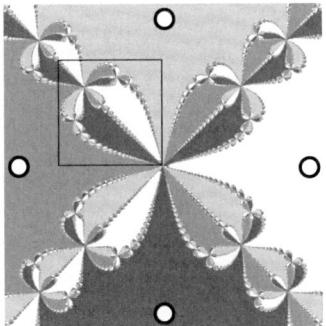

Fig. 2. Dynamics of Newt.—The four points near the edges of the figure mark the attractors. The four gray tones distinguish the basins of attraction corresponding to each attractor. Note the complex structure of the boundaries between each quadrant of the state space

The boundaries between the main triangular regions are complex. The basin of attraction for the attractor at (1+0i), colored in white, includes not only the large triangular region contiguous with the attractor, but also numerous other discontinuous regions, which ramify fractally as one examines the state space more closely (Fig. 3). This complexity has serious implications for Newt.

Fig. 3. Fractal Boundary.—The plot shows the region with upper-left corner (–.7+.7i) and lower-right corner (0+0i), outlined in Fig. 2

First, consider the case when Newt starts its explorations at (1+i). If its sensors and actuators are perfect, it will never leave the diagonal, but will oscillate erratically. However, few perfect sensors and actuators are available in the real world, and any small deviation from the diagonal will lead to a basin of attraction, thus initiating a *decision* by Newt to pursue a selected *goal*. Fig. 4 shows the trajectory of noisy Newt, starting at (1+i). (This trajectory, as well as the desire finally adopted, depends sensitively on the noise in Newt's environment; successive runs of noisy Newt can follow many different trajectories and end up at any of the four attractors with equal probability.) The first two steps lie very close to the diagonal: points on the diagonal have complex argument $\pi/4 = 0.785$ radians, while the arguments of the first two steps are 0.794 and 0.810, respectively.

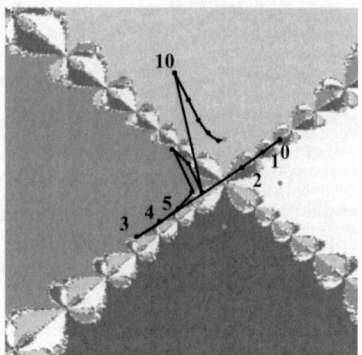

Fig. 4. Newt's trajectory with noise.—At each step, the real and imaginary parts of the location are corrupted with a random number uniformly selected between –0.01 and 0.01

Table 1. Noisy Newt's initial steps

Step	Location	Basin for attractor at
0	1+i	+ diagonal
1	.684 + .696i	0 – i
2	.321 + .338i	1 + 0i
3	–1.580 – 1.238i	0 + i
4	–1.173 – 0.900i	0 + i
5	–0.847 – 0.596i	1 + 0i
6	–0.578 – 0.218i	0 + i

The basins of attraction are so closely intertwined in this region that the small errors in Newt's movement place it in different basins of attraction in each of these early steps. Table 1 gives the location and basin of attraction at each of Newt's first six steps. Its inherent sensory noise causes it to visit three out of the four possible basins. Only in the sixth step does it adopt a goal that persists throughout the rest of the run. Even then, its trajectory carries it back and forth between the boundary regions until step 10 finally brings it to the large triangular basin, from which it makes its way deliberately to its goal. These dynamics suggest that while Newt has an *goal* at each step, only on step 6 is its goal persistent, so that it can be said (in the terms of Phil Cohen and colleagues [7, 8]) to adopt an *intention* for a particular course of action.

Newt illustrates how cognition can be imputed to a nonsymbolic agent externally, based on its behavior, using a dynamical rather than a logical foundation. It also illustrates two principles of broader interest in the agent community.

First, noise can be virtuous. Without noise, Newt gets stuck on the unstable diagonal attractor, unable to adopt an intention. A small amount of noise is sufficient to break its symmetry and let it proceed toward a goal. We have found repeatedly in swarming systems that a stochastic element in decision-making can dramatically increase a system's effectiveness.

Second, intention and commitment cannot be divorced from the agent's interactions with the outside world. When Newt is in a basin of attraction for one attractor, it perfectly plans a next step to carry it to another portion of that same attractor, but two factors can conspire to throw it into a different basin and thus change its intentions: the highly intertwined nature of the basins of attraction in its world, and the inevitable noise in its interactions with the real world that is sufficient to move it from one basin to another. These two factors do not depend on the simplistic nature of Newt's computation. More complex agents, computing intentions logically rather than arithmetically, are not immune to complex state spaces and interaction errors, and in some cases the increased complexity of their computation may actually increase the complexity of their trajectories through state space and thus their susceptibility to influence from noise. Commitment, in the sense of persistent intention, cannot be divorced from the effect of the environment on the agent.

Path Planning

At a more practical level, and to illustrate the imputation of cognition to a community as opposed to an individual agent, we consider the ADAPTIV architecture for emergent path planning with digital pheromones [27]. The foundation of this system is a network of place agents, each responsible for a region of geographical space. The structure of the network is such that two place agents are connected just when the geographic regions they represent are adjacent. Each place agent hosts avatars, agents that represent targets, threats, and vehicles in its region. These avatars deposit primary pheromones indicating their presence and strength. The place agents maintain the levels of these pheromones using the pheromone dynamics of aggregation, evaporation, and propagation.

As vehicles move through the space, their avatars continuously emit ghost agents that swarm over the network of place agents. Ghost agents execute an ant-inspired path planning algorithm [23], seeking target pheromones and avoiding threat pheromones. When a ghost finds a target, it returns to the issuing avatar, depositing a secondary pheromone. The aggregate strength of these ghost target pheromones emergently forms a path from the vehicle to the selected target.

Management of highly distributed systems of this sort is increasingly important for military and other applications, and other agents (both silicon and carbon) would benefit greatly by being able to relate to such a system at a cognitive level. What can be said about the beliefs, desires, and intentions of such a system?

Beliefs.—As anticipated in our theoretical discussion, ADAPTIV's state can be correlated with the state of the world at several different levels, representing different degrees of knowledge on the part of different agents.

- The presence of an avatar at a place constitutes belief by that place that there is an entity of the corresponding type in the region for which the place agent is responsible. This knowledge is local to an individual place agent. The system "knows" about these entities in the sense that it contains that knowledge, but not every member of the system has access to it.
- Diffusion of primary pheromone from an avatar's place to neighboring places conveys knowledge to them of the presence of the entity nearby, and by sensing the gradient to neighboring place agents, they can estimate its direction. However, the pheromone field decays exponentially, and so extends only a limited distance.
- Ghosts that encounter targets or threats embody that information in their movements and thus in the places where they deposit secondary pheromone. This pheromone condenses into a path that reaches from the avatar that sent out the ghosts to the target. At this point the issuing avatar has long-range knowledge (the best step to take toward one particular target, in light of any threats that may exist along the way).

Desires.—Concentrations of attractive and repulsive pheromones mark place agents as desirable or undesirable, and the attractors for the system consist of states in which avatars are located at place agents marked with locally maximal deposits of attractive pheromone and locally minimal deposits of repulsive pheromone. Since pheromone concentration is strongest at the place agent occupied by the depositing avatar, these maxima correspond to the locations of the targets and threats. The dynamic model warrants the semantics that the guided avatar *wants* to reach targets and avoid threats.

Intentions.—The system's decision is manifested by the emergence of a dominant pheromone path leading around threats and to a particular target. Fig. 5 shows the secondary pheromone field laid down by ghosts (shaded from dark to light), and the resulting path around the threats and to the higher priority target on the west. A secondary peak in the pheromone field leads to the northern target. If another threat were to be discovered blocking the primary path, the secondary peak would quickly capture the path, manifesting a decision at the system level. This decision cannot be attributed to any individual agent, but is an emergent property of the system. Nevertheless, using dynamical concepts, we can legitimately describe it as a decision, and other agents (synthetic and natural) can reason about it in cognitive terms.

Fig. 6 shows a subset of the state space that illustrates basins of attraction for two decisions. (These basins can be clearly seen in this two-dimensional subspace of the highly-dimensional state space of this system, exemplifying the reduction of dimensionality discussed above.) The vehicle avatar is between two target avatars. The strength of target 1 is 10, and its distance to the vehicle is 5 units, and the strength and distance of the second target vary as indicated along the axes. On the upper left of the space, where the two targets are of comparable strength and target 1 is closer than target 2, the path forms to target 1 with probability 1. In the lower right, as target 2 grows stronger and closer compared with target 1, the path forms to target 2 with probability 1. The transition region between these two attractors is very narrow. Over

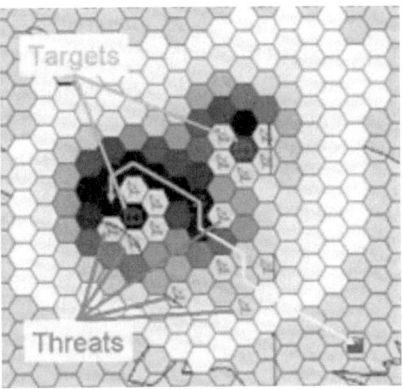

Fig. 5. Alternate Decisions.—The system has currently decided on the western target, but a backup path has concurrently formed leading to the northern one

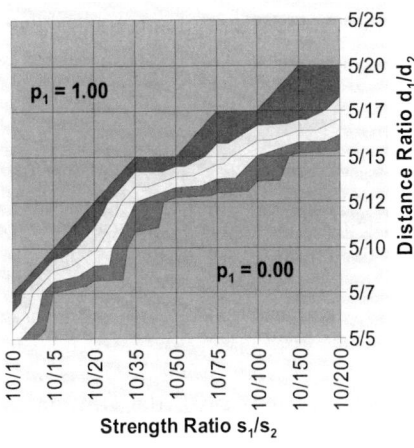

Fig. 6. Attractors for alternate desires.—The vehicle is between two targets, and selects one or the other based on their relative distances and strengths

most of the parameter space, the decision can be predicted with great certainty based on knowledge of the relative strength and distance of the different targets.

5   Conclusions

The dominant approach to cognition in agents is based on *design*, and most often on explicit representation of cognitive structures within individual agents. It is often desirable for other agents (both synthetic and natural) to be able to interact cognitively with agents and agent systems whose design is either inaccessible or else does not explicitly support cognition. We describe this approach to cognition as the *analytic* approach. This was in fact the perspective under which cognitive concepts were originally applied to artificial systems, using logical formalisms. When dealing with non-symbolic agents or swarms with emergent properties, it is more natural to base this imputation on the observed behavior of the system, using concepts from dynamical systems theory. This approach was adumbrated in the situated automata community, but for single agents and from a design rather than an analysis perspective. By extending it appropriately, we obtain a disciplined semantics for discussing communities of agents and nonsymbolic or black-box individual agents in cognitive terms.

Acknowledgements

This work was supported in part by DARPA (JFACC under contract F30602-99-C-0202, NA3TIVE under contract N00014-02-C-0458). The views and conclusions in this document are those of the authors and should not be interpreted as representing the official policies, either expressed or implied, of the Defense Advanced Research Projects Agency or the US Government.

References

1. R. Abraham and C. D. Shaw. *Dynamics--The Geometry of Behavior*. 2nd edition ed. Addison-Wesley, 1992.
2. E. Bonabeau, M. Dorigo, and G. Theraulaz. *Swarm Intelligence: From Natural to Artificial Systems*. New York, Oxford University Press, 1999.
3. B. Browning, G. A. Kaminka, and M. M. Veloso. Principled Monitoring of Distributed Agents for Detection of Coordination Failures. In *Proceedings of Distributed Autonomous Robotic Systems (DARS-02)*, 2002.
4. S. Brueckner. *Return from the Ant: Synthetic Ecosystems for Manufacturing Control*. Dr.rer.nat. Thesis at Humboldt University Berlin, Department of Computer Science, 2000.
5. S. Brueckner and H. V. D. Parunak. Information-Driven Phase Changes in Multi-Agent Coordination. In *Proceedings of Autonomous Agents and Multi-Agent Systems (AAMAS 2003)*, 950–951, 2003.
6. S. Brueckner and H. V. D. Parunak. Resource-Aware Exploration of Emergent Dynamics of Simulated Systems. In *Proceedings of Autonomous Agents and Multi-Agent Systems (AAMAS 2003)*, 781–788, 2003.
7. P. Cohen and H. Levesque. Intention is Choice with Commitment. *Artificial Intelligence*, 42:213–261, 1990.
8. P. R. Cohen and H. J. Levesque. Persistence, intention and commitment. In P. R. Cohen, J. Morgan, and M. E. Pollack, Editors, *Intentions in Communication*, 33–69. MIT Press, Cambridge, MA, 1990.
9. D. C. Dennett. Intentional Systems. *Journal of Philosophy*, 8:87–106, 1971.
10. D. C. Dennett. *The Intentional Stance*. Cambridge, MA, MIT Press, 1987.
11. J. Ferber and J.-P. Müller. Influences and Reactions: a Model of Situated Multiagent Systems. In *Proceedings of Second International Conference on Multi-Agent Systems (ICMAS-96)*, 72–79, 1996.

12. Haddadi and K. Sundermeyer. Belief-Desire-Intention Agent Architectures. In G. M. P. O'Hare and N. R. Jennings, Editors, *Foundations of Distributed Artificial Intelligence*, 169–185. John Wiley, New York, NY, 1996.

13. R. Hegger and H. Kantz. Embedding of Sequences of Time Intervals. *Europhysics Letters*, 38:267–272, 1997.

14. Hewitt. Open Information Systems Semantics for Distributed Artificial Intelligence. *Artificial Intelligence*, 47:79–106, 1991.

15. M. Jonker, J. Treur, and W. d. Vries. Temporal Analysis of the Dynamics of Beliefs, Desires, and Intentions. *Cognitive Science Quarterly*, (forthcoming), 2002.

16. G. A. Kaminka, M. Fidanboylu, A. Chang, and M. Veloso. Learning the Sequential Behavior of Teams from Observations. In *Proceedings of RoboCup Symposium*, 2002.

17. H. Kantz and T. Schreiber. *Nonlinear Time Series Analysis*. Cambridge, UK, Cambridge University Press, 1997.

18. G. Kiss. Autonomous Agents, AI and Chaos Theory. In *Proceedings of First International Conference on Simulation of Adaptive Behavior (From Animals to Animats)*, MIT Press, 1991.

19. G. Kiss and H. Reichgelt. Towards a semantics of desires. In E. Werner and Y. Demazeau, Editors, *Decentralized AI 3 --- Proceedings of the Third European Workshop on Modelling Autonomous Agents in a Multi-Agent World (MAAMAW-91)*, 115–128. Elsevier Science B.V., Amsterdam, Netherland, 1992.

20. Lux and D. Steiner. Understanding Cooperation: An Agent's Perspective. In *Proceedings of First International Conference on Multi-Agent Systems (ICMAS'95)*, 261–268, MIT and AAAI, 1995.

21. M.Wooldridge. *Reasoning about Rational Agents*. Cambridge, MA, MIT Press, 2000.

22. J. McCarthy. Ascribing Mental Qualities to Machines. In M. Ringle, Editor, *Philosophical Perspectives in Artificial Intelligence*, Harvester Press, 1979.

23. H. V. D. Parunak. 'Go to the Ant': Engineering Principles from Natural Agent Systems. *Annals of Operations Research*, 75:69–101, 1997.

24. H. V. D. Parunak. From Chaos to Commerce: Practical Issues and Research Opportunities in the Nonlinear Dynamics of Decentralized Manufacturing Systems. In *Proceedings of Second International Workshop on Intelligent Manufacturing Systems*, k15–k25, K.U.Leuven, 1999.

25. H. V. D. Parunak, S. Brueckner, M. Fleischer, and J. Odell. Co-X: Defining what Agents Do Together. In *Proceedings of Workshop on Teamwork and Coalition Formation, AAMAS 2002*, 2002.

26. H. V. D. Parunak, S. Brueckner, M. Fleischer, and J. Odell. A Preliminary Taxonomy of Multi-Agent Activity. In *Proceedings of Autonomous Agents and Multi-Agent Systems (AAMAS 2003)*, 1090–1091, 2003.

27. H. V. D. Parunak, S. A. Brueckner, and J. Sauter. Digital Pheromone Mechanisms for Coordination of Unmanned Vehicles. In *Proceedings of First International Conference on Autonomous Agents and Multi-Agent Systems (AAMAS 2002)*, 449–450, 2002.

28. R. Pfeifer and C. Scheier. *Understanding Intelligence*. Cambridge, MA, MIT Press, 1999.

29. R. F. Port and T. van Gelder. *Mind as Motion: Explorations in the Dynamics of Cognition*. Cambridge, MA, MIT Press, 1995.

30. Pynadath, M. Tambe, N. Chauvat, and L. Cavedon. Toward team-oriented programming. In *Proceedings of Workshop on Agents, theories, architectures and languages (ATAL'99)*, Springer, 1999.

31. S. Rao and M. Georgeff. Decision procedures for BDI logics. *Journal of Logic and Computation*, 8(3):293–344, 1998.

32. S. J. Rosenschein. Formal Theories of Knowledge in AI and Robotics. *New Generation Computing*, 3(3):345–357, 1985.
33. S. J. Rosenschein and L. P. Kaelbling. A Situated View of Representation and Control. *Artificial Intelligence*, 73, 1995.
34. T. Sauer, J. Yorke, and M. Casdagli. Embedology. *Journal of Statistical Physics*, 65:579, 1991.
35. R. Savit and M. Green. Time series and dependent variables. *Physica D*, (50):95–116, 1991.
36. Schmitz and T. Schreiber. Testing for nonlinearity in unevenly sampled time series. University of Wuppertal, Wuppertal, Germany, 1998. URL http://xxx.lanl.gov/abs/chao-dyn/9804042.
37. R. Shalizi. *Causal Architecture, Complexity and Self-Organization in Time Series and Cellular Automata*. Ph.D. Thesis at University of Wisconsin, Department of Physics, 2001.
38. S. H. Strogatz. *Nonlinear Dynamics and Chaos: with Applications to Physics, Biology, Chemistry, and Engineering*. Reading, MA, Addison-Wesley, 1994.
39. Takens. Detecting Strange Attractors in Turbulence. In D. A. Rand and L.-S. Young, Editors, *Dynamical Systems and Turbulence*, vol. 898, *Lecture Notes in Mathematics*, 366–381. Springer, New York, 1981.

I am Autonomous, You are Autonomous

Hans Weigand[1] and Virginia Dignum[2]

[1]Infolab, Tilburg University, PO Box 90153,
5000 LE Tilburg, The Netherlands
h.weigand@uvt.nl
[2]University Utrecht, Intelligent Systems Group, PO Box 80089,
3508 TB Utrecht, The Netherlands
Virginia@cs.uu.nl

Abstract. Autonomy is regarded as a crucial notion in multi-agent systems and several researchers have tried to identify what are the agent's parts that give it an autonomous character. In this paper, we take a different approach. If we assume that agents are autonomous (and this is a quite reasonable assumption in many practical situations, such as e-commerce), the more interesting question is: how to cope with the autonomy of agents? What are the effects on the way agents have to coordinate their behavior with other agents, and on the agent design process? And what are the effects of that (secondary effects) on the architecture of agents and agent societies. We address these questions by working out the concept of "collaboration autonomy", and by describing an agent society model that respects this kind of autonomy.

1   Introduction

Why is autonomy still a poorly understood concept? Perhaps because autonomy is such a loaded term in Western culture: individual autonomy was the dream of the Enlightenment, and it is still a cornerstone of most of the Western political and social theories, although it is not a univocal concept at all. It might be argued (but not here) that our problems in understanding agent autonomy are symptomatic of the dilemmas that we face daily in dealing with human autonomy in our societies.

In his seminal article on agent autonomy [4], Castelfranchi separated autonomy from agenthood. In his view, not all agents are autonomous. So he asked himself the question what makes an agent autonomous. In his view, the most interesting kind of autonomy is goal-autonomy, which itself is dependent on the "non-negotiability of beliefs". This means "we cannot believe a certain assertion just because there would be some advantage for us in believing it". Unfortunately, Castelfranchi does not motivate this axiom, and he also has no argument for why this property that he states about humans would also hold for agents. It seems that human autonomy at some point cannot be further analyzed.

The contribution of this paper is twofold. First, we present our position on the agent autonomy issue, and secondly, we provide a short overview of the Oper A model for agent societies and how it meets autonomy requirements.

M. Nickles, M. Rovatsos, and G. Weiss (Eds.): AUTONOMY 2003, LNAI 2969, pp. 227–236, 2004.

2   Towards a Transactional Analysis of Autonomy

Traditionally, agent autonomy is viewed as a *property* of some piece of software, that is one of the features that the software must fulfil to be considered an agent. This descriptive property can also be interpreted as a design requirement: you can build an agent in any way you want, but it must exhibit the autonomy property in the end. In this section, we make a few remarks about this software property approach, and then propose another way of looking at autonomy, inspired by the psychological approach of transactional analysis.

2.1   Autonomy as Self-governance

Abdelkader [1] discusses two interpretations of autonomy: *self-governance* (the agent is steered in selecting what goals have to be achieved by a set of motivations), and *independence* (the agent is independent from other agents) – and he prefers the first one, as independence is not a sufficient criteria. Carabelea et al [3] identify five forms of autonomy: user-autonomy, social autonomy, norm-autonomy, self-autonomy and environment autonomy. They rightly state that autonomy is a relational property (in line with [4]): X is autonomous *from* (other agent or object) Y *for* p (the object of autonomy), and they define this further as: X's behaviour regarding p cannot be imposed by Y. They call this an external perspective, which should be supplemented by an internal perspective. They suggest that one should be able to identify what are the agent's parts that give it an autonomous character (e.g. a goal maintenance module).

We agree that autonomy is not the same as independence. If autonomy is interpreted as *independence*, then the whole enterprise of agent societies is doomed to failure. After all, what makes a society interesting, for its members or for others, is that there are certain dependencies [4]. What is often overlooked in the discussions, is that these dependencies work to ways: not only the agent is dependent on its environment, but also the environment (other agents) will depend on the agent., to some extent. In other words, the agent will perform a certain role, a function with an added-value.

A controversial assumption is that the agent autonomy should be reflected in its architecture. One could argue against this assumption that if autonomy means anything, the agent is "master in his own house", so it seems odd to pose any requirement on the internal architecture. Of course, it is nice if a goal maintenance module, obligations, norms etc can be identified in the architecture, but the fact that these modules can be identified does not prove anything yet about autonomy, because it does not tell how the agent deals with them. If the "should" is taken as a design principle – if the intended meaning is that agent designers can better use such an architecture – then the assumption may be valid, but it must be clear that this is in no way a guarantee of autonomy nor an absolute requirement.

It should be noted that the notion of self-governance (behavior cannot be imposed by another party) assumes an intensional stance. Let s be a stimulus from another object (we will restrict ourselves to other agents) that aims at behavior x (for example, a REQUEST(x) message). The software receiving such a message will decide somehow whether to perform x or not (we may assume that it is capable of performing x). In traditional systems, this typically depends on authorizations and user roles, but the decision procedure can be made much more subtle, for example, by including a computation of the utility of doing x. If no stimulus has effect, we again have some "independence" situation that is not very interesting, so we can assume that at least some stimuli do have effect. In that case, the other agent can choose exacty these stimuli, in that way impose the required behavior, and there seems be no room for autonomy. So even if the agent performs only behavior congruent with its goals or utility function, and one may think that therefore it behaves autonomously, it can still be manipulated, and to say whether the behavior is self-chosen or imposed is all in the eyes of the beholder. Therefore, we think that also the criterion of self-governance is not useful as a litmus test for autonomy. However, the question is whether such a litmus test is what we need.

2.2   A New Approach

Rather than *assuming* autonomy to be a *required* property of agents, and from there infer some architecture that would *guarantee* this property, we propose a radically different approach: namely, to *require* autonomy as an *assumed* property of agents, and to infer from there some architecture that *respects* this property.

Why would we assume autonomy?
Because software programs are deployed by human users and human users are deemed to be autonomous. In most of the practical applications of agents, e.g. in e-commerce, the agent is an agent *of* some human user. For example, the agent has to buy an item on the Internet, or has to sell it in virtual shop. The same is true for a robot in a production plant or on the battle field. If the software we meet (or our software meets) is the agent of another human, then we cannot assume that it simply performs every request we make, and that it will always live up to its commitments. Hence we better not assume that to be the case, in other words, *assume* its autonomy. We don't have to make this assumption in a simulated world of programs that we control ourselves. But as soon as we enter the real world, we are dealing with software belonging to other humans or human organizations.

Why would we require *our agents to assume autonomy?*
We argued that it is a reasonable assumption that the piece of software you encounter on the Internet or anywhere in the real world behaves autonomously, but what if you do not? Is it not possible to manipulate this software, as it was sketched above? Yes, we agree that it is possible, and even if the human owner of the sofware is closely monitoring and controlling its behavior, one might attempt to manipulate software and owner together. But the question is not whether it is possible, but whether it is desirable. There are two important reasons why it is not. First, manipulating the agent quickly becomes manipulating the human owner, and this violates the ethical principle that humans should always be treated as a subject, not as an means (Kant). Secondly, you can at most try to manipulate, but you are not sure that you will succeed. Therefore, it is also safer to assume that the piece of software is autonomous, because you may very well miss your own target if you do not make that assumption. Let us get more practical: if you deploy a shopping agent, then you better build it in such a way that it does not naively assume that when something is advertized in a web shop, it is also available. Or that when the shop has promised to deliver the item, this will always happen. Instead, it is better to design your agent in such a way that it can

cope with contingencies caused by the other agents autonomous decisions. That is, to ask first whether the item is available. To ask for a firm commitment in the case of a risky transaction. To look for trust-enablers. Trust is only relevant in a situation of potential risk, so by using trust-enabling mechanisms you already assume a certain autonomy of the other party. This second argument (that it is safer) is not only valid in the case of cooperative agents, but also if you want them to be opportunistic and adversarial – even then, you don't want your agent to be naive about its opponents. As it is simply *better* to assume autonomy, we require our agents to do so. This holds for agents operating within some society (and their designers), but also for agent governing an agent society (and its designer).

How do we respect agent autonomy?
By turning around the question and requesting our agents to assume the autonomy of other agents, the next question is: what are the consequences for the agent architecture? The question is not the old one: what requirements do we put on a certain piece of software before we call it an autonomous agent? Rather the question becomes: how do we cope with the autonomy of agents? Or, in other words, how do we build agents that respect (other) agent autonomy?

In section 3, we will try to answer these questions in the context of OperA, a framework for building agent societies. The requirements that we formulate there do not have the goal of *guaranteeing* autonomy but rather of *coping with* the given autonomy. In our view, agent contracts become highly relevant then. Similarly, we can imagine requirements on the individual agents within a society that deal with other agents. To respect autonomy means at least to be prepared that any request (or assertion) *can* be rejected. From there it follows that you must also be prepared that your request can be accepted. Hence your agent must support a protocol that includes accepts and rejects. This can be done at different levels of sophistication.

2.3 I am Autonomous, You are Autonomous

"I'm OK, you are OK" is a popular expression originating from the Eric Berne's theory of transactional analysis [2]. Berne abandoned psychoanalytic theory in favor of a theory centered on communication. He focused on the information that is exchanged between people and conceptualized and categorized it in terms of transactions. By isolating transactional stimuli and responses he provided us with a method with which to study how people influence each other, and made possible the fine-grained analysis of person-to-person communication. People often adopt socially dysfunctional behavioral patterns called "games". For Berne, *individualism* means that a person is acting within a dysfunctional life script, maintaining a belief that others are not OK. In contrast, *autonomy* means in this theory that the individual is in tune with himself, others and the environment and therefore is acting freely (script free).

We refer to transactional analysis not because we think it is directly applicable to the agent autonomy question, but because of the interesting paradigm shift it offers. Autonomy is not viewed as an individual property, not even a relationship, but as a something that governs interactions. We just argued that autonomy is not something that must be required from agents, but that must be assumed. "You are autonomous" – that means that I will treat you as a subject, which does not follow my requests slavishly, but only may do so when I am able to provide good reasons. "I am autonomous" – that means also that I will not slavishly obey your requests, it means that I am able to enter meaningful relationships – call it contracts, which I expect you to respect. In other words, what we propose is that agent autonomy is first of all viewed as a norm governing agent interactions in an agent society. Agents don't need to prove that they are autonomous, but they (including their designers and agent society designers!) should live up according to this norm. This has certain consequences for the architecture of the agent society. It also imposes a norm on that piece of software representing an individual agent, although it does not specify how its designer is going to fulfill this norm.

Transactional analysis can help to see the goal autonomy in perspective: the agent has its own goals (as it is a piece of software, these goals are ultimately delegated to it by a human actor), but it also has to perform a role, deliver a function within the society it finds itself. This function is to the benefit of other agents. How can this paradox be solved? Basically, there are two choices: either the agents goals are completely congruent with the function it offers, leading to so-called benevolent agents that can hardly be called autonomous – or the paradox is solved by organizing agent transactions as *exchanges*: because in the exchange both the goal of the agent himself and the benefit of the other agent can be equally respected. The institutional way of organizing exchanges is by means of contracts, and contracts play a central role in our agent society model.

3   Autonomy in OperA

In this section, we briefly describe the model for agent societies **OperA** (**O**rganizations **per** **A**gents) [8][1]. This framework emerges from the realization that in organizations interactions occur not just by accident but aim at achieving some desired global goals, and that participants are autonomous, heterogeneous and not under the control of a single authority.

The purpose of any society is to allow its members to coexist in a shared environment and pursue their respective goals in the presence or in co-operation with others. A collection of agents interacting with each other for some purpose and/or inhabiting a specific locality can be regarded as a society. Societies usually specify mechanisms of social order in terms of common norms and rules that members are expected to adhere to [5]. An organization can be defined as a specific solution created by more or less autonomous actors to achieve common objectives. Organizational structure can therefore be viewed as a means to manage complex dynamics in (human) societies. This implies that approaches to organizational modeling must incorporate both the structural and the dynamic aspects of such a society.

From an organizational perspective, the main function of an individual agent is the enactment of a role that contributes to the global aims of the society. That is, society goals determine agent roles and interaction norms. Agents are actors that perform role(s) described by the society design. The agent's own capabilities and aims

---

[1] The name illustrates the dual relation between organizations and agents, the fact that organizations are outmost dependent on its agents, but, as in a musical opera, a script is needed that guides and constrains the performance of the actors, according to the motivations and requirements of the society designer.

determine the specific way an agent enacts its role(s), and the behavior of individual agents is motivated from their own goals and capabilities, that is, agents bring in their own ways into the society as well. However, a society is often not concerned about which individual agent will actually play a specific role as long as it gets performed. Several authors have advocated role-oriented approaches to agent society development, especially when it is manifest to take an organizational view on the application scenario [6] [11].

The above considerations can be summarized in the recognition that there is a clear need for multi-agent frameworks that combine and use the potential of a group of agents for the realization of the objectives of the whole, without ignoring the individual aims and 'personalities' of the autonomous participant agents. That is, in order to represent interactions between agents in such an open context, a framework is needed that meets the following requirements:

- **Internal Autonomy Requirement**: interaction and structure of the society must be represented independently from the internal design of the agents.
- **Collaboration Autonomy Requirement**: activity and interaction in the society must be specified without completely fixing in advance the interaction structures.

The first requirement relates to the fact that since, in theory, an open society allows the participation of multiple, diverse and heterogeneous entities, the number, characteristics and architecture of which are unknown to the society designer, the design of the society cannot be dependent on their design. With respect to the second requirement, fundamentally, a tension exists between the goals of the society designer and the autonomy of the participating entities. On the one hand, the more detail the society designer can use to specify the interactions, the more requirements are possible to check and guarantee at design time. This allows, for example, to ensure the legitimacy of the interactions, or that certain rules are always followed [10]. On the other hand, there are good reasons to allow the agents some degree of freedom, basically to enable their freedom to choose their own way of achieving collaboration, and as such increase flexibility and adaptability.

The OperA approach consists of a 3-layered model that separates the concerns of the organization from those of the individual. The top layer, called the Organizational Model, describes the structure and objectives of a system as envisioned by the organization, and the bottom layer, the Interaction Model, the activity of the system as realized by the individual agents. In order to connect individual activity with organizational structure we add a middle layer, the Social Model that describes the agreed agent interpretation of the organizational design. Fig. 1 depicts the interrelation between the different models. In the following subsections, we will describe each of the three models in more detail.

An OperA model can be thought of as a kind of abstract protocol that governs how member agents should act according to social requirements. Interaction is specified in contracts, which can be translated into formal expressions (using the logic for contract representation, described in [7]), and therefore ensure that compliance can be verified.

The actual behavior of the society emerges from the goal-pursuing behavior of the individual agents within the constraints set by the Organizational Model. From the society point of view, this creates a need to check conformance of the actual behavior to the desired behavior, which has several consequences. Firstly, we have to make explicit the commitments between agents and the society. An actor is an agent

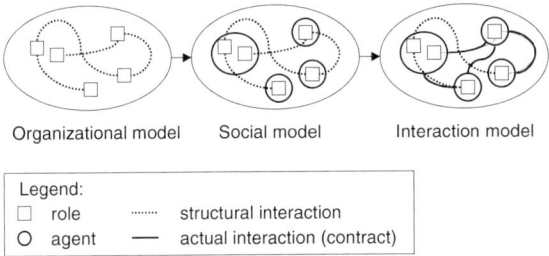

Organizational model   Social model   Interaction model

Legend:
☐   role   ⋯⋯⋯   structural interaction
O   agent   ——   actual interaction (contract)

Fig. 1. Organizational framework for agent societies

performing one or more roles in the society. The objectives and capabilities of the role are described in the OM and the actual agreements of the actor concerning its interpretation of the role are explicitly described in a social contract. We use the term Social Model to refer to the social contracts that hold at a given moment in a society.

Secondly, actual interaction between actors must also be made explicit, which can be done through (bilateral) contracts as well. We call this the Interaction Model (IM). Checking the conformance of actual behavior to the desired behavior can now be realized in two steps:

- Checking whether the contract set up by two or more agents conforms to the OM. The OM can be more or less elaborate, depending on the type of society. Typically, it does not provide more than a few "landmarks" that describe the main features (conditions, obligations, and possibly partial plans) of interaction between roles.
- Monitoring whether the actual messaging behavior conforms to the contract. This is primarily a responsibility of the agents themselves. Depending on the type of society, the OM can constraint the monitoring or even assign it to the society.

3.1   The Organizational Model

Starting point of the Agent Society Model is the organizational model (OM) that describes the structure and global characteristics of a domain from an organizational perspective. That is, from the premise that it is the society goals that determine agent roles and interaction norms. The organizational model is based on the analysis of the domain in terms of the coordination and normative elements. The OM specifies the global objectives of the society and the means to achieve those objectives.

The OM specifies an agent society in terms of four structures: social, interaction, normative and communicative. The social structure specifies objectives of the society, its roles and the model that governs coordination. The global objectives of an organization are represented in terms of objectives of the roles that compose the organization. Roles are tightly coupled to norms, and roles interact with other roles according to interaction scripts that describe a "unit" of activity in terms of landmarks. The interaction structure gives a partial ordering of the scene scripts that specify the intended interactions between roles. Society norms and regulations are specified in the normative structure, expressed in terms of role and interaction norms. Finally, the communicative structure specifies the ontologies for description of domain concepts and communication illocutions. The way interaction occurs in a society depends on the aims and characteristics of the application, determines the relations between roles, and how role goals and norms are 'passed' between related roles. For example, in a hierarchical society, goals of a parent role are shared with its children by delegation, while in a market society, different participants bid to the realization of a goal of another role.

3.2   The Social Model

We assume that individual agents are designed independently from the society to model the goals and capabilities of a given entity. In order to realize their own goals, individual agents will join the society as enactors of role(s) described in the organizational model. This means that several populations are possible for each organizational model. Agent populations of the organizational model are described in the social model (SM) in terms of commitments regulating the enactment of roles by individual agents. In the framework, agents are seen as autonomous communicative entities that will perform the society role(s) according to their own internal aims and architecture. Because the society designer does not control agent design and behavior, the actual behavior of the society instance might differ from the intended behavior. The only means the society designer has for enforcing the intended behavior is by norms, rules and sanctions. That is, when an agent applies, and is accepted, for a role, it will commit itself to the realization of the role goals and it will function within the society according to the constraints applicable to its role(s). These commitments are specified as social contracts that can be compared to labor contracts between employees and companies. The society can sanction undesirable (wrong) behavior as a means to control how an agent will do its 'job'.

The Social Model is defined by the **role enacting agents** (**reas**) that compose the society. For each agent, the rea reflects the agent's own requirements and conditions concerning its participation in the society. Depending on the complexity of the implemented agents, the negotiation of such agreements can be more or less free. However, making these agreements explicit and formal, allows the verification of whether the animated society behaves according to the design specified in the OM. The SM specifies a population of agents in a society, which can be seen as an instantiation of the OM. When all roles specified in the OM are instantiated to agents in the SM, we say that the SM provides a full instantiation of the society; otherwise, it is a partial instantiation.

3.3   The Interaction Model

Finally, interaction between agents populating a society is described in the interaction model (IM) by means of interaction contracts. This model accounts for the actual (emergent) behavior of the society at a given moment. Interaction agreements between agents are described in interaction contracts. Usually interaction contracts will 'follow' the intended interaction possibilities specified in the organizational model. However, because of the autonomous behaviour of agents, the interaction model must be able to accommodate other interaction contracts describing new, emergent, interaction paths, to the extent allowed by the organizational and social models.

OperA provides two levels of specification for interactions. The OM provides a script for interaction scenes according to the organizational aims and requirements and the IM, realized in the form of contracts, provides the interaction scenes such as agreed upon by the agents. It is the responsibility of the agents to ensure that their actual behaviour is in accordance with the contracts (e.g. using a monitoring agent or notary services provided by the society for that). However, it is the responsibility of the society, possibly represented by some of its institutional roles, to check that the agents fulfill these responsibilities.

The architecture of IM consists of a set of instances of scene scripts (called scenes), described by the interaction contracts between the role enacting agents for the roles in the scene script. An interaction scene results from the instantiation of a scene script, described in the OM, to the reas actually enacting it and might include specializations or restrictions of the script to the requirements of the reas.

4   Conclusion

The OperA model integrates a top-down specification of society objectives and global structure, with a dynamic fulfilment of roles and interactions by participants. The model separates the description of the structure and global behaviour of the domain from the specification of the individual entities that populate the domain. This separation provides several advantages to our framework above traditional MAS models. On the one hand, coordination and interaction in MAS are usually described in the context of the actions and mental states of individual agents [9]. In open societies, however, such approach is not possible because agents are developed independently from the society and there is therefore no knowledge about the internal architecture of agents, nor possibilities to directly control or guide it. Furthermore, conceptual modeling of agent societies (based on the social interactions) requires that interaction between agents be described at a higher, more abstract level, that is, in terms of roles and institutional rules. On the other hand, society models designed from an organizational perspective reflect the desired behaviour of an agent society, as determined by the society 'owners'. Once 'real' agents populate the society, their own goals and behaviour will affect the overall society behaviour, that is, such social order as envisioned by the society designer is in reality a conceptual, fictive behaviour. From an organizational perspective, the main function of individual agents is the enactment of roles that contribute to the global aims of the society. That is, society goals determine agent roles and interaction norms. Agents are actors that perform role(s) described by the society design. The agent's own capabilities and aims determine the specific way an agent enacts its role(s).

The OperA model can be viewed as an attempt to design agent societies that respect agent autonomy. We do not claim that this is achieved exhaustively in its present form. But it illustrates what a transactional analysis of autonomy could mean in practice.

References

1. Abdelkader, G.: Requirements for achieving software agents autonomy and defining their responsibility. *Proc. Autonomy Workshop at AAMAS 2003*, Melbourne, 2003.

2. Berne, Eric: *Games People Play*. New York, Grove Press, 1964.
3. Carabelea, C. O. Boissier, A. Florea: Autonomy in multi-agent systems: a classification attempt. *Proc. Autonomy Workshop at AAMAS 2003*, Melbourne, 2003.
4. Castelfranchi, C.: Guarantees for Autonomy in Cognitive Agent Architecture. *Proc. ATAL'94*. LNAI 890, Springer 1995.
5. Davidsson, P.: Emergent Societies of Information Agents. Klusch, M, Kerschberg, L. (Eds.): *Cooperative Information Agents IV*, LNAI 1860, Springer, 2000, pp. 143–153.
6. Dignum, V., Meyer, J-J., Weigand, H.: Towards an Organizational Model for Agent Societies Using Contracts. In: *Proc. of AAMAS, the 1st International Joint Conference in Autonomous Agents and Multi-Agent Systems*, Bologna, Italy, 2002.
7. Dignum, V., Meyer, J.-J., Dignum, F., Weigand, H.: Formal Specification of Interaction in Agent Societies. In: Hinchey, M., Rash, J., Truszkowski, W., Rouff, C., Gordon-Spears, D., (Eds.): *Formal Approaches to Agent-Based Systems (FAABS'02)*. LNAI 2699, Springer-Verlag, 2003.
8. Dignum, V.: *A Model for Organizational Interaction: Based on Agents, Founded in Logic*. PhD thesis, Utrecht University, 2004.
9. Ferber, J., Gutknecht, O.: A meta-model for the analysis and design of organizations in multi-agent systems. *Proc. of ICMAS'98*, IEEE Press. 1998.
10. Weigand, H., Dignum, V., Meyer, J-J., Dignum, F.: Specification by Refinement and Agreement: Designing Agent Interaction Using Landmarks and Contracts. In: Petta, P., Tolksdorf, R., Zambonelli, F. (Eds.): *Engineering Societies in the Agents World III: Proceedings ESAW'02*, LNAI 2577, Springer-Verlag, 2003, pp. 257–269.
11. Zambonelli F., Jennings, N., Wooldridge, M.: Organisational Abstractions for the Analysis and Design of Multi-Agent Systems. In: Ciancarini P., Wooldridge, M. (eds.): *Agent-Oriented Software Engineering*, LNCS 1957, Springer-Verlag, 2001, pp. 235–251.

Agents with Initiative: A Preliminary Report

Raymond So and Liz Sonenberg

Department of Information Systems, The University of Melbourne
Melbourne, VIC 3053, Australia
w.so2@pgrad.unimelb.edu.au
l.sonenberg@unimelb.edu.au

Abstract. This paper introduces a notion of forward thinking agents. Recent research in situation awareness and anticipatory behavior provides useful insights and the relevant cognitive underpinnings that can be used to design adaptive software agents that capitalize on the environmental cues and their repercussions in the context of motivational autonomy. In particular, we are interested in building software agents that exhibit non-trivial anticipatory behaviors – exploiting knowledge about the current situation and the future to improve their current behavior. Some of the key issues pertaining to the realization of forward thinking, anticipatory behavior and situation awareness in agency will be briefly discussed. A conceptual extension to the generic BDI framework is also presented as part of our initial efforts in tackling some of these issues.

1 Introduction

The Eastern Airlines Flight 401 that crashed into the Florida Everglades in December 1972 is one of the classic examples that many cognitive scientists often cited when illustrating the concept and significances of situation awareness in dynamic and complex task environments. The crew of the flight demonstrated surprisingly deficient situation awareness when they became preoccupied with a faulty landing gear indicator light and failed to be aware of the fact that the autopilot had became disengaged, which eventually sent them to the ground and to their deaths [17]. Intelligent software agents, in particular for those that are designed to work in highly dynamic and time-pressure environments, must be engineered to ensure sufficient attention is dedicated to both goal-driven and reactive activities at the appropriate time. More importantly, software agents should have the *forward thinking* capability to comprehend the repercussions of the environmental cues perceived through their sensors, and to anticipate and proactively prepare (through planning or adaptive behavior) for foreseeable situations, in addition to executing sequences of actions that are hardwired during design time. We believe forward thinking, and situation awareness in general, are essential qualities, which (a) enable agents to exhibit nontrivial awareness and initiative[1], and result in high level of autonomy [9] and (b) act as an alternative

---

[1] Also refer to "Types and Limits of Agent Autonomy" by Beavers and Hexmoor in this volume.

M. Nickles, M. Rovatsos, and G. Weiss (Eds.): AUTONOMY 2003, LNAI 2969, pp. 237–248, 2004.

meta-level control mechanism for hybrid agent architecture that consists of both a deliberative and reactive component.

In this paper, we aim to set the scene for a larger research effort that in search of an alternative meta-level control mechanism, which is based on existing theories and research in *situation awareness* (SA) and *anticipatory behavior*, for a hybrid agent architecture [20]. The TouringMachine [7] and InterRRap [12] are examples of hybrid agent architectures. The focus on situation awareness reflects our beliefs that (a) SA offers highly relevant theoretical underpinnings for our idea of forward thinking, and ultimately bring about high degree of goal or motivational autonomy [4], and (b) SA is, in general, concerned with achieving the optimal balance between the top-down (deliberative and goal-driven) and bottom-up (reactive or reflective) information processing models [5], which also can be used to implement an alternative control mechanism for hybrid agent architectures. As part of our initial efforts, we begin our investigation by studying the IRMA [2] and the standard BDI model [15], with an objective to identify the additional features that are imperative to the realization of forward thinking and SA in general.

Our approach is partially inspired by the considerable body of work in human factors engineering associated with situation awareness, and while we seek to build artificial agents, not constrained by many of the emotional (and related) issues that form the focus of much human factors work, we have identified a number of general organizing principles from this literature that have value for the design of artificial agents.

One of our starting points is the OODA (Observe, Orient, Decide, Act) Loop, formulated by John Boyd [1], an American scholar of military tactics to capture the individual agent's decision cycle, and specifically the interaction between situation awareness and action. While the OODA loop is only an informal conceptualization, it has been shown to have wide applicability across many domains, and proves to be a useful (conceptual) tool. In some sense, our proposal exploits both the similarity and the differences between the OODA loop and the perceive-deliberate-act cycle of the popular BDI execution model for artificial agents. Importantly, the standard BDI model offers little formalised support for the "Orient" phase in Boyd's loop. We argue that traditional plan libraries in a BDI agent only account for some of the dynamics of the environment. Our proposal seeks to fill this gap by introducing a new element, the *situation model*, to capture (from the domain expert at system design time) other sources of influence that are imperative to understanding the dynamics of the environment.

Technically, our initial challenge is twofold. To find an appropriate representation of environmental dynamics that (a) facilitates recognition of possible future patterns from past events and also (b) enables the system to project forward efficiently, so that the outcome is useful as an input to (real time) deliberative reasoning. Subsequently, we need to formalize, by leveraging existing theories and prior work in situation awareness and anticipatory behavior, the appropriate control mechanism that coordinate between the reactive, goal-driven and proactive activities in agency.

The remainder of this paper is organized as follows: we first furnish a brief overview of situation awareness, followed by a discussion of how situation awareness fits in the general agent theories. We conclude this paper by presenting a conceptual extension to the standard BDI framework, through which we illustrate how situation awareness and forward thinking can be realized and incorporated in deliberative agency, and highlight some of concerns and technical challenges.

2   A Brief Overview of Situation Awareness

Good situation awareness lies in making use of the available information. Or, as Flach would have it[2], "a situation is not a stimulus, but a set of affordances. Humans process 'meaning' not information. SA alerts the researcher to consider meaning with respect to both the objective task constraints (i.e., the situation) and the mental interpretation (i.e., awareness). SA can contribute to a renewed appreciation for the role of perception in problem solving and decision making and for the intimate and dynamic coupling between perception and action."

With high level situation awareness, you project your actions into the future [5]. Simply seeing and being aware of the potential problem at hand is not enough. What you do when you are able to recognise a problem ready to unfold is also very important. Just as predictive displays have been shown to improve human air traffic controller performance [6], we argue that providing corresponding information to a deliberative artificial agent also has the potential to improve performance.

Endsley's definitions of SA is "the perception of the elements in the environment within a volume of time and space, the comprehension of their meaning, and the projection of their status in the near feature" [5]. This definition of SA breaks down into three seperate levels:

- Level 1: *perception* of the elements in the environment,
- Level 2: *comprehension* of the current situation, and
- Level 3: *projection* of future status

The first step in achieving SA is to ensure the capability to perceive status, attributes and dynamics of relevant elements in the environment. Level 1 SA requires awareness of the relevant elements in the environment. Level 2 SA goes beyond being aware of the existence of the relevant elements by incorporating the significance of those elements, in the context of some operator goals. One of the specific capabilities in this level is to consolidate the knowledge in Level 1 elements and to form patterns, or to form a holistic perspective of the current environment. This is an important capability that allows a higher level of comprehension, which is well beyond understanding the individual or isolated pieces of information. The third level of SA refers to the capability to anticipate

---

[2] www.iav.ikp.liu.se/hfa/courses/Flach__documents.htm

future possible events or any foreseeable situations, which is based on a better understanding of the current situation.

2.1   Situation, Situation Model and Forward Thinking

The notion of *situation* in this paper refers to a specific state (or configuration) of the environment. Our primary interest is the dynamics of the environment that bring about some situations. A situation is a sequence of events, in particular those induce changes, that occurred in the environment over some time interval (Figure 1a). In general, situation can be considered as a trimmed down version of the branching temporal network [19]. There are two important attributes in our notion of situation. First, situation is concerned with environment features that are abstract in nature, or Level 2 SA concepts. Second, situation is concerned with history or evolvement of those abstract environment features over time, as opposed to snapshot view. Figure 1b illustrates these two important characteristics of situation.

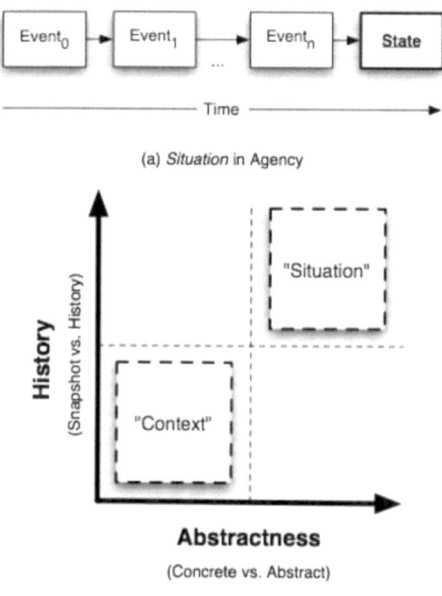

(a) *Situation* in Agency

(b) Taxonomy of Environment Dynamics

Fig. 1. The notion of situation and environment dynamics in agency

We propose to use schematic representations to capture the dynamics in the environment that is associated with situations. This paper seeks to introduce a new construct known as *situation model* that captures the environment dynamics. Situation model aims at enriching the descriptions of situations and contexts to be attended by software agents. We argue that only when this additional knowledge becomes available, agents can then engage in anticipatory and proactive behaviors that deal with foreseeable situations (Level 3 SA). In essence, situation model attempts to capture (a) the relevant environmental cues that leads to states to be attended by agents and (b) the desirable and appropriate proactive actions that deal with these situ ations as they unfold. Situation model, throughout this paper, can be considered as partial predictive model of the environment, which is a crucial element in anticipatory systems [16,3]. An anticipatory system is "a system containing predictive model of itself and/or of its environment, which allows it to change state at an instant in accord with the model's predictions pertaining to a later instant [16, p. 339]". Using predictive model, current decision-making in typical anticipatory systems can be based on future predictions, which objective is to improve performance.

There are three major justifications (or hypothesis) of introducing situation model: (1) Traditional plans are, in nature, insufficient in capturing most of the environmental dynamics that are required to realize Level 3 SA. (2) There are subtle and fundamental differences between the knowledge captured in traditional plans and situation model: plans map situations to external actions; situation model maps situations to future beliefs. (3) There are significant differences in the nature of reactive actions (as prescribed in traditional plans) and proactive actions (as in situation model) that handle the same state of the world (see figure 2).

From a software design and engineering perspective, situation model plays an important role in reducing the complexity of plans by offloading knowledge that is not directly relevant to the means-end analysis. For instance, *internal* actions that only assert beliefs and desires. In most agent architecture, the complexity of plans tend to grow as the sophistication of agents increases. The management of plans, in the long term, becomes a nontrivial task. The idea to separating knowledge that is required for deliberation and means-end analysis encourages the idea of modularization and reuse.

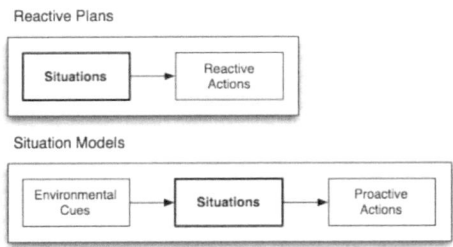

Fig. 2. Taxonomy of situation model

In the abstract BDI architecture, Level 1 SA was achieved by the presence of sensors; Level 2 SA was achieved by the pattern matching of environment status vectors to context conditions; and Level 3 SA, to a certain extent, was achievable in an ad hoc way through the use of meta-level plans. And to put our work into perspective, we argue that the use of situation model in standard BDI's deliberation and planning processes improves the ability to anticipate foreseeable situations, and as a result, to dynamically adapt their behaviors accordingly and appropriately. We loosely refer to this collection of capabilities as *forward thinking*.

2.2   An Example: The Tileworld

In this section, we further elaborate the various concepts that have been introduced so far with the help of Pollack's Tileworld [14]. Figure 3 shows an interesting scenario whereby two agents attempt to pursue the same piece of tile.

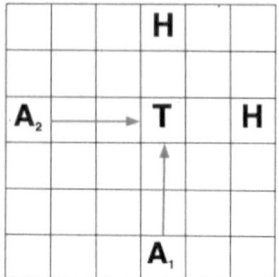

Fig. 3. Multi-agent tileworld example

In figure 3, a collision situation is about to unfold. A collision situation materializes when two (or more) agents are on collision courses, typically caused by having targeted at the same piece of tile.

The schematic representation (aka situation model) of the collision situation, for instance, consists of the followings:

1. Identification of Level 1 SA element(s): history of each agent's physical locations and their movements (aka trajectories)
2. Identification of Level 2 SA concept(s): collision, whereby two (or more) trajectories collide
3. Identification of Level 3 SA element(s): proactive action(s) that deal with *collision* - for instance, to avoid collision by aborting agent's current pursue, to pursue other tiles or to do nothing.

Let agent A_1 be our forward thinking agent and it keeps track of A_2's trajectory. Given the situation model, A_1 will be able to actively observe the positions of other agents in its vicinity over time, compare them with its own trajectory (inferred based on its current goal to pursue the tile in the middle of the grid) and to watch out for the likelihood of collision. When A_1 concludes that a collision situation is about to unfold, proactive actions can then be taken to avoid the undesirable circumstance and any futile efforts[3].

At first glance, it appears that the detection of collision, for instance, can be implemented through the use of subgoals. In the previous Tileworld example,

---

[3] In some of the tileworld's implementations, for instance, agents are given subgoal to conserve fuel, and it is not a rational choice to pursue non-achievable targets.

avoid collision can be considered as a subgoal to the high-level goal *pursue target*. This particular view implies that situations, their detection and awareness are tightly coupled with certain high-level goals. The significant contribution of situation model, however, lies in its role as a meta-level control mechanism in hybrid agent architecture that typically involves both bottom-up and top-down model of processing. For instance, awareness of the collision situation (especially at the early stage of detection) can influence the decision behavior of an agent in (a) selecting an alternative path approaching the target (b) considering other viable targets in its vicinity or (c) coordinate or negotiate with the competing agent(s) and to come up with some sort of arrangements.

We will look at the role of situation awareness and its significancy in agency in the next section.

3   Situation Awareness and Agency: How do they Connect?

Situation awareness is a set of constructs (perception, comprehension and projection) that primarily concerns with acquisition and interpretation of information gathered in the environment, which is highly relevant to how software agents perceive and process events in their situated environment. In general, SA is concerned with the awareness of what is happening around us and understanding of what that information means to us now and in the future. This awareness is usually defined in terms of what information is important for a particular job or goal. In our tileworld example, agents should watch out for the collision and competition situations when they are in the process of pursuing tiles.

3.1   Information Processing Models: Bottom-Up vs Top-Down

Situation awareness offers a description of the intimate interactions between agents and environment, and it underlies the intimate coupling between process stages perception, decision and action [8]. Human information processing in operating complex systems has been seen as switching between bottom-up (data-driven) and top-down (goal-driven) processes. The attention of goal-directed agents is directed across the environment in accordance with their active goals. In other words, they actively seek information that is required for goal attainment, and goals acts as a filter in interpreting the information perceived through their sensors [5]. There is a potential danger that goal-directed agents develop tunnel vision – when they are pre-occupied with the activities that are imperative to achieving their goals and failed to dedicate the efforts and attentions required for unfolding critical situations. The bottom-up alternatives include reactive and recognition-primed decision making. The reactive approach is primarily event-driven behavior and usually does not involve high-level comprehension of the current situation or context. Recognition-primed decision making (such as Naturalistic Decision Making) involves situation recognition and pattern matching to mental models, which implies comprehension of the current situation. A critical mechanism in achieving situation awareness is to combine both the bottom-up and top-down processing, and develop a strategy to intelligently switch between the two approaches [5].

We envision forward thinking, and SA in general, providing an appropriate framework that can be used to design a novel meta-level control mechanism that manage the switching between the top-down and bottom-up approach (Figure 4). Forward thinking capability (with its implicit predictive capability and the use of situation model) supports two types of anticipatory behavior that can be used to control and coordinate bottom-up and top-down information processing models typically coexists in hybrid agent architecture: sensorial and state anticipation [3]. Based on prediction (or anticipation) of future states, sensory anticipatory mechanism directs an agent's attention to certain (usually critical) sensorial data and hence influence sensory (pre-)processing; and state anticipatory mechanism directly influence an agent's behavioral decision making.

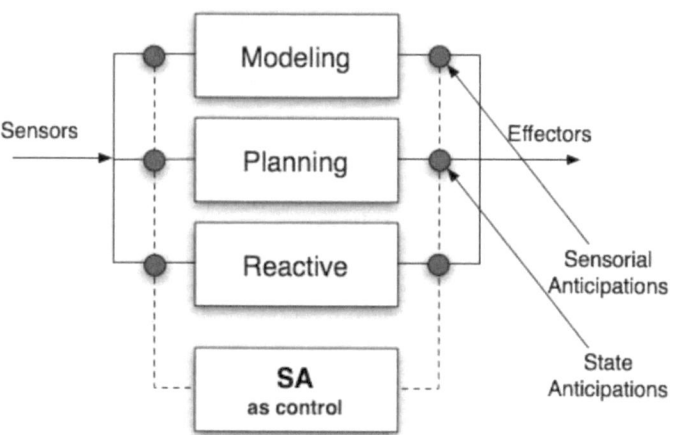

Fig. 4. Using SA as control mechanism in hybrid agent architecture (adapted from [7])

4   Operationalizing Situation Awareness and Forward Thinking

4.1   A Brief Review of Related Work

In the domain of computer-generated force (CGF) domain, a previous attempt has been made to incorporate Endsleys model of SA in a war-fighter simulation project (SYNTHER) using cognitive agents [10]. A SYNTHER (Synthetic Team Member) is a cognitive agent that simulates a human war-fighter capable of interacting with human war-fighters, and capable of participating as a member of a team of war-fighters. A blackboard representation of long-term working memory was used to consolidate the Level 1 SA elements perceived through SYNTHER agents Sensation and Perception Process. The Level 1 SA elements can then be used by the various cognitive processes to guide formulation of new courses of action, or modifications to existing courses of action. The Level 2 SA capability is implemented by making connections, or semantic links, between the disparate pieces of Level 1 SA information, which help the agents to refine and elaborate their understanding of the current situation. However, the project was not able to observe how and to what extend the cognitive process was influenced (via Level 3 SA).

As part of a larger effort to create synthetic characters for computer games with human-level intelligence, an enhanced Soar-based Quakebot was developed that incorporates anticipation of its opponents actions [11]. In order to achieve the anticipation capability, the Quakebot creates an internal representation that mimics what it thinks the enemy's internal state is, based on its own observation of the enemy. It then predicts the enemy's behavior by using its own knowledge of tactics to select what it would do if it were the enemy. Using simple rules to internally simulate external actions in the environment, the softbot forward projects until it gets a prediction that is useful or reaches a solution where there is uncertainty as to what the enemy would do next. The prediction is used to set an ambush or deny the enemy a weapon or health item.

The enhanced Quakebot has demonstrated very constrained forward thinking that based on observation and reasoning using its plan library. A major problem with this approach is that the ability to anticipate what the opponent will do next is greatly constrained by the Quakebot's own capabilities. In fact, Quakebot presumed that its opponents will act in exactly the same way as it does in all known situations. In other words, Quakebot denies all *other* possible maneuvers, which are not in its plan library, can be adopted by its opponents. A better approach will be supplying Quakebot situation models that describe, for instance, the *possible* maneuvers (in terms of spatial information of how an opponent navigate in a certain type of landscape, its speed and direction over time) that seize a weapon at the end of an hidden corridor. The Quakebot can then utilize this additional information to observe and anticipate what its opponents will do when a weapon is available at a nearby hidden corridor (or in a similar situation).

We note further potential lessons for us from other studies in human factors research – this time the stream of work on *ecological interface design* [18]. In this approach, the relationship between people and their environment is taken as the focus of study, and demands that we explicitly analyze the constraints that the environment imposes on behavior, and the ways these constraints are presented to the (human) agent (in addition to investigating the characteristics of the actors). This provides a contrast to the information processing approach to human factors, where in many cases, the environment is not even explicitly represented (except perhaps in a very spartan manner by a feedback loop from motor control to sensation).

4.2   Realizing Forward Thinking in BDI Agents

As part of our initial effort to investigate the use and implementation of forward thinking in agency, we are prepared to present a conceptual extension to the generic BDI framework (Figure 5). This is an important first step to begin our investigation of the practicality and the technical challenges pertaining to the realization of forward thinking (and SA in general) in agency.

The original belief revision function (*brf*) models an agents belief update process [19]. In other words, this function determines a new set of beliefs on the basis of the current beliefs and current percepts. In order to realize situation awareness, and eventually the forward thinking capability, the *brf* must be re-engineered to better incorporate Level 2 SA (Figure 6):

1. Incorporate model-based, bottom-up information processing - organize and structure perceptual efforts using environmental dynamics encoded in situation model.
2. Identify emerging and unfolding situations by keeping track of the relevant events and their repercussions, and comparing internal representations of

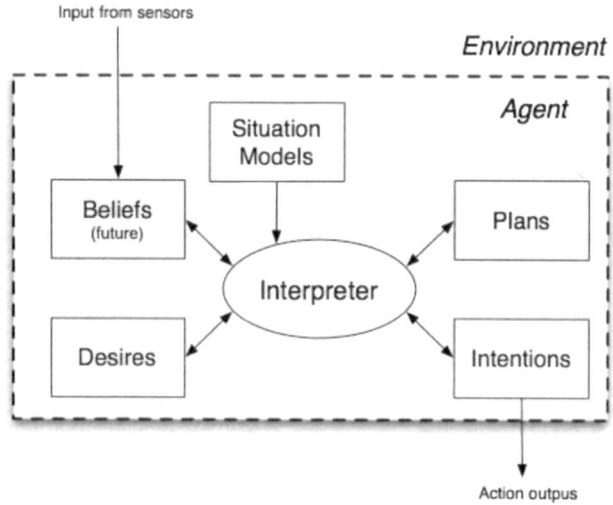

Fig. 5. Incorporating situation awareness into generic BDI framework

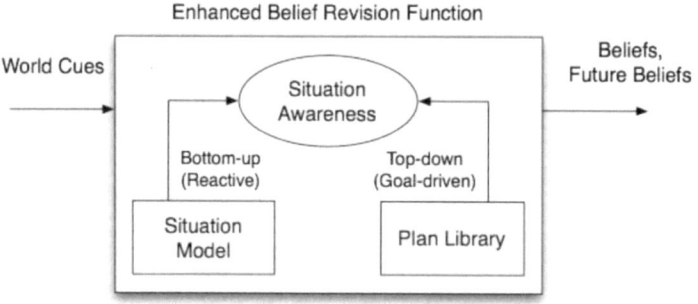

Fig. 6. Enhanced belief revision function

the world against the knowledge encapsulated in situation models and the anticipated effects of the committed plans (actions in the pipeline).

3. Generate *future beliefs* [13] for foreseeable situations.

In order to enable proactive behavior and to ultimately achieve higher degree of motivational autonomy, a number of changes are to be made in the original BDI interpreter to cater for the followings:

1. Preempt deliberation and goal-driven activities and attend to unfolding situations in respond to future belief(s) - a control mechanism for switching between top-down and bottom-up model of processing.
2. Update current desires and intentions to reflect proactive behaviors that deal with unfolding situations.
3. Resume to goal-driven mode of operation after attended to the critical situations.

5 Conclusion

Awareness and purposefulness in interaction are required before autonomy can be considered [9]. This paper sets the scene for a larger research effort that investigates the various issues of incorporating and implementing situation awareness in deliberative agent systems. We have introduced the notion of forward thinking, a capability that we believe plays a crucial role in improving ways autonomous agents interact with their situated environment through purposeful and proactive behavior. In addition, we also highlighted the use of situation awareness as an alternative control mechanism in hybrid agent architecture that consists of both the top-down and bottom-up information processing models.

We need to point out that our work is still in its infancy stage and additional efforts are required further to formalize the ideas introduced in this paper. Our future work includes formalizing the situation model construct, the algorithm of forward thinking in the BDI computational model, and more importantly, to investigate the practicality and technical challenges associated with incorporating situation awareness in the context of agent autonomy.

We have discussed, at the conceptual level, of how situation awareness can be used as a framework to improve our understanding of the complex interactions between agents and their situated environment. We have also discussed the notion of forward thinking, a capability that we believe play a crucial role in improving ways autonomous agents interact with the environment through proactive behaviors.

References

1. John R. Boyd. Pattern of Conflict. Technical Report Accessed, April 2003, http://www.d-n-i.net/boyd/pdf/poc.pdf, December 1986.
2. Michael E. Bratman, David J. Israel, and Martha E. Pollack. Plans and resource-bounded practical reasoning. *Computational Intelligence*, 4(4):349–355, 1988.

3. Martin V. Butz, Siguard Olivier, and Gerard Pierre. Internal models and anticipations in adaptive learning systems. In *The seventh international conference on the Simulation of Adaptive Behavior(SAB 2002). Workshop Proceedings Adaptive Behavior in Anticipatory Learning Systems(ABiALS 2002)*, Edinburg, Scotland, 2002.

4. Cristiano Castelfranchi. Modeling social interaction for AI Agents. In Martha E. Pollack, editor, *IJCAI-97. Proceedings of the Fifteenth International Joint Conference on Artificial Intelligence*, Pages 1567–1576. Morgan Kaufman Publishers, San Francisco, CA, 1997. Section 4 Principles of coordination Section 4.1 Reactive vs. anticipatory: Coordination among cognitive agents.

5. Mica R. Endsley. Theoretical underpinnings of situation awareness: A critical review. In Mica R. Endsley and Daniel J. Garland, editors, *Situation Awareness Analysis and Measurement*, pages 3–32. Lawrence Erlbaum Associates, Publishers, Mahwah, New Jersey, 2000.

6. Mica R. Endsley, R. Sollenberger, and E. Stein. The use of predictive displays for aiding controller situation awareness. In *Proceedings of the 43rd Annual Meeting of the Human Factors and Ergonomics Society*, 1999.

7. Innes A. Ferguson. Touringmachines: Autonomous agents with attitudes. *IEEE Computer*, 25(5):51–55, 1992.

8. John M. Flach. Situation awareness: Proceed with caution. *Human Factors*, 37(1):149–157, 1995.

9. H Hexmoor, C Castelfranchi, and R Falcone. A prospectus on agent autonomy. In H Hexmoor, C Castelfranchi, and R Falcone, editors, *Agent Autonomy*, pages 1–8. Kluwer Academic Publishers, 2003.

10. James H. Hicinbothom. Maintaining situation awareness in synthetic team members. In *The Tenth Conference on Computer Generated Forces*, pages 231–241, Norfolk, Virginia, 2001.

11. John E. Laird. It knows what you're going to do: Adding anticipation to a quake-bot. In *The fifth international conference on autonomous agents*, pages 385–392. ACM Press, 2001.

12. J. P. Muller and Markus Pischel. The agent architecture InteRRap: Concept and application. Technical Report Technical Report RR-93-26, DFKI Saarbrucken, 1993.

13. Timothy J. Norman and Derek Long. Goal creation in motivated agents. In Michael Wooldridge and Nicholas R. Jennings, editors, *Intelligent Agents: Theories, Architectures, and Languages*, volume 890 of LNAI, pages 277–290. Springer-Verlag, Heidelberg, Germany, 1995.

14. Martha E. Pollack and Marc Ringuette. Introducing the tileworld: Experimentally evaluating agent architecture. In Thomas Dietterich, editor, *Proceedings of the Eighth National Conference on Artificial Intelligence*, pages 183–189, Menlo Park, CA, 1990. AAAI Press.

15. A. Rao and M. Georgeff. BDI agents: From theory to practice. In *The First International Conference on Multiagent Systems*, San Francisco, 1995.

16. Robert Rosen. *Anticipatory Systems*. Pergamon Press, Oxford, UK, 1985.

17. Renee J Stout, Janis A Cannon-Bowers, and Eduardo Salas. The role of shared mental models in developing team situational awareness: Implications for training. *Training Research Journal*, 2:85–116, 1996/1997.

18. K J Vicente and J Rasmussen. Ecological interface design: Theoretical foundations. *IEEE Transactions on Systems, Man, and Cybernetics*, 22(4):589–606, 1992.

19. Michael Wooldridge. *Reasoning about rational agents*. MIT Press, 2000.

20. Michael Wooldridge and Nicholas R. Jennings. Intelligent agents: Theory and practice.*The Knowledge Engineering Review*, 10:115–152, 1995.

A Teamwork Coordination Strategy Using Hierarchical Role Relationship Matching

Susannah Soon[1,2], Adrian Pearce[1], and Max Noble[2]

[1] Department of Computer Science and Software Engineering
The University of Melbourne, Victoria, 3010, Australia
{susannah, pearce}@cs.mu.oz.au
[2] ADI Limited
20-22 Stirling Highway, Nedlands, WA, 6009, Australia
max.noble@adi-limited.com

Abstract. We present a teamwork coordination approach called the Rolegraph coordination strategy. It is used to dynamically coordinate agent teams that perform distributed collaborative tasks. Current agent coordination approaches recognise team activity by considering team member mental states using explicit communication, plan recognition, and observation. Such information is sometimes incomplete or unavailable. Our strategy recognises team activities by analysing role relationships in hierarchical teamwork structures. At runtime, graph representations of team structures, called Rolegraphs, are extracted, capturing collaborative team activity. Rolegraphs are approximately matched against predefined templates of known behaviour to infer agent mental states, such as intention, in order to recognise team activities. A case base is then referenced to retrieve suitable examples of coordinated team action.

1 Introduction

Multiagent teamwork [4, 7, 12] involves autonomous agents performing cooperative coordinated action to achieve common goals. Coordination is required when agents are interdependent [16], for example, when agents share tasks or avoid resource conflicts. Teamwork is based on theoretical Belief-Desire-Intention (BDI) [19] teamwork frameworks such as [3, 4, 7, 12, 25]. These frameworks use communications between agents to build mental models of one another, in order to coordinate actions. For instance, in *joint intentions* theory [4] team members commit to a joint intention to perform a team goal.

Current multiagent teamwork coordination approaches perform dynamic coordination in changing environments by recognising agent mental states. This generally relies on one or more sources of information, including observation [9], plan recognition [9–11, 21, 23], and communication [23]. These approaches are limited when communications are incomplete, conflicting, or costly [9, 10, 23], for example, when plan failure is not communicated [23]; when full plan knowledge is unavailable; or when observations are partial.

One successful coordination approach to recognising agent activity is to examine plan execution. An example is the general purpose teamwork model STEAM [23], in which paths through reactive plan hierarchies are hypotheses of possible plan

M. Nickles, M. Rovatsos, and G. Weiss (Eds.): AUTONOMY 2003, LNAI 2969, pp. 249–260, 2004.

execution. STEAM also provides other features to support coordination such as selective communication, role-monitoring constraints to monitor for team failure, and replanning on failure. In execution monitoring, Kaminka and Tambe [11] use a combination of plan recognition through temporal understanding of plan execution, team coherence relationships, and communications to overcome unreliable, and insufficient communications and sensors.

Unfortunately, such plan-based approaches suffer when plan information is unavailable. Knowledge abstraction is often used to overcome insufficient plan information [10, 21]. One approach is Sonenberg and Tidhar's [21] conceptual model for plan-based BDI mental state team modelling. Agent mental states are inferred using partial knowledge of observed action, and hypothesis testing of possible teams and structure, while abstracting lower level detail when full plan knowledge is unavailable. They identify a main limitation as the overheads and inaccuracies associated with reasoning over numerous team mental states and structure hypotheses.

This paper presents a practical distributed coordination approach for autonomous agent behaviour, the Rolegraph coordination strategy (RCS). The approach considers situations where plan knowledge of other teams is inaccessible. Instead, team behaviour is approximated by recognising hierarchical role relationships. At runtime, agent teams use their recursive *reasoning agent* to identify relevant hierarchies representing cooperating team member activities. The *reasoning agent* enables teams to analyse other teams' activities from various individual perspectives. Graph representations of the selected hierarchical structures, called Rolegraphs, are extracted, and matched against templates of known behaviour to infer team mental states, and approximate current team activity. The coordinated actions to be performed are then selected from a case base.

Hierarchical structures contain important team relationship information whilst abstracting away lower level detail. Coordination effort is reduced when agents are represented hierarchically [5, 24], as agent responsibilities, beliefs, and actions are limited by their organisational relationships. Therefore, fewer Rolegraph matching iterations are required to identify agent mental states, reducing the number of individual hypotheses to be tested. In addition, these simpler hierarchies allow graph techniques to be practically applied to assist reasoning of their relationships.

Kok et al. [13] present a role-based coordination approach by extending coordination graphs (CG) [8], which contain agent nodes, separated by edges representing coordination dependencies. Relative roles are assigned to agents at particular time instances, discretising the continuous environment. These local role relationships are coordinated using *a priori* discrete environment coordination rules [8]. Our coordination approach differs as we consider local and global dynamic team hierarchy role relationships, to which agents are assigned at runtime. In addition, we perform coordination by recognising team activities using graph matching of roles and their characteristics.

Section 2 discusses dynamic teamwork structures. Section 3 presents the Rolegraph coordination strategy. Section 4 concludes.

2   Dynamic Teamwork Structures

We define the teamwork terminology used to describe hierarchical teamwork structures. A *team* is an agent reasoning entity performing and requiring roles. *Roles* specify the relationship, including behaviour and responsibilities between teams and their subteams [1]. Roles can be filled by subteams, or individual agents that have the capability to perform the role. *Teamwork structures* specify hierarchical team-role-team relations. Initially, teams are allocated to roles. At runtime, environmental changes cause *dynamic team formation* to occur. Subteams then change in applicability, and are reallocated to suitable roles. In some instances, roles may be unallocated when no teams satisfy the role's requirements.

The JACK Teams[1] programming language [1] is used in our teamwork implementations. It adopts the BDI reasoning model, and is based on earlier work on the Procedural Reasoning System [6]. JACK Teams is an example of a centralised command model where team coordination is performed by role-tendering teams that execute team plans coordinating high level subteam goals. Individual role-filling subteams are then responsible for autonomously achieving their specific goals.

2.1 Mission Planning Team Structure Example

Our coordination strategy is motivated by the desire to improve distributed desktop decision support systems for collaborative mission planning. Mission planning processes are enhanced through coordination of customised agent assistance. At runtime, human mission planners perform tasks using agent decision support services. Evolving Rolegraph representations of these agent services model the mission planning process. Rolegraphs are matched against templates to recognise agent actions, and intentions, reflecting the human mission planner's needs. This knowledge is then used to provide further coordinated agent assistance.

We briefly describe the three levels of teamwork structure used to model collaborative mission planning processes. Levels 1 and 2 represent the military organisational hierarchy. Level 1 considers strategic level personnel, such as a tasking authority. Level 2 represents operational and tactical personnel, such as mission planners. Level 3 represents teams performing decision support services to assist the level 2 personnel to carry out their tasks. Filled roles within this collection of teamwork structures model the collaborative mission planning process at a specific time [22]. For example, the combination of all level 2 military personnel's level 3 expansions provides information about each user's decision support activities with respect to others.

Fig. 1 shows a simple mission planning team structure example, at a particular time instance. It includes the tasking authority, mission planning teams 1 and 2, and their associated level 3 teams. Each mission planning team contains an Interface Agent role filled by applicable Interface Agent teams. The Interface Agent teams provide agent services such as recognising user goals, responding to queries, making suggestions, and automating analysis and information distribution. Particular team structure configurations model mission planner activities, such as flight planning or surveillance.

At runtime, human mission planners use the agent system to suit their decision support requirements, desired situation awareness, and expertise. To reflect these needs, and in response to the changing environment, the team structures dynamically change configurations. For example, a user sets an agent's analysis ability to high,

---

[1] Agent Oriented Software's JACK Teams.

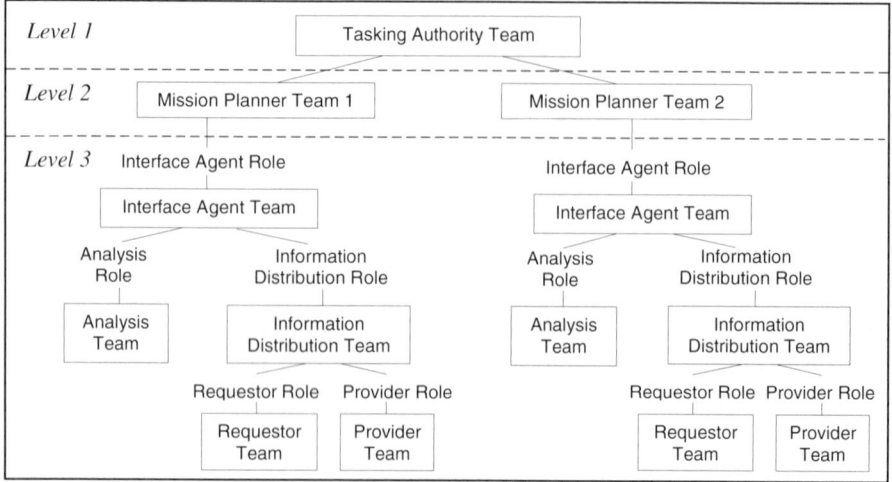

Fig. 1. A mission planning team structure hierarchy showing a level 1 tasking authority, level 2 mission planners and their associated level 3 agent services teams

forcing high quality analysis. A higher capability Analysis subteam having complex calculation capabilities, and additional knowledge, attaches to the Analysis role replacing the standard Analysis team.

A simple motivating scenario involves mission planners 1 and 2 (MP1, and MP2), jointly planning a helicopter reconnaissance mission. The mission planning systems provide customised and efficient support when MP1's Interface Agent team (IA) receives an enemy helicopter sighting. Before MP1's IA automatically warns MP2's IA of the helicopter sighting, MP1's system first establishes MP2's current activity to determine if it requires this information. Should MP2 be conducting flight planning for imminent execution, this information is immediately required. Otherwise, should MP2 be conducting general surveillance, this information is non-critical.

3   The Rolegraph Coordination Strategy

The Rolegraph coordination strategy is based on recognising team intention using hierarchical teamwork structures. The coordination we examine in this paper involves recognising the potential for agents to interact [25]. We illustrate this using a scale of one team hierarchy interacting with another team hierarchy of similar magnitude, as shown in Fig. 1. Although, the strategy can be applied to recognise team activity at various levels within the hierarchy, this paper does not explore teamwork coordination concerning team structures of different scale or abstraction.

3.1   Rolegraphs

Graph theory has been used to represent self organisation amongst agents in dynamic environments. Examples include dependence graphs [20] that form dependencies between agents where graph vertices represent agents, goals, actions, and plans; and

MAGIQUE [17], where graph vertices represent agents within a dynamic organisation hierarchy, and form acquaintances with one another to provide services.

Our Rolegraph graph representations symbolise hierarchical organisation relationships. Rolegraphs are directed acyclic graphs extracted from dynamic teamwork structures at runtime, representing role relationships between teams. Our use of the term Rolegraph is independent of Nyanchama and Osborn's [18] Role Graph in computer security.

A Rolegraph $G_{R_i}(t) = (V, E)$, contain vertices V, connected by directed edges E, where the edge (u, v) is different from (v, u), and no paths through the graph have the same start and end node. Rolegraphs have labelled vertices, x_i, which are unary label unique identifiers representing role-tendering or role-filling teams. Rolegraphs also contain labelled and attributed edges, $r_{i,j}(\lambda_1,..,\lambda_n)$ representing binary role relationships between the tendering and filling teams. At time t, the Rolegraph $G_{R_i}(t)$ in Fig. 2(b) is extracted from the team structure part in Fig. 2(a).

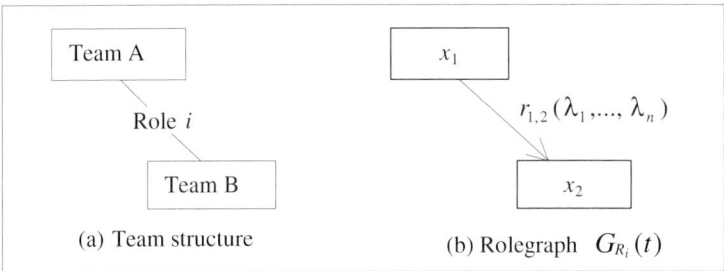

(a) Team structure   (b) Rolegraph $G_{R_i}(t)$

Fig. 2. (a) A team structure part and (b) The extracted rolegraph, containing unary label unique identifiers, x_1, and x_2, and a binary role relationship, $r_{1,2}(\lambda_1,...,\lambda_n)$

Fig. 2(b)'s Rolegraph can be represented as a tertiary relationship, involving the unary labels, and the binary role relationship:

$$G_{R_i}(t) = (x_1, x_2, r_{1,2}(\lambda_1,...,\lambda_n)) \ . \tag{1}$$

where, x_1 : role-tendering team;
          x_2 : role-filling team;
          $r_{1,2}(\lambda_1,...,\lambda_n)$: binary role relationship between tendering and filling teams.

3.1.1 Binary Role Relationships

A binary role relationship $r_{1,2}(\lambda_1,...,\lambda_n)$ specifies the unique relationship and responsibilities between teams and their subteams, where $r_{1,2}$ is the unique identifier for the role. The role relationships consist of multiple attributes, $\lambda_1,...,\lambda_n$. Role attributes closely map team relationships within corresponding team structures, and apply directionally between teams, as signified by the Rolegraph's directed edge. We now consider key attributes that define a role relationship within a teamwork structure:

Attribute 1, λ_1, is the role name. An equivalence comparison based on this attribute dramatically prunes the search space, as role names having different semantics can be discarded.

Attribute 2, λ_2, are capabilities, the actions that a subteam must be able to perform in order to fill the role. Capabilities are an important attribute for approximating teams that may be attached to roles. However, they are not an absolute representation of runtime relationships that are formed, as capabilities express possible rather than actual activities being conducted at a point in time. In JACK Teams, capabilities are represented as communications or events posted and handled between team members [1]. Capabilities can be of two types, those which the filling team must exhibit to fill the role, and non-essential team capabilities.

Attribute 3, λ_3, are role specific beliefs, these are the beliefs that the filling team must possess in order to fill the role. These beliefs may be domain specific, for example, the coordinates of an area of interest in the mission planning domain.

Attribute 4, λ_4, concerns teamwork relationships in the organisational hierarchy. For instance, particular relationships may specify that a team can only work effectively with a specific team, when role relationships are active, or when delegation of activities are feasible between particular team members. Such information can be expressed in the attribute using joint beliefs, commitments to joint intentions, team and subteam plans, potential subteams, and active roles. Kaminka and Tambe also use organisational information [11], termed social relationships, representing relative agent information such as beliefs, goals, plans and actions.

3.2   Reasoning Agent

Teams recursively possess a BDI *reasoning agent* to reflect over their own and other teams' structures. The agent recognises team activities by selecting Rolegraph matching evaluations to conduct, appropriate templates to match against, and when matching should take place. The agent uses matching results to build mental models of agent intentions relative to their goals in the environment. This information is used to select possible coordinated action.

The *reasoning agent* reasons to select Rolegraph evaluations that provide useful information using minimal processing resources. During matching, the agent uses its beliefs of events that should occur to search for patterns to trigger recognition, and then inferentially reasons about the observation [9].

In the helicopter reconnaissance mission, MP1's Interface Agent team (IA) tests hypothesis H_1: that MP2 is developing a flight plan for imminent execution rather than conducting general surveillance. MP1's IA performs this by matching relevant Rolegraphs against flight planning and surveillance templates. Flight planning involves determining enemy location and capability, and observing the environment relevant to the mission's flight path and time. Surveillance involves determining enemy location and capability in the area of interest over an extended period of time. Although, flight planning and surveillance activities are similar, they are differentiated by different times, areas of interest, and urgency. These differences are reflected in the Rolegraph representations specifying how agent teams perform decision support activities. These differences trigger the *reasoning agent's* recognition of the activity. The agent then performs inferential reasoning to identify the activity as flight planning or surveillance.

3.2.1   Restricting Evaluation by Selecting Appropriate Rolegraphs

Rolegraphs of specific activities can be simple, or be combinations representing complex activities. For example, a team activity is represented by the Rolegraph set $G_{R_i}(t) = \{G_{R_a}(t), G_{R_b}(t), G_{R_c}(t)\}$. Table 1 shows Rolegraph set examples from Fig. 1's team structures, and the activities they represent.

Table 1. Rolegraphs and activities

Rolegraph	Activity
$G_{R_{IA}}(t)$: Interface Agent Activity	Interface agent activities
$G_{R_{ID}}(t)$: Information Distribution Activity	Information distribution activities
$G_{R_A}(t)$: Analysis Activity	Analysis activities

The *reasoning agent* constrains matching by selecting relevant Rolegraphs. The agent tests H_1 by performing reasoning over Rolegraph sets representing surveillance and flight planning agent decision support services. The flight planning Rolegraph is $G_{R_F}(t) = \{G_{R_{IA}}(t), G_{R_{ID}}(t), G_{R_A}(t)\}$. It comprises of Rolegraphs representing the Interface Agent team for user interaction, the Information Distribution team for distributing information to coordinate military activities, and the Analysis team for waypoint and altitude analysis. The surveillance activity Rolegraph requires similar Rolegraphs and is expressed as $G_{RS}(t) = \{G_{R_{IA}}(t), G_{R_{ID}}(t), G_{R_A}(t)\}$.

3.2.2   Example: The Information Distribution Activity Rolegraph

We consider an expansion of the Information Distribution Activity Rolegraph $G_{R_{ID}}$ (from Fig. 1). It contains an Information Distribution team responsible for providing and requesting information for the higher level Interface Agent team. The Information Distribution team consists of a Provider team and a Requestor team. The Information Distribution Activity Rolegraph is illustrated in Fig. 3, and is defined as:

$$G_{R_{ID}}(t) = \{G_{R_I}(t), G_{R_P}(t), G_{R_R}(t)\} \ . \tag{2}$$

where, $G_{R_I}(t)$: Information Distribution Rolegraph;
$G_{R_P}(t)$: Provider Rolegraph;
$G_{R_R}(t)$: Requestor Rolegraph.

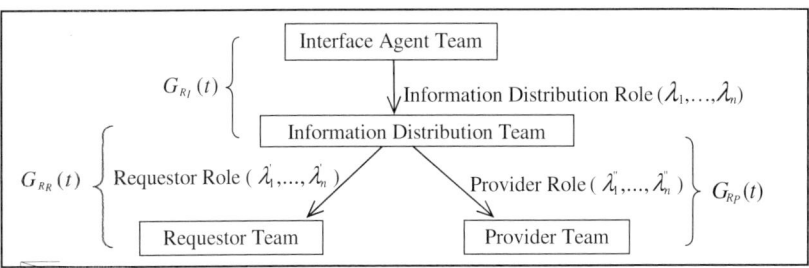

Fig. 3. Information distribution activity Rolegraph $G_{R_{ID}}(t)$

3.3   Rolegraph Matching

Sonenberg and Tidhar [21] identify team and structural relationship knowledge as a requirement in recognising team mental state. Rolegraph matching examines role and structure relationships to approximate team role filling. Possible team activities are determined by matching Rolegraphs against templates of known activity or mental states, using role attributes as matching parameters. Mental models are inferred using matching results, and then team activities are recognised. Such role relationship evaluation is approximate, as actual role filling at a particular time instance is unknown.

Although the graph matching or subgraph isomorphism problem [2] is considered to be NP complete, we use the Kuhn-Munkres algorithm [15], a polynomial time bipartite graph matching algorithm for approximate Rolegraph matchings. Such polynomial time algorithms are sufficient for the scale of Rolegraphs we consider.

A bipartite graph $G_b = (V_b, E_b)$ contains the set of vertices V_b, which can be divided into two partites V_1 and V_2 such that no edge E_b, connects vertices in the same set. We assign Rolegraph vertices to partite V_1, and template vertices to partite V_2. The weightings of the edges between V_1 and V_2 are an approximation of the degree the vertices correspond to one another. We describe how these bipartite graph weightings are derived in section 3.3.2. The matching algorithm is then used to find the maximal weighted matching of the bipartite graph. That is, the largest weighted set of edges from E_b such that each vertex in V_b is incident to at most one edge.

3.3.1   Templates

Templates represent particular team activities by specifying Rolegraph configurations including teams, subteams, and role relationship attribute values. In the military domain, template knowledge acquisition is obtained from sources such as military standard operating procedures, mission planner experience, and mission briefs. Templates can represent high level activities such as surveillance, or low level activities such as a particular calculation analysis. Templates are expressed as in equation 3, and can be schematically represented in the same way as Rolegraphs.

$$G_{T_i} = (y_1, y_2, r_{1,2}(\lambda_1, ..., \lambda_n)). \tag{3}$$

where, y_1 : role-tendering team;
              y_2 : role-filling team;
              $r_{1,2}(\lambda_1, ..., \lambda_n)$: binary role relationship between tendering and filling teams.

Templates enable generalisation by detailing the range of allowable values an attribute can hold. Generalisation specifies the degree of matching required for a Rolegraph to be classified as a particular activity. Some templates contain specialised attribute values representing precise activities. These can be generalised at runtime by being formed into larger sets of attributes sourced from multiple templates.

3.3.2   Determining Weightings

The weighting w, between the vertices of Rolegraph $G_{R_i}(t)$ and template G_{T_i}, is determined by the summation of attribute comparisons between corresponding roles (equation 4). The attribute comparisons are conducted using an attribute specific equivalence function. Equation 4 is illustrated in Fig. 4. The function f_k is the comparison function for attribute k.

$$w = \bigvee_{\lambda_k \in G_{R_i}(t), \lambda_k' \in G_{T_i}} \sum f_k(\lambda_k, \lambda_k'), \text{ where } f_k \text{ is an equivalence function} \tag{4}$$

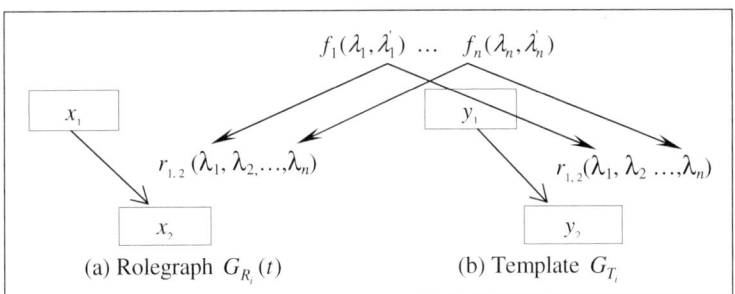

Fig. 4. Summing the comparisons of the role attributes of the Rolegraph G_{R_i} and template G_{T_i} to determine the weighting w between Rolegraph team x_2 and template team y_2

The comparison function $f_k(\lambda_k, \lambda_k')$ can be an equivalence function or a generalisation, testing whether attributes belong to sets, or ranges of numeric, symbolic, or syntactic values. Rolegraph matching can be performed using sufficient attribute subsets.

3.3.2.1 Example: Rolegraph versus Surveillance and Flight Planning Templates

Testing H_1 involves comparing MP2's Information Distribution Activity Rolegraph (Fig. 3) against those of the flight planning and surveillance templates. The Information Distribution Activity Rolegraph includes the Information Distribution, the Requestor, and the Provider teams, represented by partite $V_1 = \{x_1, x_2, x_3\}$. The corresponding teams for the flight planning template are represented by $V_2 = \{y_1, y_2, y_3\}$, and those for the surveillance template are represented by $V_2 = \{y_4, y_5, y_6\}$.

To illustrate, we compare only the Requestor roles of the Rolegraph $G_{R_i}(t)$ and the flight planning template G_{T_F}. This comparison is similar to that in Fig. 4. We determine the weighting w, between the teams filling the Requestor role in the Rolegraph and the flight planning template. For each Requestor role, we compare the role specific beliefs attribute λ_3, using the equivalence function, $f_3(\lambda_3, \lambda_3')$.

> For $G_{R_i}(t)$, $\lambda_3 = \{$Area: (115E 31S, 116E 32S); Time: 02:00–03:00;...$\}$
> For G_{T_F}, $\lambda_3' = \{$Area: (114E 30S, 116E 32S); Time: 01:40–03:00;...$\}$

The λ_3 values are similar, we calculate a normalised value of 4 using the equivalence function $f(\lambda_3, \lambda_3')$ (details omitted). This value is added to those obtained for comparison calculations for other attributes in the role, giving a total of 7. Since this is the comparison of teams x_2 and y_2 which both fill the Requestor role, the value of 7 is inserted into a weightings matrix W_F at position (x_2, y_2). The weightings for all possible matches between the Rolegraph and the flight planning template teams are calculated using equation 4 to populate the weightings matrix W_F.

Fig. 5(a) shows the weightings matrix W_F derived by comparing the Rolegraph $G_{R_I}(t)$ and the flight planning template G_{T_F}. Fig. 5(b) shows the weightings matrix W_S derived by comparing the Rolegraph $G_{R_I}(t)$ and surveillance template G_{T_S}. Using the matching algorithm, the matching between $G_{R_I}(t)$ and G_{T_F} is $M_F = \{x_1y_1, x_2y_2, x_3y_3\}$ with a weighting of 23. The matching between $G_{R_I}(t)$ and G_{T_S} is $M_S = \{x_2y_5, x_3y_6, x_1y_4\}$ with a weighting of 13. Since the teams of the Rolegraph $G_{R_I}(t)$ match those of the template G_{T_F} with the maximum weighting, the Rolegraph activity is classified as flight planning.

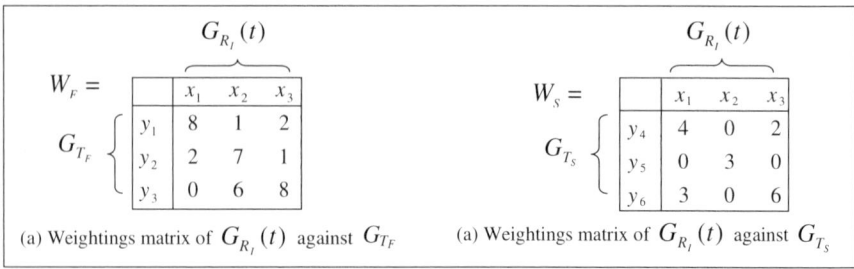

Fig. 5. Weightings matrix between Rolegraph $G_{R_I}(t)$ and templates G_{T_F} and G_{T_S}

3.4 Coordination Using Case-based Reasoning

Case-based reasoning [14] models the human reasoning process, solving new problems by referring to past solutions. Following the Rolegraph matching, the *reasoning agent* uses case-based reasoning to select generic team coordination suitable in the current circumstance. It achieves this by examining similar structured Rolegraph cases, whilst ensuring agent intentions, and beliefs are consistent with those held currently. Coordination then involves reusing past Rolegraph characteristics such as team plans, and subteam choices. Teams are notified of recommended coordination through team structure modifications, or locations of Rolegraphs exhibiting desired coordination.

In the motivating scenario, coordination cases are stored for Rolegraphs classified as flight planning or surveillance. The context condition of the plan to send the sighting to MP2, is satisfied for flight planning, and remains unsatisfied for surveillance.

4   Conclusions and Future Work

The Rolegraph coordination strategy eliminates full plan knowledge requirements by reasoning over Rolegraphs extracted from dynamic teamwork structures, generated as teams perform collaborative activities. *Reasoning agents* recognise team activities by approximately matching Rolegraphs against templates of known activities. The approach relies on meaningful role relationships, and is limited by the available predefined templates, and coordination scenarios, restricting recognisable activities and possible coordination.

The benefits of our approach include that team activities can be recognised efficiently using approximate graph matching. The recognition process is easily controlled as the abstraction at which the graphs are compared is limited by the role attributes considered. The recursive *reasoning agent* allows team activity recognition to be performed from different perspectives in the hierarchy, enabling robustness. Our future work considers the applicability of this approach to heterogenous and large-scale team structures, where full plan knowledge is unavailable. This involves developing suitable team structure abstraction techniques. Further work involves formalising the *reasoning agent* to describe its BDI reasoning process in selecting hypotheses and templates, and inferencing intentions using matching results.

Acknowledgments

This research is supported by an Australian Research Council APA (I) grant, and the industry partner ADI Limited.

References

1. Jack Teams User Guide, Agent Oriented Software Pty Ltd, 2002.
2. Barrow, H.G. and Burstall, R.M. Subgraph Isomorphism, Matching Relational Structures and Maximal Cliques. *Information Processing Letters, 4* (4). 83–84.
3. Bratman, M. Shared Cooperative Activity. In *Faces of Intention: Selected Essays on Intention and Agency*, Cambridge University Press, 1999, 93–108.
4. Cohen, P. and Levesque, H. Teamwork. *Nous, Special Issue on Cognitive Science and AI, 25* (4). 487–512.
5. Durfee, E.H. Practically Coordinating. *AI Magazine, 20* (1). 99–116.
6. Georgeff, M. and Lansky, A., Reactive Reasoning and Planning. In *Proceedings of the Sixth National Conference on Artificial Intelligence (AAAI-87)*, (Seattle, WA, 1987), 677–682.
7. Grosz, B.J. and Kraus, S. Collaborative Plans for Complex Group Action *Artificial Intelligence*, Elsevier Science B.V., Amsterdam, Netherlands, 1996, 269–357.
8. Guestrin, C., Koller, D. and Parr, R. Multiagent Planning with Factored MDPs. *Advances in Neural Information Processing Systems, 14.*
9. Heinze, C., Goss, S. and Pearce, A., Plan Recognition in Military Simulation: Incorporating Machine Learning with Intelligent Agents. In *Proceedings of the International Joint Conference on Artificial Intelligence (IJCAI99) Agent Workshop on Team Behaviour and Plan Recognition*, (Stockholm, 1999), 53–64.
10. Huber, M. and Durfee, E., On Acting Together: Without Communication. In *Proceedings of the Conference on Uncertainty in Artificial Intelligence (UAI'96)*, (1996).
11. Kaminka, G.A. and Tambe, M. Robust Agent Teams Via Socially-Attentive Monitoring. *Journal of Artificial Intelligence Research, 12.* 105–147.
12. Kinny, D., Ljungberg, M., Rao, A., Sonenberg, E., Tidhar, G. and Werner, E., Planned Team Activity. In *Artificial Social Systems. 4th European Workshop on Modelling Autonomous Agents in a Multi-Agent World, MAAMAW '92. Selected Papers. Springer–Verlag. 1994, pp.227–56. Berlin, Germany.*

13. Kok, J., Spaan, M. and Vlassis, N., An Approach to Noncommunicative Multiagent Coordination in Continuous Domains. In *Proceedings of Benelearn '02 Annual Machine Learning Conference of Belgium and The Netherlands*, (Utrecht, The Netherlands, 2002).

14. Kolodner, J. *Case-Based Reasoning*. Morgan Kaufmann, San Francisco, California, 1993.

15. Kuhn, H.W. The Hungarian Method for the Assignment Problem. *Naval Res. Logist. Quart*, 2. 83–97.

16. Lesser, V. Reflections on the Nature of Multi-Agent Coordination and Its Implications for an Agent Architecture. *Autonomous Agents and Multi-Agent Systems*, 1. 89–111.

17. Mathieu, P., Routier, J.C. and Secq, Y., Dynamic Organization of Multi-Agent Systems. In *Proceedings of the 1st International Joint Conference on: Autonomous Agents and Multiagent Systems, Jul 15–19 2002*, (Bologna, Italy, 2002), Association for Computing Machinery, 451–452.

18. Nyanchama, M. and Osborn, S. The Role Graph Model and Conflict of Interest. *ACM Transactions on Information and Systems Security, vol. 2* (no. 1). 3–33.

19. Rao, A. and Georgeff, M., Modeling Rational Agents within a Bdi-Architecture. In *Proceedings of the Second International Conference on Principles of Knowledge Representation and Reasoning*, (1991), Morgan Kaufmann Publishers, San Mateo, CA, pages 473–484.

20. Simao Sichman, J. and Conte, R., Multi-Agent Dependence by Dependence Graphs. In *Proceedings of the 1st International Joint Conference on: Autonomous Agents and Multiagent Systems, Jul 15–19 2002*, (Bologna, Italy, 2002), Association for Computing Machinery, 483–490.

21. Sonenberg, E. and Tidhar, G., Observations on Team-Oriented Mental State Recognition. In *Proceedings of the International Joint Conference on Artificial Intelligence (IJCAI99) Agent Workshop on Team Behaviour and Plan Recognition*, (Stockholm, 1999).

22. Soon, S., Pearce, A. and Noble, M., Modelling the Collaborative Mission Planning Process Using Dynamic Teamwork Structures. In *Proceedings of the 2nd International Joint Conference on: Autonomous Agents and Multiagent Systems, Jul 14–18 2003*, (Melbourne, Australia, 2003), Association for Computing Machinery, 1124–1125.

23. Tambe, M. Towards Flexible Teamwork. *Journal of Artificial Intelligence Research, 7*. 83–124.

24. Tidhar, G., Rao, A.S. and Sonenberg, E.A., Guided Team Selection. In *ICMAS–96 Proceedings. Second International Conference on Multi-Agent Systems. AAAI Press. 1996, pp.369–76. Menlo Park, CA, USA.*

25. Wooldridge, M. and Jennings, N.R. Cooperative Problem-Solving Process. *Journal of Logic and Computation, 9* (4). 563–592.

A Dialectic Architecture for Computational Autonomy

Mark Witkowski[1] and Kostas Stathis[2]

[1] Intelligent and Interactive Systems Group, Department of Electrical and Electronic
Engineering, Imperial College, Exhibition Road, London SW7 2BT, U.K.
m.witkowski@imperial.ac.uk
[2] Intelligent Computing Environments, Department of Computing,
School of Informatics, City University, London EC1V 0HB, U.K.
kostas@soi.city.ac.uk

Abstract. This paper takes the view that to be considered autonomous, a
software agent must possess the means by which to manage its own motivations
and so arbitrate between competing internal goals. Using the motivational
theories of Abraham Maslow as a starting point, we investigate the role that
argumentation processes might play in balancing the many competing aspects
of a whole agent's motivational agenda. This is developed into an Agent
Argumentation Architecture (AAA) in which multiple "faculties" argue for
different aspects of the total behavior of the Agent. The overall effect of these
internal arguments then defines which actions the agent will select for
expression, and so define the overt and observable "personality" of the agent.

1 Introduction

In this discussion paper we consider the nature of autonomy, what it means to be
autonomous and how a greater degree of autonomy might be achieved in independent
machines. We shall use the notion of a software agent as the exemplar type of system
to discuss the nature of autonomy, but also draw on ideas from psychological theory.

Autonomy should be considered as separate from automatic or independent
operation [1], [13], [14], [17], [28]. Luck and d'Inverno [13] have proposed the view
that an object can be regarded as an agent once it has goals and the means to effect
them, and an autonomous agent one that has motivations, and so the ability to create
goals according to some internal, hidden and changeable agenda. By this definition
the overwhelming majority of "autonomous agents" of the type characterized by [8]
and [15] would be more properly defined as independent. In this view autonomous
systems cover a wide spectrum of degrees of autonomy. The earliest obvious
precursors, Cybernetic or homeostatic [30] machines use pre-defined (but possibly
self-adjusting) control strategies to maintain a pre-defined set point. As with these
automatic systems, the reactive or behaviorist model [3] is certainly independent of
the programmer, but tied to the strategies laid down by its program. In the standard
BDI agent model [24], for instance, immediate pre-programmed strategies are
replaced by a goal driven approach in which decisions about which specific actions to
be taken are deferred until the goals are activated.

This paper investigates the situation where a software agent has control of its goals
and decision-making process, but also the ability to set and maintain its own agenda

M. Nickles, M. Rovatsos, and G. Weiss (Eds.): AUTONOMY 2003, LNAI 2969, pp. 261–273, 2004.

of goals and select actions for expression according to an individualized class of arguments and internal priorities. Full autonomy and thus complete freedom from the programmer requires that the agent can learn new activities, adopt new goals and, for complete autonomy, devise new learning strategies also.

The term "autonomy" is derived from the ancient Greek Αυτονομια meaning self-regulation by having self-laws. Formal models of Deontics, the formalization of duties, obligations and prohibitions, seem counterproductive in this context. At their heart, the paradox of the notion of something that is obligatory, yet coupled to sanction to be applied if that obligation is not fulfilled ("contrary to duty"). Note that this paper is not a discussion of morality, of rights and wrongs, but rather of mechanism. A fully autonomous agent cannot be obliged (as in deontics) to conform to law, but must decide the consequences under its own prioritization and proceed or not (e.g. without sanction, law has no effect). An agent need not be isolated from other agents, it will need to interact and cooperate with peers. Furthermore, an agent does not need to act as a servant, but might elect to act as such.

We are interested in producing a generic architecture, building on notions proposed by Kakas and Moraïtis [10], for embodied and non-embodied entities to achieve full autonomy for its own sake. But we will then consider implications for practical technology, which may in turn tell us something about the human condition. Why, for instance, do we consider ourselves autonomous? Having made the proposal for such an architecture in the body of this position paper, we return in the discussion section to these issues. The issues under discussion here are at the edge of what current techniques in reasoning can be expected to achieve, and are perhaps at the very boundary of what logic can be expected to represent, at least in its current form ([11] for a discussion). Detailed considerations of the logic formalization fall outside the scope of this position paper.

Section two presents a view of the motivational theories [16] of Abraham Maslow (1908–1970), and asks whether they can offer any insight into how a software agent may be made more completely autonomous. A software agent might do so by taking more immediate control of its own behavioral agenda, setting its own goals and determining the overall outcomes and consequences to the agent in terms of those various and varying motivating factors. We shall use the approach laid out by Maslow as a starting point, but argue that its provisions do not entirely apply in the case of Software Agents.

Section three we will look at extensions to the last class of autonomous agents, loosely based on notions of the BDI architecture ([2], [24]), in which processes ("faculties") encapsulating a number of different top-level goals (which appear as "motivations" to an observer of the agent, or to the introspective agent) must both propose new actions or goals that support the area they are responsible for, and argue that these goals should be adopted by the agent as a whole. We shall take the view this overall process can and should be viewed and modeled as one of a dialectic argumentation game, in which individual faculties must both argue for the value of their individual contribution, of which they are a part.

Argumentation has found favor recently as a way of modeling legal arguments using logic ([12], [22], and [23] for review). We note that there are significant differences between legal and internal argumentation, and that the categories of argument and that the argumentation strategies employed must therefore be different. Argumentation alone, is however, not sufficient to determine the overall outcome of a conflicted situation. Just as legal argumentation relies on a notion of the judgment of an individual or panel of judges, or reference to a superior court or legislature to resolve otherwise undecidable arguments, the autonomous agent will require a corresponding, but internal mechanism. Section five considers the role a scheme of preferences might play in achieving this end. Section six will discuss a procedural layer, providing for a game-like protocol by which the argumentation process might take place.

2   Personality and Motivations

This section takes as its starting point a discussion of the motivational theories of Maslow [16]. Maslow's work has been influential in the understanding of human drive and its relationship to the expression of personality (and thence to our notions of autonomy as individuals). It is important because it attempts to relate a model of underlying processes to observable traits, and as such stands apart from work which primarily serves to categorize and measure personality (e.g. [7], [9], and [25] for review). Carabelea *et al.* [4] attempt a personality classification system for agent systems.

Maslow describes five classes of motivation, forming a dynamic "needs hierarchy". At the base level are **Physiological needs**, the immediate requirements to maintain the function of the organism. These needs, in the living organism, will include homeostatic [30] functions, such as blood sugar or oxygen balance, or hydration levels. Physiological needs may also drive complex behaviors, such as food acquisition, that are not well modeled with a control theoretic approach. At the next level **Safety Needs** predominate. In this respect Maslow specifically refers to the search for stability and freedom from fear and anxiety in the individual's continuing context, rather than freedom from immediate danger. At the third level **Belongingness and Love Needs** emerge. These refer to the apparent human requirement to seek out and maintain immediate contact with other individuals in a caring and cared for context ("the giving and receiving of affection"). Maslow argues that failure to achieve or denial of these needs leads to a wide range of distressing psychological symptoms. At the fourth level **Esteem Needs** emerge. Maslow divides these needs into two primary categories, the need for self-esteem, "... the desire for strength, achievement ... independence and freedom" and for the esteem of others, "... status, fame and glory, dominance, recognition, attention, importance, dignity, or appreciation". It is clear that these two categories (as with the other major needs) cover a broad spectrum of possible drivers for activity. At the final level **Self-actualization Needs**, the individual is driven to develop its capabilities to the highest degree possible, "what humans can be, they must be". This level will include invention and aesthetic (musical, artistic, and, presumably, scientific) achievement.

In describing this as a needs hierarchy, Maslow postulates that until a lower level need is satisfied, the next level will not emerge. We suggest that this does not represent a true hierarchy (in the sense that one level facilitates, or is facilitated by, the next), rather that the lower, say physiological needs, are just generally more urgent than the others, so the argument for satisfying them becomes correspondingly stronger and not easily overturned by less urgent topics. Such a mechanism still appears to the observer as a "hierarchy" in the manner indicated by Maslow. Yet it is clear that, in humans at least, the higher-level "needs" can completely subsume the lower. An artist might starve in his garret to produce works of great personal aestheticism, which nobody else appreciates. An ambitious person might seek public esteem at the expense of personal relationships and happiness – yet still be a glutton.

Kakas and Moraïtis [10] describe the power structure between the basic Maslow levels (motivations M_i and M_j, for instance) using a priority relation ($h\_p(M_i,M_j)$) indicating that M_i has a higher priority than M_j. Morignot and Hayes-Roth [19] suggest a weighting vector to arbitrate between levels. These schemes have a gross effect on "personality". The levels appear to have, and need to have, a more subtle interaction, sometimes winning out, sometimes losing.

Maslow elegantly captures this idea: "sound motivational theory should ... assume that motivation is constant, never ending, fluctuating, and complex, and that this is an almost universal characteristic of practically every ... state of affairs." The model we propose divides the agent's reasoning into many independent but interacting faculties. Each faculty is responsible for arguing the case for the activity or approach for which it is responsible and any that support it, and arguing against any that would contradict it. The strength of each faculty is determined primarily by the number and applicability of arguments it has available, but ultimately on notions of implicit preference which definitively answers the question "what is it that I want more?"

Maslow argues that there is no value in trying to enumerate a fixed list of drives (despite producing long lists of need descriptive adjectives to illustrate his levels), and that drives overlap and interact. Yet to produce an equivalent "motivation theory" for a software agent, we must exactly isolate the specific factors that will motivate the behavior of the individual agent and produce some form of explanation as to why they should be incorporated into the design in addition to proposing a mechanism by which they might appear as "never ending, fluctuating, and complex".

In this part of the paper we consider a partial equivalence between the human interpretations of the Maslow motivations to that of a software agent. Agent faculties can be divided into several main grouping. (1) **Operational:** those relating to the immediate protection of the agent and its continuing operation. (2) **Self-benefit:** Those relating to aspects of the agent's behavior that is directly related to its ongoing protection and individual benefit. (3) **Peer-interaction:** those relating to individually identified entities, human or artificial, with which it has specific relationships. (4) **Community-interaction:** those relating to the agent's place in an electronic society that might contain both other software agents and humans, with which it must interact. This category might include both informal and institutionalized groups. (5) **Non-utilitarian:** longer-term activities, not directly related to tasks that offer an immediate or readily quantified benefit.

We assume that the key resource that a software agent must maintain is access to processor cycles, associated data storage and the communications medium reflecting the immediate protection criteria (1). Without immediate access to an active host the agent is completely ineffective and effectively dead. The situation is somewhat more apparent with an embodied agent, such as a robot, where access to uninterrupted power and avoidance of physical damage are clear criteria [19].

Category (2) above equates broadly to the safety needs. In this category an e-commerce agent might seek to accumulate financial strength to pay for a reliable infrastructure strategy, or construct a viable migration plan.

Category (3) addresses how to interact with immediate, individually identified humans and other software agents. Each will have its own personality, and the agent must tailor its interactions with them in specific ways to maintain appropriate relationships and be able to achieve its goals in the future. This broadly equates to the care and kinship needs, though it may be that an autonomous agent might take a hostile view towards a third party, perhaps arguing that "my friend's enemy is my enemy". That is, the relationship between an established ally is more important than the third party, and that it would be jeopardized by perceived collaboration with that third party. An alternative, more social, view would propose the agent has a duty of care towards them anyway. How an autonomous agent would represent or express a personal dislike remains to be explored.

In category (4) the agent accumulates arguments relating to the peer groups, identified, but not individualized and towards the greater society in which it, and any human partner with which it is associated must operate. In the case of an e-commerce assistant agent, this will almost certainly include aspects of the full legal system, and would certainly include the norms and standards of the particular trading circles in which the agent and partner choose to operate.

Level (5) remains problematic for software agents in the absence of the notion of a "feel-good" qualia for agents, but one might speculate that successful agents, those that have accumulated an excess of resources by careful management or good luck would have the opportunity to continue exploring their world for self-improvement alone. One could speculate even further that some might continue to accumulate excess resources for its own sake, or enter into philanthropic activities for less-fortunate agents or agent communities of their choosing.

3   A Game-Based Architecture for Dialectic Argumentation

The Agent Argumentation Architecture (AAA), shown in Figure 1, consists of the following components: an Argumentation State, a Knowledge Base (KB), a number of Faculties (F), an "attender" module (managing the flow of incoming information) and a "Planner/Effector" module (responsible for making plans from goals and performing actions as required). The interaction between components is organized as a complex game consisting of argument sub-games for achieving goals. We draw from the representation already available in [26] to describe games of this kind in terms of the valid moves, their effects, and conditions for when these argument sub-games terminate and when goals may be reestablished.

The **Argumentation State (AS):** A communal structure for the current state of the game, including the arguments put forward, and not yet defeated or subsumed, but accessible by the faculties and the input and output modules.

The **Knowledge Base (KB):** Acts, conventionally, as a long-term repository of assertions within the system. For the purpose of the discussion that follows we shall assume that the elements held in KB take the form of conjectures, rather than expressions of fact. To assume this implies that the knowledge of the agent is non-monotonic, credulous (as opposed to skeptical), and allows inconsistency. On the other hand, a monotonic, skeptical and consistent knowledge base hardly allows for an argumentation process, as it is always obliged to agree with itself.

We will partition the KB according to the Maslow motivation types (KB = $\{K_{m1} \cup K_{m2} \cup K_{m3} \cup K_{m4} \cup K_{m5}\}$, as indicated by the solid radial lines in figure 1) and according to the faculties (KB = $\{K_{f1} \cup \ldots \cup K_{fn}\}$, as indicated by the dotted radial

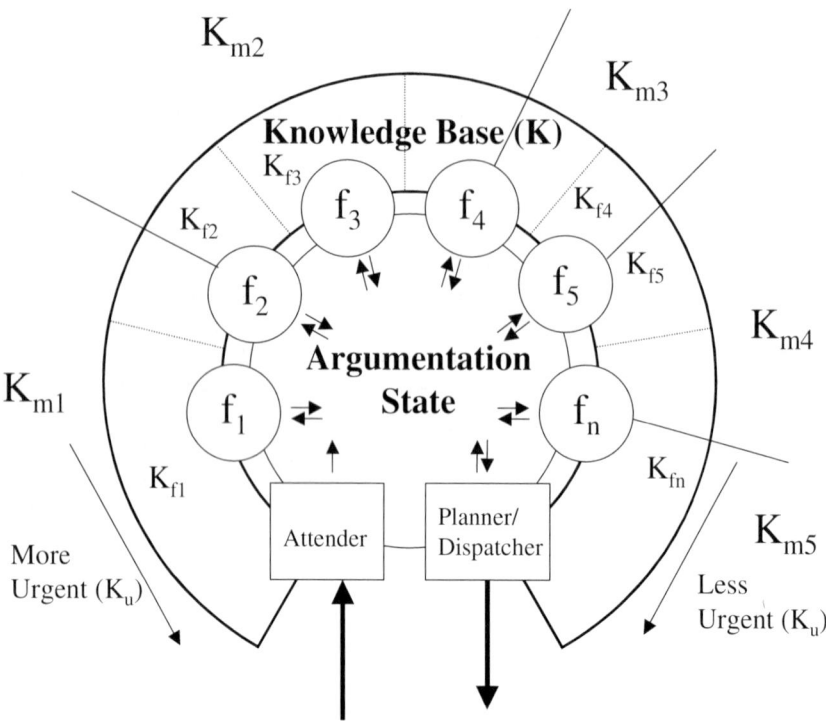

Fig. 1. The Agent Argumentation Architecture (AAA)

lines in figure 1). We assume that the number of faculties (n) will be greater than 5, and that some faculties will impinge on more than one motivational category.

The **Faculties:** Faculties ($F = \{f_1 \ldots f_n\}$) are responsible for particular aspects of the whole agent's possible agenda, which taken together will comprise every aspect that the agent can address. Each faculty may therefore argue that a goal should be established to actively achieve some aspect of that agenda (or avoid some situation that would be detrimental to the agenda). Equally it must monitor the goals and actions proposed by other faculties to determine whether the consequences of those goals or actions would interfere with or contradict some aspect of its own agenda. If some proposal supports the faculty's agenda, it will argue in support of the proposal, or argue against it if the agenda is contradicted. A faculty could, of course be ambivalent towards a proposal, which can have both positive and negative consequences, the faculty being supportive, unsupportive or neutral according to the relative merits of the two positions. Each faculty is arguing the whole agent's best interests are served by pursuing (or not pursuing) certain courses of action, but from its specific viewpoint. There is no winner or loser; each faculty is (or at least, should be) working towards the overall advantage of the whole agent. Essentially each faculty has an "opinion" of what is best for the agent as a whole, but from its limited viewpoint, and must successfully argue its case against other possibly competing views for this to prevail and become incorporated into the agent's overt behavior.

The **"Planner/Effector" module:** Is responsible for creating prototype plans from agreed goals and effecting actions from agreed plans.

The **"Attender" module:** We assume that there is a continuous stream of incoming information, which will comprise of at least the following types of item: requests to perform activities on behalf of other agents, broad suggestions of things the agent might like to do or adopt (for instance, adverts, suggestions by providers that the recipient would be advantaged by purchasing or doing something), "news" items, indicating the outcome of various previous actions and activities, and solicited or unsolicited assertions from other agents, which the current agent may or may not wish to adopt according to its current motivational strategy. These are delivered to the AS and discarded soon, each faculty having a brief time to inspect and adopt them into the KB if required.

4 The Role of Argumentation

In this section we consider how argumentation, [12], [22], [23] might play a role in the architecture for a fully autonomous agent described in the last section. The majority of work in argumentation has been conducted as an approach to the mechanization of the legal process, (often) seeking to model how lawyers conduct cases. In Prakken and Sartor's view [23] a (legal) argumentation system will have four notional layers, a *logical layer* defining the structure of arguments and the underlying semantics, a *dialectical layer* defining how conflicting arguments will be resolved, a *procedural layer* defining the discourse rules for the argumentation and a *strategic or heuristic layer* providing rational ways to conduct the argumentation. Much attention falls onto the second of these, as it captures the form of the argumentation process. As legal argumentation is primarily combative it is often considered as an n-ply game in which one party presents an assertion, which the other party then attempts to overturn, leading in turn to the possibility of a counterargument, and so on. The procedure terminates when there is no further effective counterargument to be made and the argument is therefore won or lost.

Legal argumentation is, in the English and US tradition, both combative and largely retrospective, comprising accusations and rebuttals between prosecution and defense. In the legal process there is an overall and external framework to arbitrate between the parties (the statutes) and sanctions to be applied to those found guilty. In the proposed model for autonomy, the argumentation will be about current or future events, goals and actions, and the consequences that may follow if one route or another is taken. There will be no overall guiding external principle against which a definitive decision about what to adopt as the overt behavior of the agent might be judged. Existing models of legal argumentation [23] rely on four major types of argumentation strategy (though not necessarily all in the same system): **Conflicts**, arguments reaching differing conclusions may be *rebutted*, where one aspect of an established argument is shown not to hold in the current circumstances, *assumption attack*, the assumption of non-provability is presented with a proof, and *undercutting*, where one argument challenges the rule of inference used by another. **Argument comparison**, where lines of argument are decided by recourse to a higher principle. **Argument status**, where defeated arguments may be reinstated if the rebuttal is itself defeated. In **defeasible argumentation**, general rules may be defeated by more specific ones, where a defined priority relationship exists.

The agent argumentation system is driven by the search for, and resolution of conflicts. Conflicts arise where two faculties arrive at different conclusions at the end of individual chains of consequences, or one asserts that an action by the other will be detrimental to its agenda. We identify six classes of argument appropriate to the AAA:

Goal Proposal Move: Some faculty f_n determines that prevailing circumstances imply that a goal must be asserted or will become asserted shortly.

Conflict of Interests Moves: For a goal or action proposed by faculty f_n with intended consequence x, faculty f_m asserts that some y, which it believes is also consequential leads to a conflict, either by asserting an action that f_m determines as detrimental (interest conflict) or interfering with a current plan of f_m (resource conflict). f_n may retract, propose an alternative or a covering action.

Alternative Argument Move: Faculty f_m, having detected a conflict of interests, proposes an alternative solution to achieve f_n's original goal in a manner that does not conflict with its agenda (cooperative).

Retraction Move: Faculty f_n retracts a proposed action or goal, because it is no longer applicable to f_n because of changed circumstances (including conflict with another faculty).

Undercut Move: Faculty f_n challenges the assertion that there is a conflict, and attempts to undercut f_m's consequence chain, by arguing f_m has included an invalid step or overestimated its significance.

Covering-action Move: Some action a has a positive outcome for f_n, but gives rise to a potential liability according to f_m, f_m may propose a prior action that would avoid the liability. This is a "duty of care" argument, complicating the behavior of the agent to ameliorate possible (but not certain) negative outcome. This argument is particularly valuable where the undesirable consequence is rare, but highly damaging.

In this model, the agent is required under normal circumstances to completely exhaust the argumentation process, i.e. to explore all the rational routes to a decision [2]. A hung decision would, in an isolated system, represent a considerable dilemma with the agent unable to act. In a connected system, new information is always arriving on the AS, and this new information may be sufficient to tip the balance one way or the other. This is, perhaps, the ideal, but likely the agent cannot wait.

5   The Role of Preferences

Underlying much of human decision-making would appear to be a complex web of preferences, allowing us to choose between the otherwise rationally un-decidable alternatives. In the first instance the system should determine whether there are any preference relations specific to the two items in conflict (prefer(x,y)). This being so the situation is resolved. Failing that, more general classes of preference should be invoked.

The agent will have a current preference ordering relationship between, say, the urgency 'U' of rules in the knowledge ($k_n \subseteq K$) base. For example the preference ordering of U (U_p) indicates the agent's current rank ordering of K ($k_1 .. k_n$), expressed as: U_p: prefer($u(k_{34})$, $u(k_{12})$, $u(k_{100})$ ..., $u(k_m)$). Similarly the agent can place a rank ordering on the roles of different faculties (F_p: prefer(f_n, f_q, ..., f_m)), or on different motivational areas (M_p: prefer(m_2, m_1, m_3, m_4, m_5), as in [10]). The agent may further make explicit the rank ordering of these classes of rankings (e.g. prefer(U_p, F_p, M_p)), and so be in a position to change them.

Traditionally, most formalizations of preference [20] have made an assumption of *ceteris paribus*. It is clear that, in humans at least, preference orderings are quite variable, modified by mood, emotion and recent events, and need not be applied consistently. The manner in which preferences might be dynamically updated, while interesting and perhaps central to notions of full autonomy, falls largely outside the scope of this paper. Witkowski [31] describes a scheme by which an agent's preference for selecting a rule (and so an action associated with it) may be tied closely to the confidence an agent has in it, based on internal expectations established by the rule. Witkowski *et al* [32] investigate the dynamic effects of trust relationships between agents in a trading community, where each agent's preference or willingness to trade with another agent is determined by past experience.

The outcome of the argumentation process equates to Von Wright's [29] notion of an *extrinsic* preference, subject to reason and rationality, and the preference rank ordering to *intrinsic* preference, hidden and apparently outside rationality. Each has a role to play in the full definition of agent autonomy.

6   The Procedure

This section outlines the activities (the *procedural layer*) by which an agent's faculties might interact to define a goal directed strategy. Overall, the activities are essentially asynchronous, with faculties responding to items appearing on the AS, and are bounded by the computational resource available to them and the whole agent.

Activity 1) Each faculty is responsible for proposing immediate actions or goals onto the communal AS. It is expected that a goal or action will be one that maintains or enhances the whole agent, but strictly from the viewpoint of the topic related to the proposing faculty (i.e. they are expected to be topic "selfish"). This is the primary creative component of the whole agent. Proposals can be immediate, short or long term. In general, the lower numbered "motivations" will give rise to immediate and short-term proposals, the higher numbered ones mid to long-term proposals, reflecting the scale of urgency. Proposals for immediate actions will often relate to safety or highly opportunistic situations (a purely reactive faculty could only propose actions). In an agent of any sophistication there will be many specific goals, of varying duration, active at any one time. In general, an agent will have many more suggestions for actions and goals than it could reasonably service. Proposals are therefore opportunistic and made with the likelihood that they will be rejected. Any action or goal proposed at this stage represents a "desire" on the part of the whole agent.

Activity 2) Every proposed action or goal must be vetted by each of the other faculties to determine whether it, or any of its known consequences, would violate (or augment) the agenda of that faculty in the current circumstances. Arguments against a proposal can include: the proposal, or one of its known consequences, would directly contradict an aspect of the vetter's agenda; that the proposal would, if enacted, drain resources from a previously agreed course of action, disrupting, or making that course of action untenable. To not respond to a proposal at this stage is surely to tacitly accept it. One might suppose that this, and the other, vetting stages would be subject to a policy-based [2] strategy to reduce computational load and delay.

Activity 3) Those goals that pass through stage 2, are passed to a conventional means-ends planner. The planner might, at this stage, find no reasonable instantiation of the goal in the current circumstances, and it would be (temporarily) abandoned. Otherwise the planner will deliver to the communal AS a proposed sequence of actions for all the faculties to consider. At this stage the planner is only required to produce a viable sequence of proposed actions, perhaps based on a least cost heuristic or some other (perhaps simplistic, perhaps sophisticated) "optimal strategy" metric. It may contain considerable consequential risks and liabilities to the agent.

Activity 4) With the extra detail of the instantiated plan, each faculty is required once again to review the plan, raising arguments or objections if any action proposed as part of it would cause immediate or consequential violation of any faculty's primary motivations. At this stage a plan might be rejected outright or be returned to the planner for modification, to avoid or amend the contended step or steps. Again, to not respond at this stage is to tacitly accept the plan and its consequences. The agreed sequence of actions become part the whole agent's intentions at this stage, and are passed to the plan effector module.

Activity 5) At the allotted time, or under the planned conditions, the effector module will briefly present the action as instantiated at the moment of execution to the AS, giving the assembled faculties one last chance to argue that it should be suppressed, primarily due to changes in circumstances since it was previously agreed. Cotterill [5] refers to this as "veto-on-the-fly", arguing that it brings a significant advantage to an agent; it is also reminiscent of Bratman's [2] rational reconsideration step. If the action is challenged this will appear as a hesitation, even if the action finally goes ahead.

7   Discussion

This model presents an autonomous agent as a peer group of faculties, each with a role within, and responsibility to, the whole agent. This is not a "democracy". There is no voting, but argumentation in which the "last man standing" [6] (the faculty with the most effective arguments, or whose arguments are preferred) becomes the winner, and whose suggestions become part of the overt behavior of the whole agent. Acceptance of this criterion effectively guarantees that each argumentation game will terminate in a finite time. Additional urgency constraints may, of course, be required.

The observable "personality" of the agent is therefore defined by the balance and effectiveness of the individual faculties. In a well-balanced agent system based on these principles, the range of faculties will cover all the aspects of the whole agent's individual and social requirements. Equally, each faculty will have an equitable share of the overall resources and only propose a reasonable number of goals, deferring if the arguments placed against it are stronger (i.e. not continue to assert weaker arguments, once a stronger one is presented, or to repeat defeated arguments unless a material change to circumstances has occurred).

It will be interesting to speculate as to the effects of various failings or imbalances in the relative strengths of each component faculty. If a faculty has few, or only weak, arguments, the whole agent will appear weak, or even oblivious, to that part of the normal social interaction. If "individual" completely dominates over "society", the agent shows symptoms of "pathological behavior", ignoring social convention and the needs of others. If the argumentation process that vets the initial plan (activity 4) or the release of actions the agent's behavior is inadequate (activity 5), the agent appears *akratic* (mentally incontinent) in the social context [17]. If a specific faculty were to persistently propose goals, particularly the same one, even if repeatedly rebuffed the whole agent would appear distracted. If these were not successfully rebuffed, then the agent would appear to be obsessive. This could, of course, take many forms, depending on the specific faculties involved.

The model and architecture presented here has much of the flavor of Minsky's view of a society of mind [18], with individual, focused, component parts each contributing some specialization or expertise in the service of the whole. There is further a clear analogy between this society of mind within an agent and a society of individual agents. A strongly autonomous agent makes a significant impact on that society by being able to express its internal arguments and preferences to influence those of others. Every action made contributes to the agent's perceived personality. It remains unclear whether "motivation" in the sense used by Maslow should be interpreted as reflecting a *post-priori* explanation of observable behavior arising from competitive individual "interest groups" (faculties) or a genuine drive mechanism. Maslow puts high emphasis on motivation as a sensation overtly available to the (human) agent; we place emphasis on the mechanism, giving rise to apparently complex motivated behavior within a deterministic framework.

Whether fully autonomous agents of this type will ever find a place in the world of e-commerce remains to be seen, but the outlook is not encouraging, Luck and d'Inverno [14] are certainly skeptical. Almost certainly not in the immediately foreseeable future, as most people want and expect their computers and software to behave in a (reasonably) consistent manner and perform the actions as requested. For instance work to date ([27], [33]) on agents in the connected community, presupposes that software agents have restricted flexibility and are primarily subservient. By definition a fully autonomous agent may refuse, and may even perform actions specifically contrary to the wishes or intentions of a "user".

All this could easily be seen as a recipe for apparent erratic and unpredictable behavior. On the other hand it may be that the extra flexibility offered by full autonomy will play a part in generating a whole new class of interesting cooperative activities and the development of independent communities of such agents, literally trading on their own account and purchasing the computational and support resources they need from dedicated suppliers. After all the richness and diversity of human society is the product of the interactions between uncounted millions of apparently autonomous entities.

However, e-commerce is not the only application for software agents, and true autonomy may find unexpected value in a range of applications from entertainment to providing companionship for the lonely. In each case the unexpected and unpredictable may prove to be an added bonus, otherwise missing from the pre-programmed.

For the engineer or computer scientist with a philosophical leaning, discovering what would make a software agent autonomous would in turn have the most profound implications for our understanding of how humans come to view themselves as autonomous and in possession of apparent "freewill". True autonomy in artificial agents is worth studying for this reason alone. Our suspicion is that we have hardly scratched the surface of this problem – or begun to perceive its full potential.

Acknowledgements

This research was supported in part by the European Union as part of the SOCS project (Societies Of ComputeeS), IST-2001-32530.

References

[1] Beavers, G. and Hexmoor, H. (2003) Types and Limits of Agent Autonomy, in: Rovatsos, M. and Nickles, M. (eds.) Proc. 1st Int. Workshop on Computational Autonomy – Potential, Risks and Solutions, AAMAS03, Melbourne, Australia, pp. 1–9

[2] Bratman, M.E. (1987) *Intention, Plans, and Practical Reason*, Cambridge, MA: Harvard University Press

[3] Brooks, R. (1991) Intelligence without Representation, *Artificial Intelligence*, Vol. 47, pp. 139–159

[4] Carabelea, C., Boissier, O. and Florea, A. (2003) Autonomy in Multi-agent Systems: A Classification Attempt, in: Rovatsos, M. and Nickles, M. (eds.) Proc. 1st Int. Workshop on Computational Autonomy – Potential, Risks and Solutions, AAMAS03, Melbourne, Australia, pp. 11–21

[5] Cotterill, R. (1998) *Enchanted Looms Conscious Networks in Brains and Computers*, Cambridge University Press

[6] Dung, P.M. (1995) On the Acceptability of Arguments and its Fundamental Role in Nonmonotonic Reasoning, Logic Programming and N-person Games. *Artificial Intelligence*, Vol. 77, pp. 321–357

[7] Eysenck, H. (1991) Dimensions of Personality: 16, 5 or 3? Criteria for a Taxonomic Paradigm, *Personality and Individual Differences*, Vol. 12(8), pp. 773–790

[8] Jennings, N.R., Sycara, K.P. and Wooldridge, M. (1998) A Roadmap of Agent Research and Development, *J. of Autonomous Agents and Multi-Agent Systems*, Vol. 1(1), pp. 7–36

[9] John, O.P. (1990) The "Big Five" Factor Taxonomy: Dimensions of Personality in the Natural Language and in Questionnaires, in Pervin, L.A. (ed) *Handbook of Personality: Theory and Research*, New York: Guildford, pp. 66–100

[10] Kakas, A. and Moraïtis, P. (2003) Argumentation Based Decision Making for Autonomous Agents, proc. 2nd Int. Joint Conf. on Autonomous Agents and Multiagent Systems (AAMAS-03), pp. 883–890

[11] Kakas A.C., Kowalski, R.A. and Toni, F. (1998) The role of abduction in logic programming, Gabbay, D.M. et al (eds.) Handbook of Logic in Artificial Intelligence and Logic Programming 5, Oxford University Press, pp. 235–324

[12] Kowalski, R.A. and Toni, F. (1996) Abstract Argumentation, *Artificial Intelligence and Law Journal*, Vol. 4(3–4), pp. 275–296

[13] Luck, M. and d'Inverno, M. (1995) A Formal Framework for Agency and Autonomy, in Proc. 1st Int. Conf. On Multi-Agent Systems (ICMAS), pp. 254–260

[14] Luck, M. and d'Inverno, M. (2001) Autonomy: A Nice Idea in Theory, in Intelligent Agents VII, Proc. 7th Workshop on Agent Theories, Architectures and Languages (ATAL-2000), Springer-Verlag LNAI, Vol. 1986, 3pp.

[15] Maes, P. (1994) Agents that Reduce Work and Information Overload, Communications of the ACM Vol. 37(7), pp. 31–40

[16] Maslow, A.H. (1987) *Motivation and Personality*, Third edition, Frager, R., *et al.* (eds.), New York: Harper and Row (first published 1954)

[17] Mele, A.R. (1995) *Autonomous Agents: From Self-Control to Autonomy*, New York: Oxford University Press

[18] Minsky, M (1985) *Society of Mind*, New York: Simon and Schuster

[19] Morignot, P. and Hayes-Roth, B. (1996) Motivated Agents, Knowledge Systems Laboratory, Dept. Computer Science, Stanford University, Report KSL 96–22

[20] Moutafakis, N.J. (1987) *The Logics of Preference: A Study of Prohairetic Logics in Twentieth Century Philosophy*, D. Reidel Publishing Co.

[21] Norman, T.J. and Long, D. (1995) Goal Creation in Motivated Agents, in: Wooldridge, M. and Jennings, N.R. (eds.) *Intelligent Agents: Theories, Architectures, and Languages*, Springer-Verlag LNAI Vol. 890, pp. 277–290

[22] Prakken, H. (1997) *Logical Tools for Modelling Legal Argument. A Study of Defeasible Reasoning in Law.* Dordrecht: Kluwer Academic Publishers

[23] Prakken, H. and Sartor, G. (2002) The Role of Logic in Computational Models of Legal Argument: A Critical Survey, in: Kakas, A. and Sadri, F. (eds), *Computational Logic: From Logic programming into the Future (In Honour of Bob Kowalski)*, Berlin: Springer-Verlag LNCS Vol. 2048, pp. 342–381

[24] Rao, A.S. and Georgeff, M.P. (1991) Modeling Rational Agents within a BDI-Architecture, Proc. Int. Conf. on Principles of Knowledge, Representation and Reasoning (KR-91), pp. 473–484

[25] Revelle, W. (1995) Personality Processes, *The 1995 Annual Review of Psychology*

[26] Stathis, K. (2000) A Game-based Architecture for Developing Interactive Components in Computational Logic, *Journal of Functional and Logic Programming*, 2000(1), MIT Press.

[27] Stathis, K., de Bruijn, O. and Macedo, S. (2002) Living Memory: Agent-based Information Management for Connected Local Communities, *Journal of Interacting with Computers*, Vol. 14(6), pp. 665–690

[28] Steels, L. (1995) When are Robots Intelligent Autonomous Agents? *Journal of Robotics and Autonomous Systems*, Vol. 15, pp. 3–9

[29] Von Wright, G.H. (1963) *The Logic of Preference: An Essay*, Edinburgh: Edinburgh University Press

[30] Wiener, N. (1948) *Cybernetics: or Control and Communication in the Animal and the Machine*, Cambridge, MA: The MIT Press

[31] Witkowski, M. (1997) Schemes for Learning and Behaviour: A New Expectancy Model, Ph.D. thesis, University of London

[32] Witkowski, M., Artikis, A. and Pitt, J. (2001) Experiments in Building Experiential Trust in a Society of Objective-Trust Based Agents, in: Falcone, R., Singh, M. and Tan, Y-H. (eds.) *Trust in Cyber-societies*, Springer LNCS 2246, pp. 111–132

[33] Witkowski, M., Neville, B. and Pitt, J. (2003) Agent Mediated Retailing in the Connected Local Community, *Journal of Interacting with Computers*, Vol. 15(1), pp. 5–33

Author Index